Pitfalls of Shoreline Stabilization

Coastal Research Library

VOLUME 3

The aim of this book series is to disseminate information to the coastal research community. The Series covers all aspects of coastal research including but not limited to relevant aspects of geological sciences, biology (incl. ecology and coastal marine ecosystems), geomorphology (physical geography), climate, littoral oceanography, coastal hydraulics, environmental (resource) management, engineering, and remote sensing. Policy, coastal law, and relevant issues such as conflict resolution and risk management would also be covered by the Series. The scope of the Series is broad and with a unique crossdisciplinary nature. The Series would tend to focus on topics that are of current interest and which carry some import as opposed to traditional titles that are esoteric and non-controversial. Monographs as well as contributed volumes are welcomed.

For further volumes:
http://www.springer.com/series/8795

Pitfalls of Shoreline Stabilization

Selected Case Studies

Editors

J. Andrew G. Cooper
University of Ulster, Northern Ireland, UK

Orrin H. Pilkey
Duke University, Durham, NC, USA

Springer

Editors
J. Andrew G. Cooper
School of Environmental Sciences
University of Ulster
Coleraine, Northern Ireland, UK

Orrin H. Pilkey
Division of Earth and Ocean Sciences
Nicholas School of the Environment
Duke University
Durham, NC, USA

ISSN 2211-0577 ISSN 2211-0585 (electronic)
ISBN 978-94-007-4122-5 ISBN 978-94-007-4123-2 (eBook)
DOI 10.1007/978-94-007-4123-2
Springer Dordrecht Heidelberg New York London

Library of Congress Control Number: 2012939622

Printed on acid-free paper

Springer is part of Springer Science+Business Media (www.springer.com)

Contents

Contributors

Giorgio Anfuso Dpto. Ciencias de la Tierra, Facultad de Ciencias del Mar y Ambientales, Universidad de Cádiz, Puerto Real, Spain

Edward J. Anthony Aix-Marseille Univ, Institut Universitaire de France, Cedex, France

Steven Brayshaw School of Engineering, Griffith University, Gold Coast, QLD, Australia

Simon C. Bundy Sustainable Development Projects CC, Ballito, South Africa

David M. Bush Department of Geosciences, University of West Georgia, Carrollton, GA, USA

Andrew S. Coburn Program for the Study of Developed Shorelines, Western Carolina University, Cullowhee, NC, USA

J. Andrew G. Cooper School of Environmental Sciences, University of Ulster, Coleraine, Northern Ireland, UK

Charles W. Finkl Department of Geosciences, Florida Atlantic University, Boca Raton, FL, USA

Charles H. Fletcher Department of Geology and Geophysics, University of Hawaii at Manoa, Honolulu, HI, USA

Helena Granja CIIMAR/Earth Sciences department, University of Minho, Braga, Portugal

Chester W. Jackson Jr Department of Geology and Geography, Georgia Southern University, Statesboro, GA, USA

Derek W.T. Jackson School of Environmental Sciences, University of Ulster, Coleraine, Northern Ireland, UK

James W. Jordan Department of Environmental Studies, Antioch University New England, Keene, NH, USA

Paul S. Kench School of Environment, The University of Auckland, Auckland, New Zealand

Charles Lemckert School of Engineering, Griffith University, Gold Coast, QLD, Australia

Leanne Lestak Institute of Arctic and Alpine Research, University of Colorado, Boulder, CO, USA

Gonzalo C. Malvárez Area of Physical Geography, Decano, Facultad de Humanidades, Universidad Pablo de Olavide, Sevilla, Spain

William F. Manley Institute of Arctic and Alpine Research, University of Colorado, Boulder, CO, USA

José Ángel Martínez-del-Pozo Instituto de Arqueología de Mérida – CSIC, Mérida (Badajoz), Spain

Owen K. Mason Institute of Arctic and Alpine Research, University of Colorado, Boulder, CO, USA

Andrew A. Mather Ethekwini Municipality, Durban, South Africa

William J. Neal Department of Geology, Grand Valley State University, Allendale, MI, USA

Michael R. Phillips School of Built and Natural Environment, Faculty of Applied Design and Engineering, Swansea Metropolitan University, Swansea, Wales, UK

Orrin H. Pilkey Nicholas School of the Environment, Duke University, Durham, NC, USA

José Luís Pinho Department of Civil Engineering, University of Minho, Braga, Portugal

Nelson Rangel-Buitrago Department of Earth Sciences, School of Marine and Environmental Sciences, University of Cadiz, Puerto Real, Spain

Bradley M. Romine Department of Geology and Geophysics, University of Hawaii at Manoa, Honolulu, HI, USA

François Sabatier CEREGE, UMR CNRS 6635, Aix-Marseille Université, Aix en Provence, France

Andrew D. Short Coastal Studies Unit, School of Geosciences, University of Sydney, NSW, Australia

Alan M. Smith School of Geological Sciences, University of KwaZulu-Natal, Durban, South Africa

Hilary Stevens Department of Geological Sciences, University of Delaware, Newark, DE, USA

Arthur Trembanis Department of Geological Sciences, University of Delaware, Newark, DE, USA

Introduction

J. Andrew G. Cooper and Orrin H. Pilkey

The global rush to the beach continues, and with it come many challenges, not least of which is how to live with a changing shoreline. The rush is a recent phenomenon, having taken place in large part after the Second World War, when the citizens of many western societies became sufficiently wealthy to take seaside holidays in large numbers and build or rent second homes at the beach. The closer to the beach and/or the better the view of the sea, the more sought-after the property. The problem of course is that if you can see the sea, the sea can see you.

In northern Europe over the last two decades, seashore tourist towns were being abandoned in favor of seashore tourist towns in Spain, France and Portugal where warmer water and more sunshine prevailed. During the same time frame in the US, especially in the state of Florida, hundreds of miles of sandy shorelines were lined with high-rise condos, apartments and hotels. The trend was repeated in Australia (Fig. 1), South Africa and Brazil as residents sought beachfront vacations and taken to its maximum extension in Dubai (Fig. 2) as the available beachfront land was exhausted and the shoreline was extended into the sea to maximize the sandy shoreline for development. As air travel became more affordable, the desire for beachfront living has seen the development of tourist resorts and holiday homes all around the world's shorelines during the past two decades, even in islands that were once considered remote from the main tourist source areas. Although tourist developments are perhaps the main areas to spring to mind when thinking of shoreline stabilization, there are many villages, towns and cities whose coastal location poses challenges in relation to shoreline change.

J.A.G. Cooper
School of Environmental Sciences, University of Ulster, Cromore Road,
BT52 1SA Coleraine, Northern Ireland, UK
e-mail: jag.cooper@ulster.ac.uk

O.H. Pilkey
Nicholas School of the Environment, Duke University, Durham, NC 27708, USA
e-mail: opilkey@duke.edu

Fig. 1 Beachfront high-rise development along the Gold Coast, Queensland, Australia

Fig. 2 High-rise beachfront development in Dubai

Shoreline stabilization issues have been around since antiquity, but the scale of the challenge has been multiplied by several orders of magnitude by the rush to the shore. It is now a major societal challenge to work out a way to live with an ever-changing coast and what to do with the buildings, roads and other infrastructure we have placed there. Historically and, for the most part, currently the response has been to defend wherever we can. Now we are learning that this isn't the only way and indeed it is a way littered with pitfalls. In this book we bring together accounts of efforts from around the world to stabilize the shoreline. The accounts all show that from the traditional hard engineering (Anthony and Sabatier, Kench, Jackson, Romine and Fletcher), through various kinds of soft engineering (Coburn, Finkl, Brayshaw and Lemckert) to innovative 'alternative' approaches (Anfuso et al., Pilkey and Cooper), efforts to stabilize the shoreline always encounter pitfalls. Indeed the findings point to the conclusion that shoreline stabilization cannot be achieved at any meaningful timescale.

As the rush continues, inlets are deepened for navigation (Finkl), jetties are built to protect inlets (Anfuso et al., Brayshaw and Lemckert) and seawalls are constructed for storm protection (Malvarez, Granja and Pinho, Stevens and Tembanis, Phillips, Anthony and Sabatier). Simultaneously streams and rivers are dammed, thousands upon thousands of farm ponds are excavated and larger rivers are leveed all of which lead to loss of sand supply to the beach and of course to increased shoreline erosion. But the unkindest cut of all for beach front development is the rising sea level caused by the thermal expansion of the oceans and the melting of ice sheets and glaciers.

Most sea level rise expert panels believe that there is a good chance that sea level will rise 1 m by the year 2100. Such a rise has to mean the end of beachfront development on all sandy shores and on barrier islands in their entirety. Beach nourishment to hold the shoreline in place will be economically impossible. Only hard stabilization holds out hope of at least temporary preservation of near-beach buildings and, in the case of barrier islands and atolls, the walls must completely surround the islands, a situation already achieved in Male, the Maldives Capital (Kench). But coastal engineering structures of sufficient magnitude to withstand open ocean conditions are very costly. And the cost of hard stabilization also includes a stiff environmental price – the loss of the beach (Jackson, Anthony and Sabatier, Malvarez, Jackson et al.), the *raison d' etre* for beach front development to begin with. As several papers (Jackson, Anfuso et al., Malvarez, Anthony and Sabatier Phillips, and Smith et al.) show, stabilization structures can themselves cause additional problems by creating new nearshore dynamic regimes that lead to enhanced rates or extent of erosion.

So now we are faced with hundreds of miles of high-rise-lined beaches in tourist areas of Spain, Portugal, Australia, Brazil and Florida (to name but a few) facing a sea level rise that will cost huge amounts of money to protect, at the cost of the beach, which was the reason people chose to live or visit there to begin with. But at the same time the world's major coastal cities will be in trouble. According to a UN study, cities that are particularly susceptible to sea level rise include Miami, New York, New Orleans, Osaka, Tokyo, Amsterdam, Rotterdam and Boston.

The protection of these and dozens of other major cities from sea level rise will itself involve a huge cost. It seems certain that national priorities will favor the cities over the thousands of miles of relatively lightly developed touristic shoreline development, let alone individual beach houses. In the long term there is the realistic prospect that shoreline stabilization will only be applied to the cities and that retreat will be the order of the day elsewhere.

Coastal management programs around the world increasingly recognize the twin problems of sea level rise and environmental damage from shoreline stabilization structures. Both are difficult problems to work with. There are still those who deny sea level rise or at least minimize its impact. The impact of hard stabilization remains controversial plus there are a large number of powerful and influential individuals (beach front property owners) who argue that preservation of buildings is more important than preservation of beaches.

As time rolls on and as sea level rises and beach sand supplies are reduced, more and more pressure is applied to government at all levels to save beach houses. Coastal management regulations are diluted and compromised. A South Carolina law forbidding all hard stabilization was revised with the state legislature declaration that groins are not a form of hard stabilization. In 1985, North Carolina passed a law forbidding all hard stabilization but in 2011 the state legislature allowed construction of terminal groins (small jetties) at the ends of barrier islands having been assured by engineering firms that these structures would not cause downdrift erosion.

The conflict between the engineering and the scientific mentality is a fact of life that all coastal managers must face. In general this conflict boils down to a short-term view and a long-term view. Engineers are asked to hold the shoreline in place now, with the attitude that with more funding any unexpected problems can be fixed. Scientists, if asked, recognize the longer-term problem.

This volume is a timely one coming as it does in the context of increasing global concern about rising seas and eroding shores and ever widening societal debates about holding shorelines in place. There is a large literature concerned with how to engineer shorelines but much less is written about the pitfalls of shoreline stabilization. The 18 papers herein address stabilization experience on many types of shorelines including long sandy beaches (Granja and Pinho, Short, Brayshaw and Lemckert), embayed rocky coasts (Jackson, Smith et al., Malvarez, Phillips), estuarine (Trembanis), atolls (Kench), tropical islands (Romine and Fletcher, Jackson et al.), Arctic (Mason) and lake (Pilkey). The studies from more than ten countries illustrate the role of national priorities and politics in choice of alternative shoreline stabilization techniques. Societal debates about holding shorelines in place will be occurring in ever increasing frequency in coming years. We hope the studies herein will play a useful and helpful role in informing that debate.

Chapter 1
Pitfalls of Shoreline Stabilisation – Tweed River Mouth, Gold Coast, Australia

Steven Brayshaw and Charles Lemckert

Abstract The Tweed River mouth on the Gold Coast of Queensland, Australia, has undergone extensive modification. Initial operations involved the construction of rock training walls to stabilise the river mouth and to maintain a deep navigation channel. These modifications interrupted the natural longshore transport and ebb tidal delta processes causing accretion on Letitia Spit and contributing to the erosion of southern Gold Coast beaches. In 2001 a sand bypassing system was built to mimic the natural longshore transport, pumping sand from Letitia Spit to southern Gold Coast beaches. Initial operations brought public concern over wide beach widths, however, the sand bypassing system has been highly successful in maintaining a deep navigation channel in the Tweed River entrance and delivering sand to southern Gold Coast beaches.

1.1 Introduction

Many river entrances along the east coast of Australia have been extensively modified to stabilise the river mouth and improve safety of the entrance for small and large vessels entering and leaving local ports. These modifications consisted primarily of rock training walls, with most of the work being undertaken following Australia's Federation in 1901 (Kay and Lester 1997). However, little was known about the effects these changes would have on the river mouths and surrounding environments. Environmental aspects of Coastal Zone Management (CZM) were not generally considered important, as the import and export trades and coastal shipping transport (along with development of the economy) was a priority (Kay and Lester 1997). Thus CZM was left mainly to state and local governments,

S. Brayshaw (✉) • C. Lemckert
School of Engineering, Griffith University, Engineering Building, Engineering Drive, South Port, Gold Coast, QLD 4222, Australia
e-mail: steven.brayshaw@griffithuni.edu.au; c.lemckert@griffith.edu.au

J.A.G. Cooper and O.H. Pilkey (eds.), *Pitfalls of Shoreline Stabilization: Selected Case Studies*, Coastal Research Library 3, DOI 10.1007/978-94-007-4123-2_1,

whose interests focused on their own region (Cullen 1982). Today it is evident that understanding natural coastal processes is critical to the planning of coastal activities, in order to avoid unfavourable consequences.

Ebb tidal deltas and longshore transport – two important coastal processes that define the dynamics of coastal river entrances – are both significantly influenced by meteorological conditions. Ebb tidal deltas form at river mouths from the high current velocity of water exiting the river system during the ebb tide. This current scours the river channel and deposits sediment on marginal shoals on the ocean side of the river mouth, and it is here that the current velocities reduce and the sediment they carry settles (Özsoy 1986). During the flood tide, the velocities entering the river are insufficient to resuspend the recently-deposited sediment, resulting in a build-up of the shoals over time. Longshore transport is the natural transport of sediment along the coast, forced by *prevailing wind* and swell directions. Generally, most coastlines experience a net longshore transport in one direction; for example, the Gold Coast region (Queensland, Australia) experiences a net northward transport of 500,000 m^3/year (Hyder Consulting Pty et al. 1997).

The common practice of river mouth training inhibits the longshore transport process, trapping the sand behind on the downstream end of a training wall and preventing the natural longshore transport of sediment, and consequently eroding the coastline. It is therefore vital that engineers and geoscientists working on such coastal projects fully investigate coastal processes before any action is undertaken (Castelle et al. 2009). However, this has not always been done in the past, and there are many examples where significant remediation schemes are required.

The Tweed River mouth (an example of a trained entrance) has attracted significant media attention as it is a highly modified system initially designed with one main objective – to provide a deep and safe channel for navigation through the Tweed River entrance (Hyder Consulting Pty et al. 1997). This limited aim and the ideals of governments at the time resulted in significant environmental consequences (particularly on the southern Gold Coast). Throughout the history of the Tweed River mouth stabilisation projects, no long-term management plan had been developed – each management strategy had been implemented following impacts from previous management strategies (Hyder Consulting Pty et al. 1997).

1.2 How Did It Come About?

The Tweed River mouth is located in far northern New South Wales (NSW) just south of the NSW and Queensland (QLD) state border (see Fig. 1.1), and has undergone many alterations since the first training walls were built in 1891. Training of the Tweed River mouth came about from the need to improve the conditions of the Tweed River sand bar to vessels used for the export of the developing Cedar timber trade and the importing of groceries, clothing and hardware into the Tweed Valley (Hyder Consulting Pty et al. 1997). The Tweed River mouth was a very dynamic system, and the entrance changed frequently. Although

Fig. 1.1 Location of Tweed River, Australia (Source: Google Maps 2011)– (**A**) State of Queensland, Australia (**B**) Tweed and southern Gold Coast Region (**C**) Tweed Heads (**D**) Kirra Beach (**E**) Kirra Reef (**F**) Kirra Groyne (**G**) Coolangatta/Greenmount Beach (**H**) Rainbow Bay Beach (**I**) Snapper Rocks Beach (**J**) Duranbah Beach (**K**) Tweed River training walls after extension in 1962 (**L**) Sand pumping jetty (**M**) Initial training walls built in 1891 (**N**) Beach width prior to sand pumping in 2001 (**O**) Letitia Spit

conditions were favourable at sea, bar conditions were more often than not too treacherous for fishing and transport vessels. Early steam transport vessels were often forced to load and unload inside and outside the entrance, while smaller shallow draft vessels transported the goods to and from the transport vessel. At the time, knowledge of coastal dynamics was not greatly understood.

The 1962 building and extension of rock training walls either side of the mouth helped in the short term to maintain a deep navigation channel. However, soon after construction was completed, ongoing dredging operations were required as the river continued to shoal from the ebb tidal delta and longshore transport (Hyder Consulting Pty et al. 1997). The longshore transport of sand was completely interrupted, which caused major accretion on the southern side of the system, an area known as Letitia Spit (see Fig. 1.1). Figure 1.2 shows how longshore transport was inhibited until the shoreline at Letitia Spit accreted out to the end of the break wall, allowing sand bars to form across the entrance. Many costly dredging operations were implemented between 1974 and 1996; however, a more permanent solution was required (Hyder Consulting Pty et al. 1997).

A series of intense storm events during the late 1960s and early 1970s exacerbated the biggest impact of the training wall extension. The intense storms severely eroded the southern Gold Coast beaches (see Fig. 1.3) which could not be replenished as the training walls prevented longshore transport of sediment upward along the coast. Conflict arose between neighbouring governments, the Queensland Government taking the view that the New South Wales Government (which manages the Tweed River entrance system) was responsible for the degradation. With neither of the state governments willing to take responsibility, dredging operations during the 1980s and early 1990s were undertaken to transport the sand out of the river entrance and from Letitia Spit (see Figs. 1.2 and 1.3) to replenish southern Gold Coast beaches.

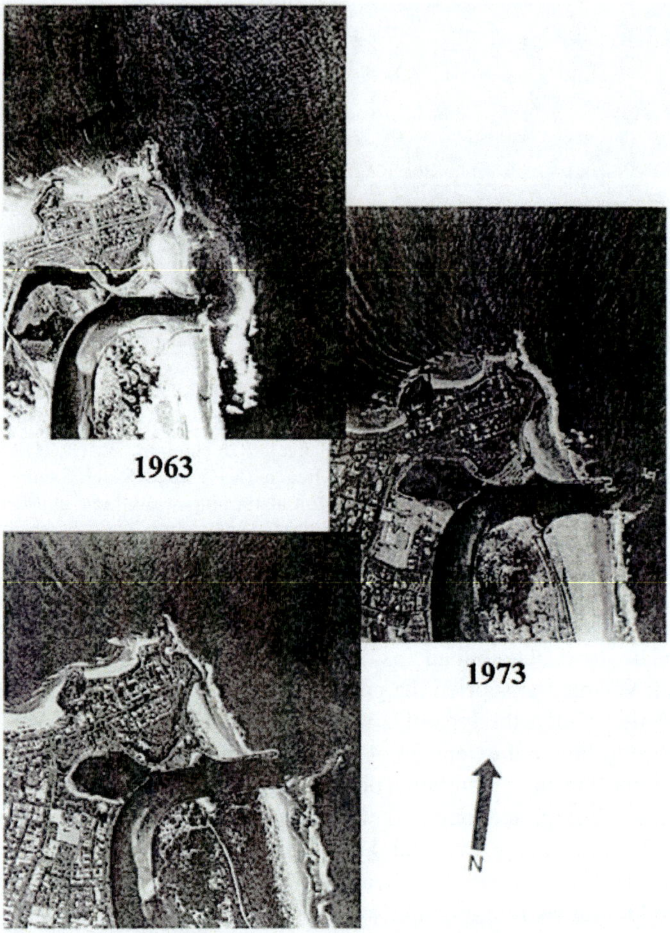

Fig. 1.2 Accretion of Letitia Spit and the shoaling of Tweed River mouth after training wall extension. The 1963 photo shows the training walls half built. The beach width reached to the end of the training walls and shoaling extended across the entrance. In 1973 (approximately 8 years after an extension was completed) the beach width had again accreted further out to the extent of the training wall, and shoaling across and within the entrance became evident once again. While dredging operations removed sediment from within the entrance, shoals continued to develop across the entrance due to longshore transport and accumulation on the ebb tidal delta (Source: Hyder Consulting Pty et al. 1997)

In 1964 the Queensland Government invited Delft Hydraulics Laboratory to investigate the ongoing erosion problems on the Gold Coast. The report outlined that longshore transport would be inhibited by the Tweed River training walls for 25 years, and that after this time the net longshore transport would be re-established, but the river training walls would not be effective in maintaining a safe entrance for navigation (Delft Hydraulics Laboratory 1970). It was then recommended that the New South Wales and Queensland Governments should

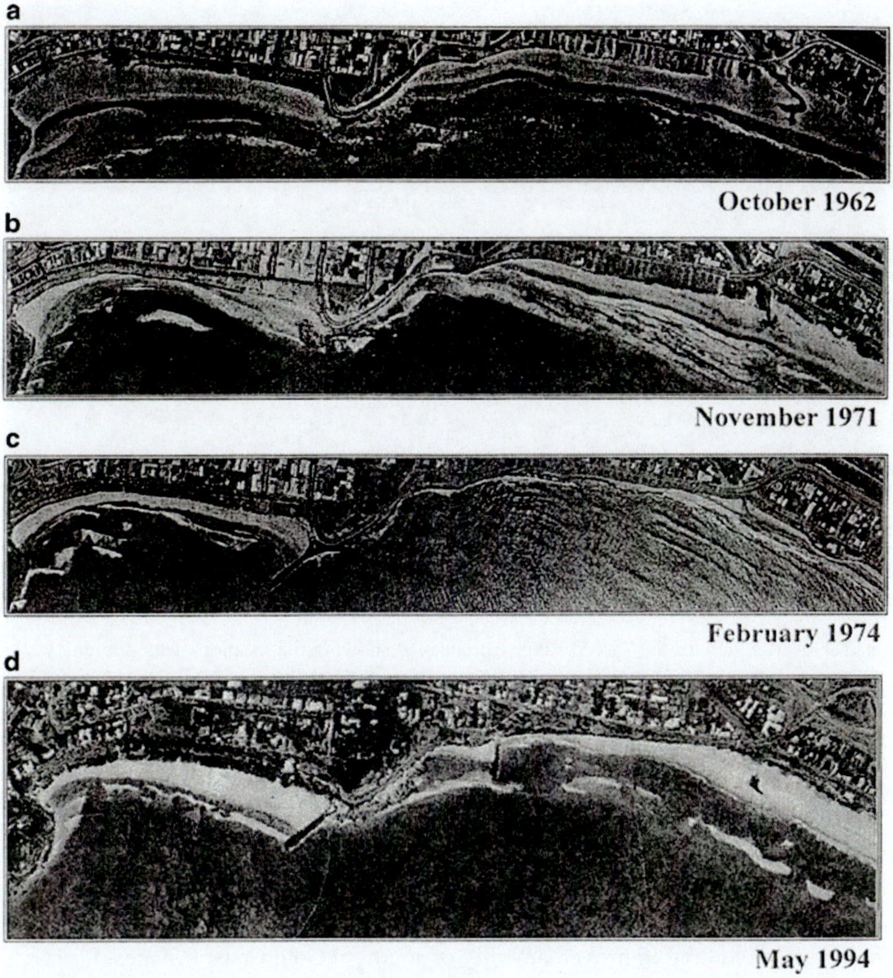

Fig. 1.3 The erosion of Coolangatta/Greenmount (*left*) and Kirra (*right*) beaches since the extension of the Tweed River training walls. In 1962 (**a**), before the training walls were extended, beach widths were considered to be in a 'natural' state. Storm events in the late 1960s severely eroded beaches (**b**). A rock groyne was built at Kirra Point to retain sediment at Coolangatta beach. However, depleted offshore beach nourishing supplies of sand could not replenish beach width after recurring storm events leading up to February 1974 (**c**), further beach width. While beach nourishment operations during the 1980s and early 1990s replenished beaches (**d**), erosion began again due to lack of natural sand supply (Source: Hyder Consulting Pty et al. 1997)

co-operate in the investigation of methods for managing the net loss of sand to southern Gold Coast beaches.

Following the Delft report and public concern over beach amenity, the New South Wales and Queensland Governments implemented a range of CZM strategies, culminating in the Beach Protection Act (QLD) in 1968 (Kay and Lester 1997) and the Coastal Protection Act (NSW) in 1979 (Watson and Lord 2005). Both state governments saw the need to act and to balance environmental aspects with

Fig. 1.4 Aerial view of the Tweed River entrance (**a**) showing the location of the training walls and sand pumping jetty (*bottom left*) and the jetty structure (**b**) (Source: NSW LPMA and QLD DERM 2011)

the need for economic growth. Detailed studies were undertaken to investigate the region's coastal processes, and it was confirmed that the Tweed River training walls were responsible for the reduction in sand supplied to southern Gold Coast beaches (Hyder Consulting Pty et al. 1997). However, each government had different objectives for the management of the river mouth – the New South Wales Government required the safe use of the Tweed River entrance, while the Queensland Government required improved amenity of southern Gold Coast beaches. It wasn't until the 1980s that both governments could agree to negotiate over implementing a permanent solution for the Tweed River entrance (Hyder Consulting Pty et al. 1997). Differing issues of concern, technical understanding and political philosophy meant the management strategies put into place by each State varied greatly (Kay and Lester 1997).

In 2001, a sand bypassing system was implemented to mimic the region's natural longshore transport process, delivering sand from Letitia Spit to the southern Gold Coast beaches while maintaining a safe entrance for navigation. The Tweed River Entrance Sand Bypassing Project (TRESBP) is a joint scheme between the New South Wales and Queensland Governments that is bound by the Tweed River Entrance Sand Bypassing Act 1998. A sand pumping jetty was built approximately 250 m south of the southern training wall (see Fig. 1.4) to collect and pump sand to four different outlets (Duranbah Beach, Snapper Rocks east and west, and Kirra Beach – see Fig. 1.1) (Boswood et al. 2001). Sand is pumped at a rate consistent with the net long-term average longshore transport rate of about 500,000 m^3 per year that would be transported north under natural circumstances (Castelle et al. 2009). A sink hole is created along the jetty where the pumps are located, trapping the sand and preventing it from reaching the southern break wall. The main objective of the system is to mimic the natural longshore transport of sediment that would occur if the Tweed River mouth was in its natural state.

To manage the effects of the ebb tidal delta, intermittent dredging operations are still required to maintain a safe water depth in the river mouth. Sediment collected from dredging is placed in dumping areas offshore from Duranbah, Rainbow Bay and Coolangatta beaches (Hyder Consulting Pty et al. 1997).

1.3 Problems That Have Arisen

TRESBP has been very successful in simulating the net northerly longshore transport of sand from Letitia Spit to the southern Gold Coast beaches. Although successful in meeting the aim of the project, the initial over-widening of the sand-receiving beaches has come under great public and media scrutiny. Rainbow Bay, Coolangatta Beach and Kirra Beach all saw dramatic rates of accretion, and initial over-pumping of sand was said to have compromised surfing, swimming, fishing, diving and beach use amenity (Castelle et al. 2009). This public perception could be considered questionable, as early photographs of Kirra and Coolangatta beaches show that at times beach width in the 1950s was comparative to that achieved after sand pumping began in 2001 (NSW LPMA and QLD DERM 2011). Littoral coasts are highly dynamic, and beach width invariably changes with cycles of weather conditions.

Operators of the project responded to public concern by stating there was a need to pump large quantities of sand to southern Gold Coast beaches to swiftly rebuild the badly eroded beaches (NSW LPMA and QLD DERM 2011). It was expected that winter storm conditions would disperse the build-up of sand quickly; however, this was not the case as the weather remained calm, which encouraged beach accretion (NSW LPMA and QLD DERM 2011; Surfrider Foundation Australia 2011).

This false anticipation had a profound effect on the Gold Coast's world-renowned southern surfing beaches. The amenity of beaches is extremely important to the Gold Coast, both socially and economically. In 2007, Gold Coast residents made 40 million trips to the beaches for recreational purposes, with annual expenditure estimated as up to $233 million (Lazarow 2009). Southern Gold Coast beaches are a part of the public domain, and public perception of beach width is what beach widths are 'now' rather than what they were in the past or will be in the future.

On a positive note, surf quality at Snapper Rocks has improved greatly since the bypass system was implemented. The 'superbank', a by-product of over-pumping, became highly regarded by recreational and professional surfers around the world (Lazarow 2006). However, surf at Coolangatta and Kirra beaches has deteriorated due to the large influx of sand and the 'superbank' (Lazarow 2006).

The erosion of Kirra Beach since the 1962 extension of the training walls made Kirra one of the world's most renowned surfing sites, with rides of up to 500 m long (Hyder Consulting Pty et al. 1997). Since the mid-1960s, the beach has not been in a natural state due to lack of longshore transport. The dramatic accretion of the beach after sand-pumping limited the surf break to a small longboard rider's wave that breaks off of the outer sand bank (Lazarow 2006). The 'Bring back Kirra' campaign organised by the local surfing community built momentum to pressure local and state governments to better manage TRESBP and return Kirra Beach to its famous profile (Surfrider Foundation Australia 2011). However, the objective of the sand by-passing system is to deliver sand to the southern Gold Coast beaches as would occur if the training walls were not there – not to maintain beaches with a favourable surfing condition.

In the early 1970s rock groynes were built to restore the beach width of Coolangatta and Kirra beaches. Now, with the implementation of TRESBP and the lack of adverse weather since, these groynes may be inhibiting the natural northward transport of the large quantities of pumped sand. This contentious topic has been left for the Gold Coast City Council to resolve (NSW LPMA and QLD DERM 2011).

An additional adverse effect of the increase of beach width at Kirra and Coolangatta beaches is the impact of these on Kirra reef, which lies a few 100 m off Kirra Beach. This was anticipated in the environmental impact study as an inherent consequence of replenishing Kirra Beach (Hyder Consulting Pty et al. 1997). Habitat diversity was compromised and benthic flora and fauna communities are now more susceptible to wave action and currents due to the shallower water depth, and the surf zone has been brought seaward and closer to the reef (FRC Environmental 2010).

With the expected impact of climate change resulting in increased cyclone activity within the region, the Gold Coast and Tweed areas are expected to experience enhanced environmental and economic damage (QLD Government 2011a, b). The implementation of the sand by-passing scheme to reduce sedimentation in the Tweed River entrance for safer navigation purposes may further expose the Tweed River and surrounding communities to increased wave and storm surge heights in the Tweed River estuary. The increased water depth of the entrance allows waves to penetrate further into the river mouth before shoaling and breaking occurs.

1.4 Present Day Circumstances – 10 Years On

In 10 years since the opening of the TRESBP, the project has transported 7,580,293 m^3 of sand from Letitia Spit to southern Gold Coast beaches (NSW LPMA and QLD DERM 2011). A deep and safe entrance for navigation has been retained and the severely eroded southern Gold Coast beaches have been widened. Although anticipated by operators, the initial operations brought community frustration over the effects on beach width from 'over'-pumping. Anticipated

Fig 1.5 Beach widths of Coolangatta Beach at 6 months (19/11/2001) (**a**), 5 years (8/3/2006) (**b**) and 10 years (1/3/2011) (**c**) after sand pumping began (Source: NSW LPMA and QLD DERM 2011)

stormy weather conditions did not occur, delaying sand dispersal and adding pressure to the local authorities.

Extensive monitoring programs have been implemented since the commissioning of the system in order to evaluate environmental impacts and to adjust the operation accordingly (Boswood et al. 2001). The pumping of sand has not incurred any major issues relating to the operation of the pumping system, nor have any major environmental impacts occurred.

The large influx of sand that entered Coolangatta Bay and Kirra Beach at the beginning of the sand pumping has now begun to disperse. During 2007 and 2008 approximately 500,000 m^3 of sand moved out of Coolangatta Bay, while 140,000 m^3 (see Fig. 1.5) moved away from Kirra Beach (NSW LPMA and QLD DERM 2011).

This trend is likely to continue under the influence of longshore transport and revised delivery rates until a dynamic equilibrium state is reached (NSW LPMA and QLD DERM 2011). The southern Gold Coast beaches appear to be returning to the variable profiles of the pre-1960s, before the training walls were extended (NSW LPMA and QLD DERM 2011). Surfing and beach amenity at Kirra and Coolangatta beaches have greatly improved, which has pleased the local surfing community behind the 'Save Kirra' campaign. However, the surfing community maintains pressure on local and state governments to ensure continuation of the trend of the southern Gold Coast beaches' return to their former (1960s) state.

The erosion of Kirra beach has also allowed for the recovery of Kirra Reef. Elevated bed levels from the greater-than-average sand transportation were chiefly responsible for the coverage (FRC Environmental 2010); sand have moved away and uncovered the rocky outcrops that form Kirra Reef. The aerial extent of Kirra Reef is still much less than before the sand pumping jetty was implemented in 2001 (FRC Environmental 2010). Although this uncovering was anticipated, the EIS did not anticipate the alterations to the ecology of the reef that increased wave and current actions would create. It has been said that fauna and flora assemblages are in decline, and that the reef has not yet reached physical or ecological equilibrium (FRC Environmental 2010).

1.5 Benefits

There has been much controversy on the implementation of TRESBP. Early modifications to the Tweed River mouth were done with one outcome in mind – to stabilise the river mouth to make the entrance safer for vessels navigating in and out of port. At the time, implications on the local coastal processes from building training walls (particularly the longshore transport of sand) was not known or considered – these effects only became evident many years later. Dredging operations took place to deepen the entrance and help replenish southern Gold Coast beaches; however, this did not provide a permanent solution (Hyder Consulting Pty et al. 1997).

The implementation of the sand pumping plant has acted to mimic the natural northerly movement of sand along the coast (Hyder Consulting Pty et al. 1997). The net long-term average longshore transport rate of sand is 500,000 m^3/year, but this varies accordingly to meteorological conditions – in low wave conditions, the transport rate could be as low as 100,000 m^3/year, while in severe storm conditions it could be as high as one million m^3/year of sand (Hyder Consulting Pty et al. 1997). The sand by-passing plant is a permanent system with greater flexibility to deliver sand at a rate closer to what would occur under calculated net longshore transport rates (Hyder Consulting Pty et al. 1997).

The location of the TRESBP jetty prevents most sediment reaching the southern training wall, which prevents sand bars forming off the end of the training wall, thereby stopping the entrance from becoming shallow and not allowing waves to break across the entrance. These improvements to the entrance mean boating accidents have rarely occurred since the project was implemented. Before this, boating accidents were frequent and often very serious (NSW LPMA and QLD DERM 2011). The entrance may still be impassable at certain times, as other aspects such as tidal flow and wind speed affect the entrance conditions (NSW LPMA and QLD DERM 2011).

Beach width and amenity of southern Gold Coast beaches have improved dramatically since sand was first pumped to them. Before the initial dredging stages of the Tweed River Sand By-passing project began, southern Gold Coast beaches were severely eroded (Hyder Consulting Pty et al. 1997). The beaches were highly susceptible to storm events, with recovery being extremely slow due to the lack of sand available for replenishment. The sand by-passing project now delivers sand to these beaches, mimicking the natural flow of sand, and the volume of sand and location of delivery can be altered in order to deliver sand where it is mostly needed (Hyder Consulting Pty et al. 1997), allowing these beaches to recover more quickly from storm events (Castelle et al. 2008). The beaches are now much wider, but not 'too wide', allowing greater use of the beach for local beach users, swimmers and surfers.

1.6 Future Management Issues

Long-term monitoring will help determine the pumping rates required to maintain beach amenity and to mimic natural supply. These may change with differing weather conditions and the number of storm events the region receives.

Many of the Gold Coast's beaches are not considered able to withstand extreme weather events (Castelle et al. 2008). A study of four Gold Coast beaches – The Spit, Narrowneck Beach, Broadbeach and Coolangatta Beach – showed that, despite numerous beach nourishment operations and coastal protection projects undertaken along the Gold Coast during the last 30 years, an increase in storm frequency and storms in quick succession could have devastating impacts (Castelle et al. 2008). It was also noted that beaches immediately north of the Tweed River Sand Pumping project are most likely to recover from extreme storm events (Castelle et al. 2008). However, dune vegetation on northern Gold Coast beaches – The Spit and Narrowneck – provided a stronger barrier to withstand against severe erosion from high wave events such as storms and cyclones. Coolangatta Beach also experienced the greatest erosion due to the wide beach width and waves not reaching the vegetated dune (Castelle et al. 2008). Accordingly, future management of TRESBP may need to include increasing vegetation on the beach dunes (of Coolangatta Beach and Kirra Beach) to further stabilise the beach and to help protect against high wave events.

Future monitoring and further studies are vital to determining whether the Kirra Beach groyne is required, and whether these areas would be better off with the groyne removed. A previous study undertaken on the Kirra groyne indicated that if longshore sand transport is kept at a natural rate, then beach widths should be similar to those of the 1960s whether the groyne is in place or not, as natural sand transport at that time was similar (Worely Parsons 2009). If the groyne was shortened, then Coolangatta Beach would become narrower and the transition zone (offshore bar) from Coolangatta Beach to Kirra Beach would also be narrower (Worely Parsons 2009). If any changes were to be made, past events indicate it is essential that in order to avoid adverse public backlash, the public should be fully informed and allowed to be part of the decision-making process.

1.7 Summary – Lessons Learnt

It is clear from the history of the Tweed River that in order to minimise environmental impact, natural environmental processes must be completely understood prior to any coastal modification. The underlying coastal processes are what define the state of the coastal system, and modifying any one of these processes can – as shown by the Tweed River mouth project – cause the greater coastline to deteriorate and become less resilient to external forcing such as adverse weather and future changes in climate.

When significant research and development have been undertaken to identify natural processes, engineering structures can be implemented within the natural environment. Modifying the environment is an expensive exercise that becomes more so when further projects are required to 'patch' up or correct existing developments. Littoral coasts are highly dynamic and understanding the effects modifications have on the coastal zone will become more apparent after each engineering development has been put into place. The success of further coastal zone management strategies will depend on what has been learned from previous practices. Thus, further monitoring and study is vital to the future management and operation of such strategies.

The current Queensland Coastal Management Plan incorporates shoreline erosion and practices that support and maintain coastal processes through its policies (QLD Government 2011a, b). Coastal regions are considered hazardous if affected by a sea level rise of 0.8 m and a 10% increase in cyclone intensity. Under the current Queensland coastal plan, future coastal protection works must demonstrate the need to protect and incorporate beach nourishment so longshore transport is not impeded. The Queensland Government now considers erosion control structures such as rock walls or groynes as last resort measures.

The lessons learned from works at the Tweed River mouth led to the successful operation of the sand by-passing project at the Gold Coast Seaway. The need for training of the Nerang River mouth in 1984 (Delft Hydraulics Laboratory 1970) (which became the Gold Coast Seaway) was much the same as with the Tweed River mouth – to provide safe entrance for navigation in and out of the Gold Coast Broadwater. From their experience with the Tweed River mouth, managers saw it was vital to consider the implications of such a project outside of the initial project objective, and the extent to which that interruption of coastal processes (such as longshore transport and ebb tidal deltas) can incur. The successful operation and management of the Gold Coast Seaway ultimately led to the Tweed River Entrance Sand By-passing Project.

References

Boswood P, McMahon J et al (2001) Environmental management of the construction and operation of a sand bypassing system at the Tweed River Entrance. In: Coasts & ports 2001: Proceedings of the 15th Australasian coastal and ocean engineering conference, the 8th Australasian port and harbour conference, Institution of Engineers, Australia, pp 322–329

Castelle B, Le Corre Y et al (2008) Can the Gold Coast beaches withstand extreme events? Geo-Mar Lett 28(1):23–30

Castelle B, Turner IL et al (2009) Beach nourishments at Coolangatta Bay over the period 1987–2005: impacts and lessons. Coast Eng 56(9):940–950

Cullen P (1982) Coastal zone management in Australia. Coast Manag 10(3):183–212

Delft Hydraulics Laboratory (1970) Gold Coast, Queensland, Australia – coastal erosion and related problems, vol 1. Delft Hydraulics Laboratory, Delft

FRC Environmental (2010) Tweed River Entrance Sand Bypassing Project (TRESBP) Kirra reef ecological monitoring 2010, Wellington Point

Hyder Consulting Pty Ltd, Patterson Britton Partners Pty Ltd et al (1997) Tweed river sand bypassing project permanent bypassing system – environmental impact statement, Southport

Kay R, Lester C (1997) Benchmarking the future direction for coastal management in Australia. Coast Manag 25(3):265–292

Lazarow N (2006) Wave quality on the Southern Gold Coast. Gold Coast shoreline management review information sheet 6. Griffith Centre for Coastal Management, Gold Coast

Lazarow N (2009) Using observed market expenditure to estimate the value of recreational surfing to the Gold Coast, Australia. J Coast Res SI 56 (Proceedings of the 10th international coastal symposium, Lisbon), pp 1130–1134

NSW LPMA and QLD DERM (2011) Tweed River Entrance Sand Bypassing Project – environmental management. Retrieved 12 Mar 2011, from http://www.tweedsandbypass.nsw.gov.au/

Özsoy E (1986) Ebb-tidal jets: a model of suspended sediment and mass transport at tidal inlets. Estuarine Coastal Shelf Sci 22(1):45–62

QLD Government (2011a) Climate change in the Southeast Queensland region. Department of Environment and Resource Management, Brisbane

QLD Government (2011b) Queensland coastal plan. Department of Environment and Resource Management, Brisbane

Surfrider Foundation Australia (2011) Kirra Point.org. Retrieved 12 Mar 2011, from http://www.kirrapoint.org/default.asp

Watson P, Lord D (2005) Coastline management in New South Wales. Over-regulation or a sound investment in the future? Coasts and ports: coastal living – living coast, Institution of Engineers, Australia

Worely Parsons (2009) TRESBP – Kirra groyne effects study. Land and Property Management Authority, New South Wales

Chapter 2
Adelaide Beach Management 1836–2025

Andrew D. Short

Abstract The 28 km long Adelaide coast consists of a near continuous sandy beach and dune system, crossed by a few small outlets and terminating at the northern Adelaide Outer Harbor. Since European settlement in 1836 the dunes have been largely leveled and developed, six jetties built across the beach-nearshore, the outlets have been trained with breakwaters, two breakwaters protect marina developments and much of the backbeach has been armored with seawalls, usually following erosion events. After more than a century of *ad hoc* beach management, a review of the coast in the 1960s lead to the establishment of a Coast Protection Board in 1972 and the beginning of coordinated management of the entire system. Sand recycling and nourishment has become the major management tool, together with better designed seawalls, dune restoration and management, improved water quality and beach monitoring. The history of management of the Adelaide coast provides an excellent example of the transformation in coastal management that has occurred in Australia and elsewhere, as solely hard engineering has given way to a range of options both hard and soft, all designed to suit the coastal system in question.

2.1 Introduction

Adelaide, the capital of South Australia, has a population of 1.3 million. It was settled in 1836 with the city well-planned on a grid pattern and radial road network that dominates the landscape today. The city centre is adjacent to the small Torrens River, 10 km east of the coast. The coast extends for 28 km from Seacliff in the south to the entrance to the Outer Harbor in the north. It consists of low to moderate energy wave-dominated beaches, grading with decreasing wave energy north into

A.D. Short (✉)
Coastal Studies Unit, School of Geosciences, University of Sydney, Sydney, NSW 2006, Australia
e-mail: andrew.short@sydney.edu.au

J.A.G. Cooper and O.H. Pilkey (eds.), *Pitfalls of Shoreline Stabilization: Selected Case Studies*, Coastal Research Library 3, DOI 10.1007/978-94-007-4123-2_2,

tide-modified beaches. These were originally backed along most of the coast by dune systems that reached 10–12 m in height and extended 200–300 m inland as two to three parallel ridges, widening to over a kilometer along the northern few kilometers. The beaches and dunes were supplied by northerly longshore sand transport and deposition throughout the Holocene.

Today the entire coastline is occupied by residential and some commercial development and beach-based recreational activities. Coincident with early development in the nineteenth century was the need for seawalls, which continued and expanded in a haphazard way well into the twentieth century. Since 1972, when the Coast Protection Board (CPB) was established, there has been an increasingly coordinated approach to the coast's management based on a much improved understanding of the coastal processes, the adverse impact of past structures, and a recognition of maintaining the public amenity values of these popular urban beaches. Today a combination of beach sand nourishment, sand recycling and sand bypassing is being used to maintain the beaches. This paper reviews evolution of and past and present approaches to managing Adelaide's beaches.

2.2 Evolution of Adelaide Coast

Adelaide is located on the eastern shore of St Vincent Gulf at latitude 34°S (Fig. 2.1), exposing it to a Mediterranean climate with warm to hot dry summers and wetter cooler winters. Although the eastern gulf shore is exposed to the prevailing westerly winds and high Southern Ocean swell dominates the open coast, the location of Adelaide 100 km into the gulf together with the blocking effect of Kangaroo Island across the gulf entrance results in only occasional low ocean swell reaching the eastern gulf coast. Consequently both wind driven westerly seas and attenuated and refracted ocean swell contribute approximately equally to the local coastal processes and sediment transport (Coastal Engineering Solutions 2004). Wind waves with a 4–6 s period arrive predominantly from the southwest. They average less than 1 m, though they can reach up to 2.6 m during storms and storm surges of up to 1.5 m have been recorded. Swell occurs with periods of 12–16 s, arriving from the southwest with heights of less than 1 m. Summer seabreezes also arrive from the west while northwesterly storms may arrive in winter. The net result is northerly longshore sand transport. Spring tides range from 1.3 m at the gulf entrance increasing northward to 2.4 m at Adelaide's Outer Harbor and 3 m in the northern gulf apex. The tides drive a 0.2–0.3 m/s alternating north–south tidal currents, which parallel the shore.

The gulf was flooded by the Holocene sea level transgression that reached around present sea level about 6.5 ka. The flooding of the shallow gulf floor reactivated siliceous sediments moving them to and northward along the coast. Bowman and Harvey (1986) dated the beach and dune ridges of the northern Adelaide coast and reconstructed the Holocene palaeo-shorelines of the LeFevre Peninsula, which occupies the northern 14 km of the coast. They observed rapid

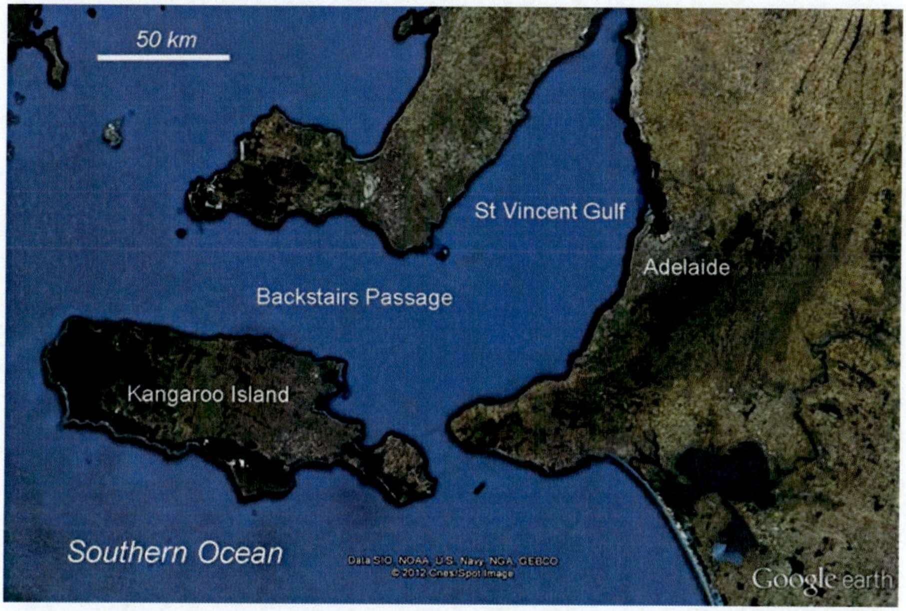

Fig. 2.1 Google image of St Vincent Gulf with Adelaide located on its eastern shore and the entrance partially blocked by Kangaroo Island (Source: Google Earth.com)

northern movement of sand between 7.5 and 5.5 ka, followed by a reduced rate of sediment supply and a change in coastline orientation contributing to spit recurvature and a flared beach-ridge pattern. Figure 2.2 illustrates the northward growth of the coast as a series of both seaward prograding dune ridges and northward prograding recurved spits and ridges. The sand has been transported into an increasingly lower energy environment and terminates in a series of recurved spits that form the northern end of the peninsula. To the north of the peninsula are extensive low energy tide-dominated tidal flats, mangroves and inner shelly beach ridges.

The Holocene produced a series of dune ridges 200–300 m wide and 10–12 m high, widening to 1–2 km along the northern several kilometre of coast. The 28 km-long beach receives its highest energy in the south gradually decreasing to the north with distance into the gulf, together with a shoaling nearshore zone from Semaphore northwards, in lee of the Wonga Shoal, and changing shoreline orientation. Along much of the southern 20 km between Seacliff and Semaphore the beach consists of a wave-dominated low tide terrace which is cut by occasional shallow north skewed rip channels, fronted by one to two shallow sand bars, the crests at times exposed at spring low tide (Short 2001). The southernmost section suffers a net loss of sand as longshore transport to the north, averaging 50,000 m^3/year, exceeds sand arriving around cliffs south of the area, resulting in shoreline retreat (Harvey and Bowman 1987). North of Semaphore, wave energy decreases and three shore parallel bars gradually dissipate into a wide low gradient inter to sub-tidal terrace, as wave energy becomes insufficient

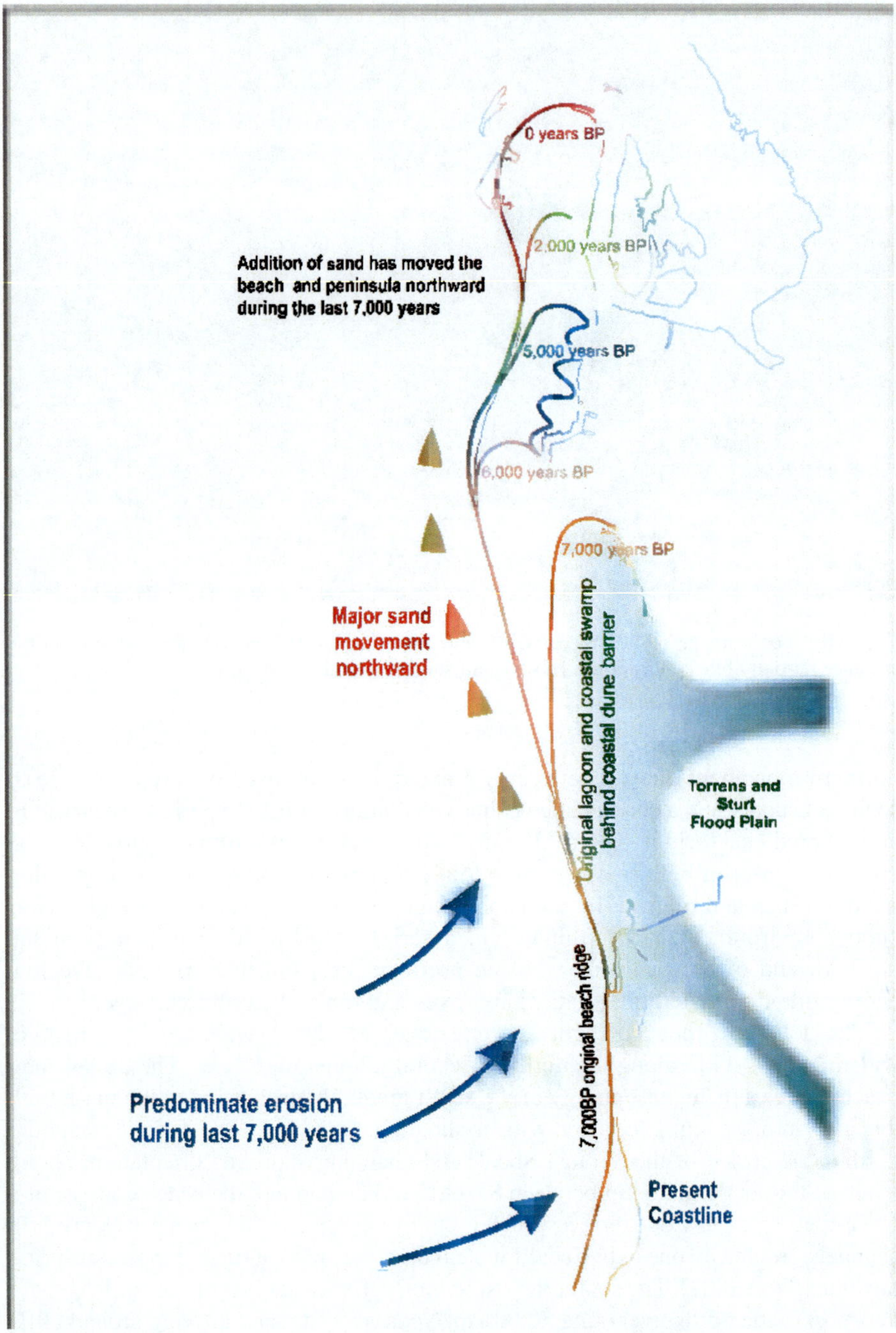

Fig. 2.2 Building out of the northern Adelaide coastline over the last 7,000 years (Based on Bowman and Harvey 1986; Source: DEH 2005)

to form and maintain the bar morphology, and tide-modified conditions dominate. Along the entire coast the individual bars run shore parallel for up to 2–3 km and are aligned eschelon to the shore. Each bar commences at a point of attachment and gradually moves offshore and dissipates as another inner bar replaces it. In the intervening shallow troughs north-trending mega ripples are maintained by both wave and tide-driven northerly currents. Seaward of the bars bare sand now extends seaward for several kilometres offshore where seagrass meadows of *Posidonia sinuosa, Amphibolis antarctica, P. angustifolia, Heterozostera tasmanica* and *Halophila australis* are encountered out to a depth of 18 m. The inshore seagrass meadows have been heavily impacted and reduced in recent decades.

Behind the beaches the incipient foredune is vegetated by a mixture of native and exotic herbs, grasses and shrubs including the exotic sea rocket *(Cakile maritime)*, dune onion grass *(Trachyandra divaricarta)* and sea wheat grass *(Thinopyrum junceiforum)*, together with the native *Spinifex hirsutus*. The foredunes are covered by pioneer woody plant species including *Olearia axillaris* and *Acacia longifolia* var. *sophorae*.

2.3 Nineteenth and Twentieth Century Shoreline Development

Usage and development of the Adelaide coast began soon after settlement. Glenelg, where the first settlers stepped ashore in 1836 was formally settled during the 1840s, with a 380 m-long jetty constructed in 1857. Major beach erosion occurred in 1943 and the old jetty was destroyed in 1948, while a shorter jetty exists today. Residential development expanded slowly along the Adelaide coast during the nineteenth century, but most of the shore developed between 1900 and 1950 (Fig. 2.3).

In most cases development, property boundaries and infrastructure extends to the rear of the beach and covers the dunes. In addition the dunes were mined, leveled for development and used to fill the swampy back barrier depression for development, including Adelaide Airport. This resulted in both removal of some dunes and quarantining of much of the remaining dunes from the beach system. Furthermore the response to property erosion was often the construction of seawalls, further alienating the dunes. From the 1850s onwards timber and concrete seawalls were built at several locations together with long jetties at Largs Bay, Semaphore, Glenelg, Henley and Brighton, with seawall bordering most of the jetties. Between 1948 and 1972 concrete, timber, gabion and rubble seawalls were constructed along sections of the coast, many replacing earlier failed walls. By the 1970s the Adelaide coast had 12 km of seawall, six jetties, six groynes and two sets of training walls-breakwaters (Fig. 2.4).

At the northern end of the system a breakwater, built in 1903–1905 to train the entrance to the Outer Harbor, trapped the northerly longshore sand transport, prograding the shoreline by 250 m. In 1974 the North Haven marina and training walls were built 1.5 km to the south, which in turn prograded the updrift shoreline by 200 m by 2005.

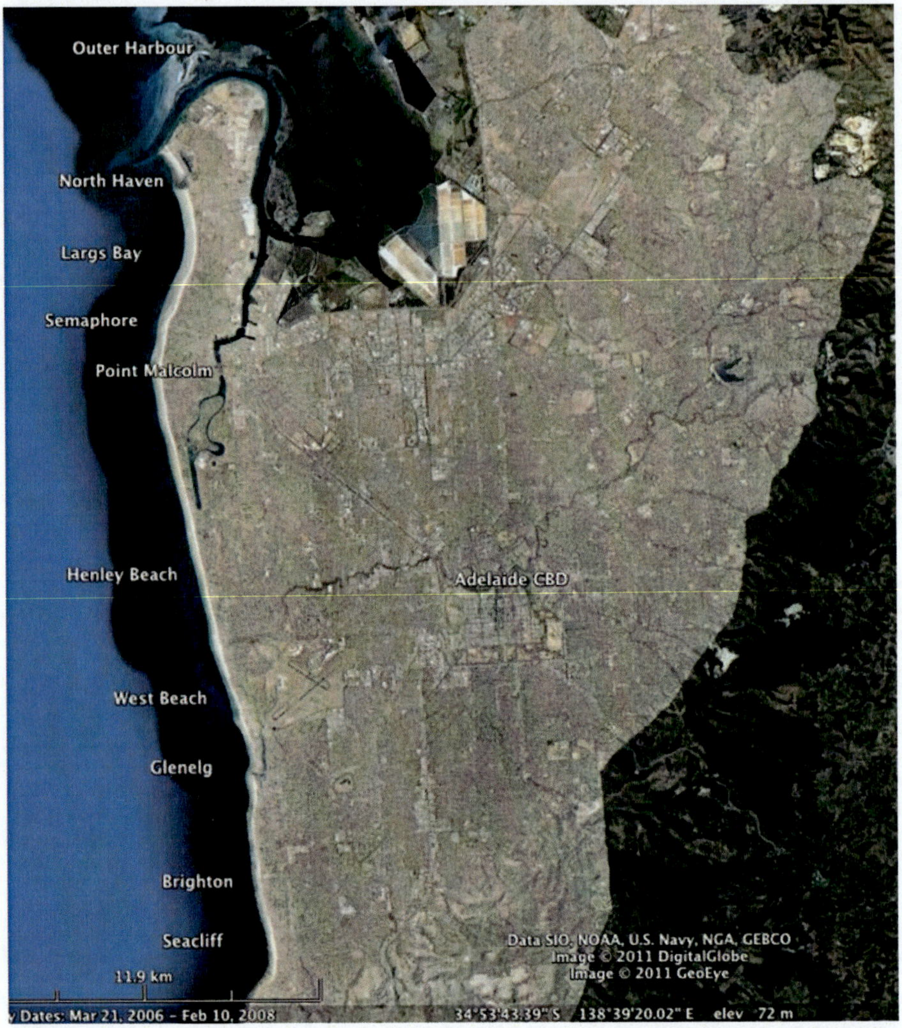

Fig. 2.3 The Adelaide metropolitan coastline (Source: Google Earth.com)

Three other structures also impede longshore sand transport. In 1936 the Torrens River outlet was constructed and the small delta it produced has acted as a hydraulic groyne, trapping sand and prograding the shoreline by 100 m. At West Beach 2.5 km to the south, a detached breakwater, built in 1998, provides shelter for a boat ramp and yacht club (Fig. 2.5a). Two km further south the mouth of the Patawalonga Creek at Glenelg has a long history of structures. Earlier damaged structures were replaced with a groyne in 1964, that was extended and augmented in 1997 as part of a marina development (Fig. 2.5b). Both the Glenelg and West Beach facilities had sand bypassing operations, now replaced by sand recycling.

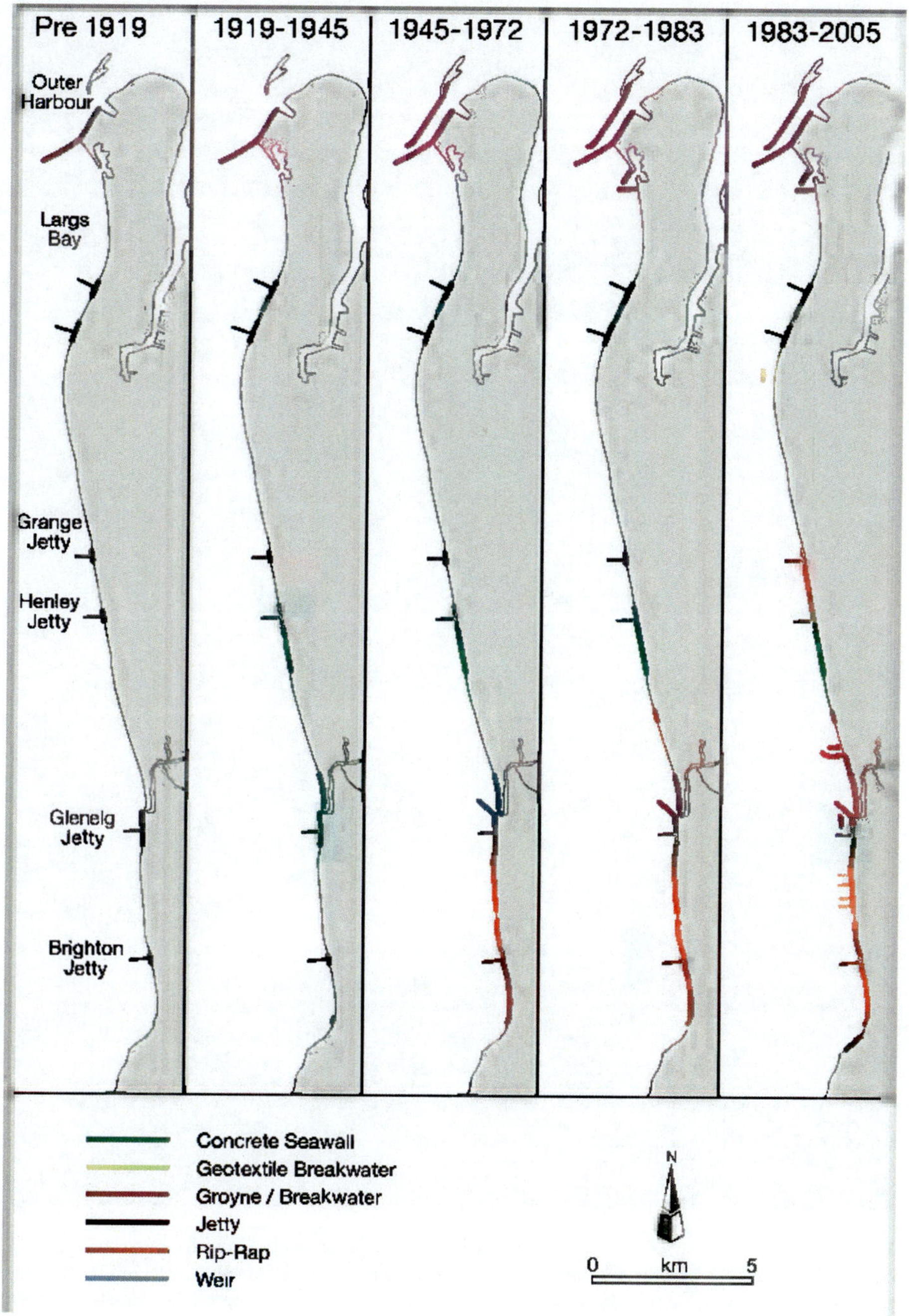

Fig. 2.4 Construction of coast protection structures up to 2005 (Source: DEH 2005)

Fig. 2.5 (**a**) West Beach breakwater and marina (*left*), with sewerage outfall under construction (*right*) and Adelaide Airport behind; (**b**) the trained mouth of the Patawalonga River at Glenelg, together with the marina development. Note the bypassing dredge in lee of the detached breakwater (Photos: S. Daw)

2.3.1 Stormwater and Wastewater Outlets and Impacts

Adelaide's stormwater is discharged though three major outlets at Barcoo, Torrens and the Port estuary, together with many smaller outlets. There is little sediment retention at the outlets. Wastewater is treated at Bolivar, adjacent to the Outer Harbor and Glenelg, and at Christies Beach 12 km south of Seacliff and discharged into the gulf at these locations. The impact of the wastewater and past sludge discharges has been to enrich nutrients along the coast, which has lead to algal growth on and dieback of the seagrass meadows. During the past 50–60 years it is estimated 25% (5,200 ha) of seagrass has been lost along the Adelaide coast. This in turn has resulted in mobilization of the once vegetated and stable near-shore sands. Currently it is estimated that 80,000 m^3/year of sand is eroded from the former seagrass beds and moved shoreward. In addition the seagrass loss has reduced wave attenuation and refraction across the former meadows so that waves have greater sediment transport potential when they reach the shore. Models have estimated that longshore sand transport has increased by 10% in affected areas.

2.3.2 Subsidence and Sea Level

Groundwater extracted from the Adelaide coastal plain has caused subsidence in the northern part of the coast, while tectonic movement is a more widespread factor. Subsidence ranges from low at Marino to 2 mm/year at Pt Adelaide. Concurrently the Pt Stanvac tide gauge has recorded a rise in sea level of 5 mm/year. Hence relative sea level is rising along parts of the coast at more than 5 mm/year.

In summary, 130 years of European settlement and occupation of Adelaide has had the following changes along its coast.

- Covering, removal, leveling and mining of the coastal dunes;
- Alienation of most of the dunes from the beach system;
- Armoring of 12 km (43%) of the back beach and frontal dune, including some sections with little or no beach fronting the seawall (Fig. 2.6);
- Construction of six long jetties across the beach and bar systems;
- Construction of three impediments to sand movement in the central coast;
- Construction of a 3 km long terminal breakwater at the downdrift sediment terminus;
- Discharge of stormwater and wastewater along the coast resulting in substantial seagrass dieback;
- Sand remobilization and shoreward transport and seabed erosion due to seagrass dieback;
- Increased northerly sand transport due to seagrass dieback;
- Active shoreline retreat at Seacliff;
- Substantial loss of public beach amenity from beach narrowing or armored backshore.

Fig. 2.6 (**a**) Seawall at West Beach looking north to Henley Jetty; and (**b**) partly sand covered seawall at Glenelg Beach (Photos: A. D. Short)

- possible increase in public risk owing to higher waves, stronger longshore currents, and deeper and narrower surf zone.
- Occurrence of both ground subsidence and rising sea level.

2.4 The Evolution of Contemporary Beach Management

For the first 130 years the Adelaide coast was developed with little knowledge of beach-dune systems and managed in response to storm damage and erosion events. This culminated in severe erosion in the 1950s and 1960s when a series of major storms occurred resulting in extensive seawall construction. In response, the University of Adelaide Civil Engineering Department was commissioned in 1965 to study and find a solution to the problems.

2.4.1 Culver Report 1970

The University of Adelaide report, known as the Culver Report (Culver 1970) was completed in 1970. Its major recommendations were:

1. Stop any further development on the beach and dune;
2. Nourish or protect the most at risk areas, particularly Brighton, Glenelg and Henley;
3. Maintain all known coastal reserves of sand for preservation for future possible use;
4. Establish a beach protection authority; and
5. The authority to begin the detailed appraisal of the best restorative measures.

2.4.2 Coast Protection Board 1972

In 1972 the *Coast Protection Act* was passed and the Coast Protection Board was established. The 4,792 km long South Australian coast was divided into seven Coastal Protection Districts, of which Adelaide the smallest, remains the most intensively managed.

The Board's work along the Adelaide coast encompassed beach replenishment as the main means of shoreline and dune protection, coupled with seawall maintenance and dune protection. Beach replenishment involved moving sand from the northern terminus areas back towards the southern end of the system. Replenishment commenced in 1973 and between 1977 and 1984 an average of 105,000 m^3/year of sand was trucked from Semaphore and Glenelg to the southern central end of the system at Brighton, Glenelg North and West Beach (Fig. 2.7).

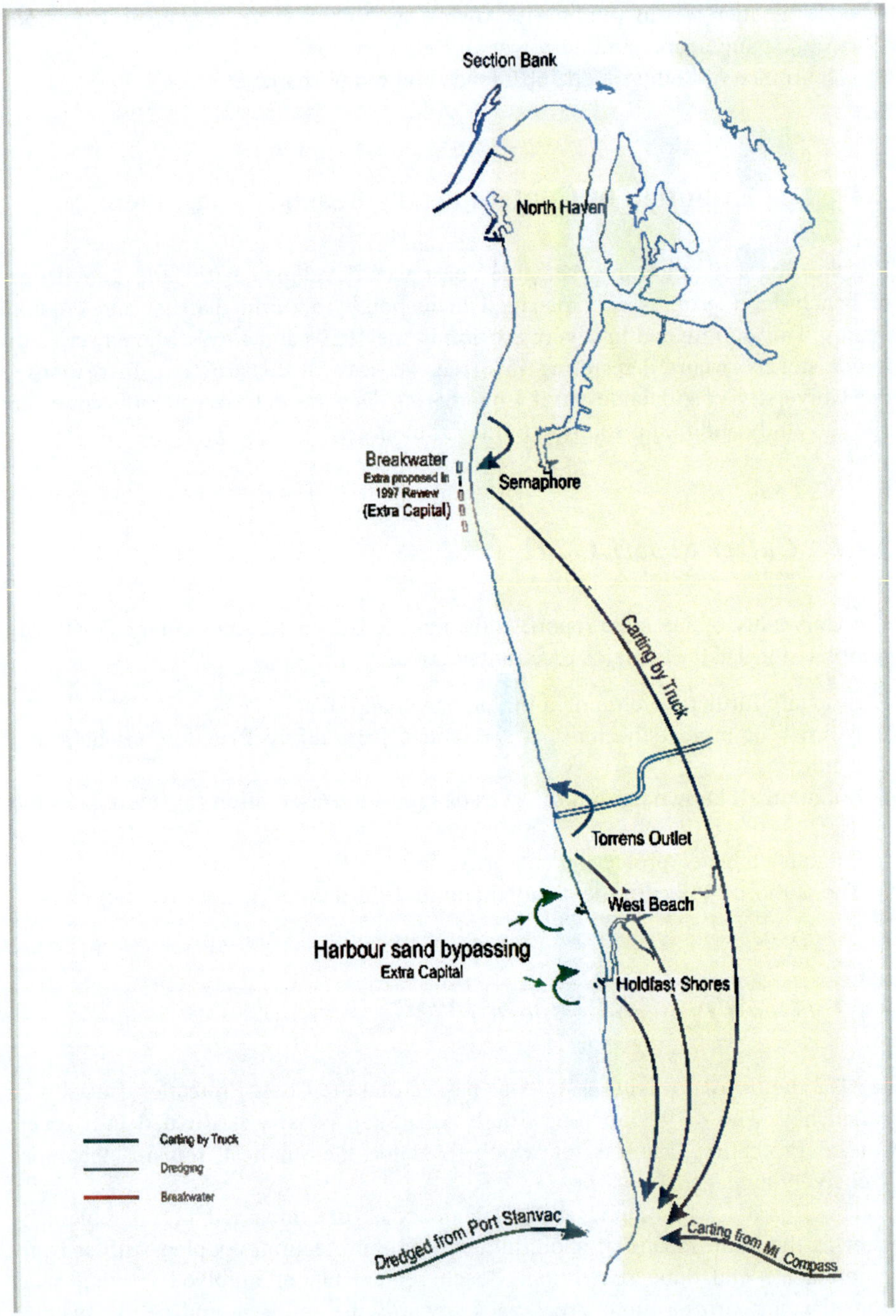

Fig. 2.7 Summary of the existing beach management strategy (Source: DEH 2005)

Seawall construction was also improved after 1970 with new sloping walls built of rip-rap that are more wave absorbent and designed to minimize undermining failure. Many of the new walls were built to improve and/or extend existing walls. Figure 2.4 illustrates the extent of seawall construction up to 2005. Dune stabilization, involving planting and fencing, was undertaken to lessen sand blowing inland, as well as to stabilise the backshore-fordune area.

2.4.3 *Adelaide Coast Protection Strategy Review 1984*

The Adelaide Coast Protection Strategy Review (Coastal Management Branch 1984; Wynne et al. 1984) reviewed the beach replenishment scheme, as well as examining potential offshore sand sources. Following a review of options for coastal protection including do nothing, seawalls, groynes and detached breakwaters, existing measures and massive nourishment, it recommended continuation of the existing strategy of replenishing 100,000 m^3 sand annually, and replacing existing seawalls where necessary.

2.4.4 *Metropolitan Coast Protection District Management Plan 1985*

In 1985 the Coast Protection Board (1985) released the *Metropolitan Coast Protection District Management Plan,* as part of its systematic planning for each of the State's coastal protection districts. This plan covered the greater Adelaide coast from Sellicks Beach in the south to Port Gawler in the north (Fig. 2.3). Two major issues identified by the plan were the pollution of coastal waters by wastewater, which had adversely affected the seagrass meadows and mangroves; and the need to maintain and improve the coastal sand dunes. To safeguard the latter all proposed developments in the beach-dune area had to be carefully assessed and in many cases prohibited.

2.4.5 *Review of Alternatives for the Adelaide Metropolitan Beach Replenishment Strategy 1992*

Coastal Management Branch (1992) undertook a review of the beach replenishment strategy including an assessment of the impacts of the existing strategy and an investigation of offshore sand sources. Based on a beach monitoring program commenced in 1975 it found up to 160,000 m^3 of beach replenishment sand was required annually in order to maintain beaches to 1977 levels. The review recommended that a 2-year replenishment cycle equivalent to 160,000 m^3 annually be initiated, with the sand from existing replenishment sources as well as dredged from offshore. This strategy was accepted and remains in place to this day.

2.4.6 Review of the Management of Adelaide Metropolitan Beaches, 1997

An independent group (of which the author was a member) was appointed in 1997 to review the management of the Adelaide beaches. It made three major recommendations (DENR 1997)

- Recreational benefits be given due regard in State Government budgeting and in providing grants to local councils;
- Further study into seagrass loss be urgently undertaken; and
- More offshore sand be found for beach replenishment as a matter of urgency.

Each of these have since been acted upon.

2.5 Adelaide's Living Beaches 2005

The most recent report on management of the Adelaide coast is the *Adelaide's Living Beaches: A strategy for 2005–2025* (Coastal Protection Board 2005). This technical and strategic report reviews the past history of management and details the ongoing management required to maintain the beaches to 2025 and beyond.

The present strategy is based around five actions. The main objective being managing sand on the beach through continued replenishment, including recycling trapped sand (from north to south using a slurry pipeline) within sand management cells, improving integration of the sand management at harbors with the replenishment program and nourishing beaches with sand from external sources as summarized in Fig. 2.7. The strategy includes maintaining seawalls as terminal structures for erosion during the extreme storms, and continuing beach monitoring and dune management activities. The following section provides a more detailed review of these five strategies.

2.5.1 Beach Replenishment

Figure 2.8 plots the past and present sediment transport potential rates along the coast determined from modelling. It shows the increase over the past 100 years owing to seagrass dieback, as well as northerly decrease as wave energy decreases.

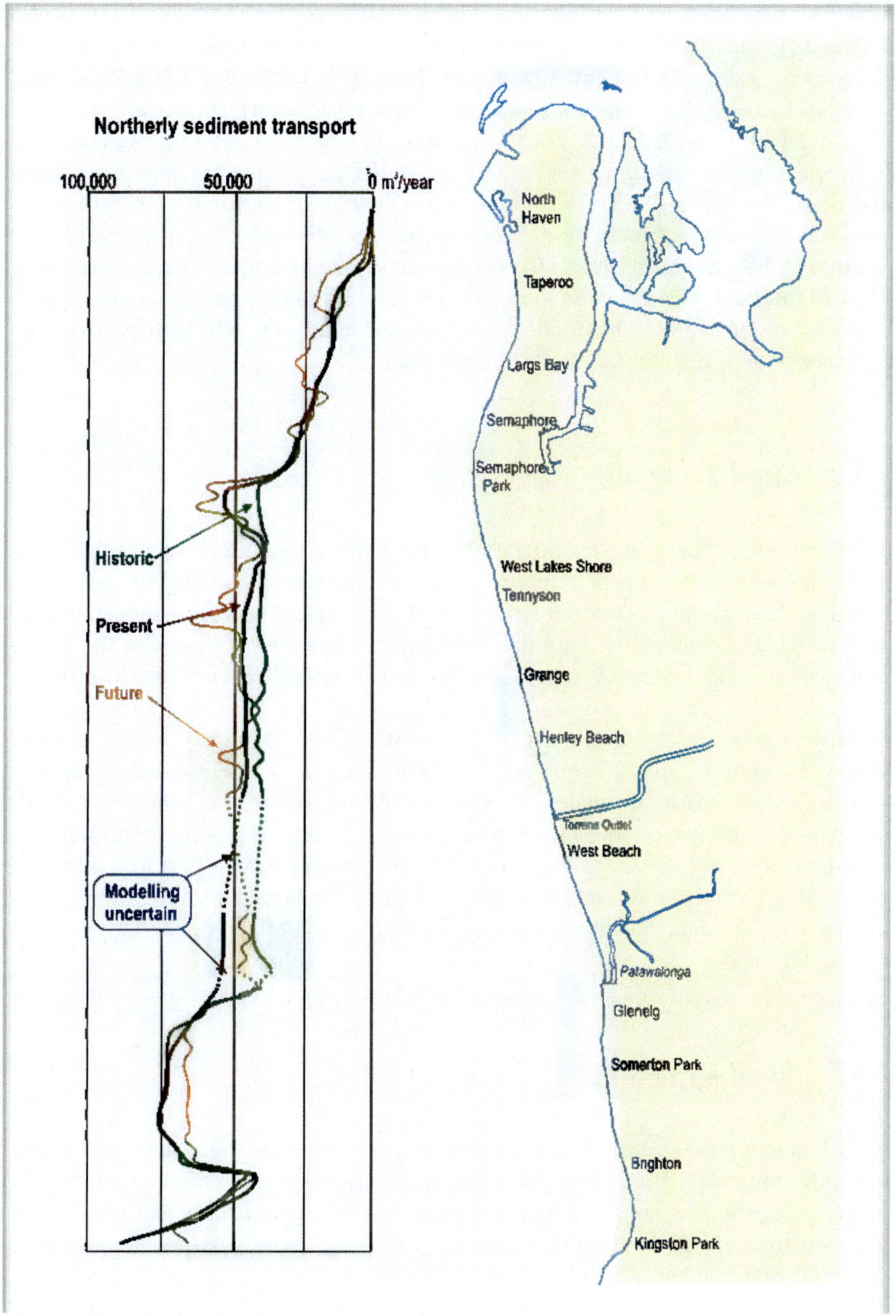

Fig. 2.8 Sediment transport rates for 100 years ago, present conditions and predicted for 2050 (Source: DEH 2005)

Whereas 100 years ago the rate was about 40,000 m^3/year, it is now estimated between 50 and 70,000 m^3/year, with the updrift system unable to supply this rate. In order to redress the negative sediment budget, between 1973 and 2005 over 2,500,000 m^3 of sand was recycled both by sand trucking (2,163,000 m^3) and dredging within the Adelaide beach system (450,000 m^3). The sand was sourced from the northern terminus of Semaphore and Point Malcolm, and the central updrift accumulations at Torrens Outlet and Glenelg. In addition 1,500,000 m^3 of sand was brought in from external sources, mostly dredged from offshore of Port Stanvac (1,140,000 m^3), while 370,000 m^3 was derived from external land sources. Most of the sand was placed towards the south at Brighton and Glenelg (Fig. 2.7). The cost of the 4,000,000 m^3 of sand replenishment was $36 million over the 32 years, averaging about $1.2 million per year.

2.5.2 *Sand Trapping*

Sand trapping occurs at several points along the coast either naturally at the downdrift terminus from Semaphore north or intentionally owing to structures. Existing sand trapping structures have been used to define seven sand management cells along the coast, within which sand is 'recycled' to provide the beach replenishment necessary to maintain beaches and coastal protection. In order to concentrate and control the downdrift trapping in 2005 a 200 m long, shore parallel detached breakwater, with a crest of 1 m AHD, was constructed 200 m seaward of the shoreline at Semaphore (Fig. 2.9). The purpose of the breakwater was to trap sand south of the normal sand terminus at Semaphore jetty, thereby providing an accessible sand store closer to the southern recycling point, resulting in a saving on trucking distances and costs. The breakwater is also part of a strategy, which if effective could see a breakwater field constructed south of the first breakwater. To date the breakwater has proved very effective in trapping and storing the sand.

2.5.3 *Sand Bypassing*

Sand bypassing was formerly undertaken at four locations along the coast where structures impede natural longshore transport. This is now mainly replaced by recycling of trapped sand to southern beaches. At Glenelg a detached breakwater was constructed off the southern training wall to purposely trap sand, which is collected and trucked south to Brighton beaches. However, sand and seagrass entering the entrance channel is dredged and pumped to the northern side of the entrance (Fig. 2.5b) at a rate of 54,000 m^3/year. Two kilometres to the north at the Adelaide Shores boat ramp at West Beach, sand is trapped against the southern detached breakwater and in the marina at a rate of about 90,000 m^3/year. This sand

Fig. 2.9 The 200 m long detached Semaphore breakwater constructed in 2005. Note the sand accumulation between the shore and breakwater, and deep rip channel exiting north of the breakwater (Source: GoogleEarth.com)

is carted south to Glenelg North beach or dredged to immediately north of the marina. In addition 6,000 m^3 of seagrass wrack accumulates each year, which is occasionally (when it interferes with sand carting) trucked north and placed on the dunes as a buffer. At the Torrens Outlet, another 2.5 km north, sand trapped on the southern side is carted to the south at a rate of about 20,000 m^3/year. Finally, sand is trapped along the southern breakwater at North Haven and eventually moves into the marina channel, where approximately 20–30,000 m^3 a campaign is periodically dredged from the channel. However, this sand is very fine and unsuitable for beach nourishment. The 2005 strategy includes the replacement of sand carting to recycling sand with a pipeline system to pump the sand in a slurry with seawater.

2.5.4 Seawalls

Seawalls have been constructed along 14 km of the coast since the 19th C (Figs. 2.4 and 2.6). While they are considered a 'last line of defence' they do impound the backing dune sand, as well as lower adjacent beach levels and exacerbate erosion at their ends where exposed to wave attack. Today seawalls are constructed to a design height of 4.2 m AHD to prevent overtopping, with a toe at -1.25 m AHD to prevent

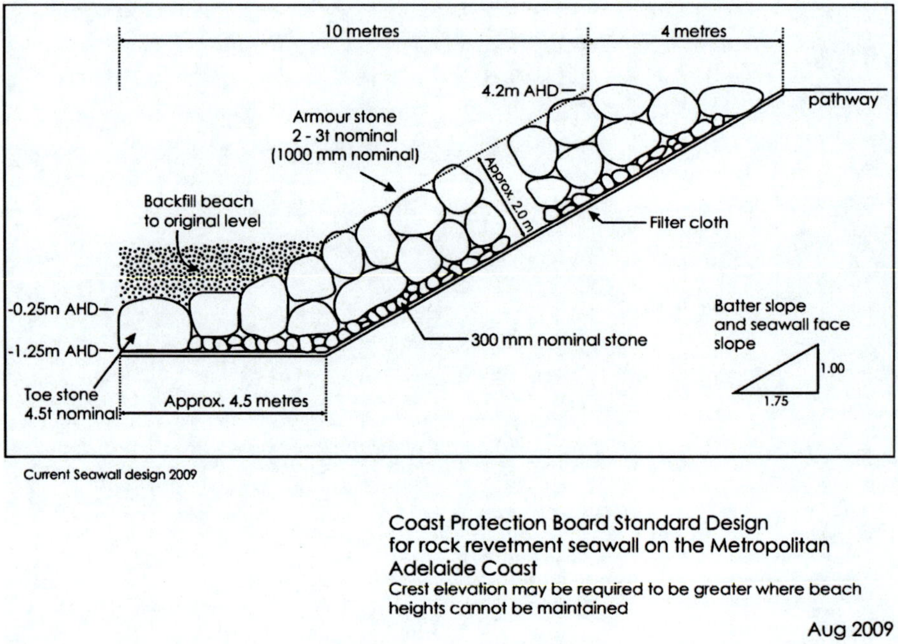

Fig. 2.10 Current Adelaide seawall design (Source: DEH 2005)

toe undermining. They are constructed of rip-rap underlain by a geotextile fabric to allow wave absorption while minimizing loss of sand through the voids (Fig. 2.10).

2.5.5 *Sand Dunes*

Maintenance of sand dunes along the coast is a critical part of the management strategy, with the dunes being valued for their natural habitat and as an essential coastal buffer. Most of the dunes are however an artifact of the management strategy, with only a few areas of remaining natural dune systems at Minda, West Beach, Tennyson, and North Haven to Outer Harbor. Where possible these natural systems will be maintained and enhanced. Elsewhere if the dunes accrete seaward of the protective buffer, the beach in front will be trimmed in the intertidal zone, and the sand recycled to areas requiring replenishment.

The local Natural Resource Management Board and local councils also use dune fencing, planting and access control to maintain dune stability and reduce the landward loss of dune sand. Finally, a number of exotic plants have invaded the dunes. While they assist in sand trapping and dune stabilization they do pose a threat to the indigenous plants and associated habitats. To monitor and control the spread of the invasive species the Urban Forest Biodiversity Program has worked with councils to develop reserve-specific vegetation management plans.

2.5.6 *Monitoring*

Since 1975 monitoring of the coast and its sand resources has been a critical part of the management strategy. Today the monitoring program consists of the following activities.

Beach profile lines were established approximately every 500 m along the shore. These are land surveyed to wading depth and then extended 1 km offshore, and more recently some lines to 2.5 and 5 km offshore (Fig. 2.11). In addition sand level rods consisting of 1.6 m-long brass rods were hammered into the seabed along 16 of the profile lines, spaced every 25–50 m and extending between 300 and 2,000 m offshore. Most were placed by 1990 and were surveyed annually for a number of years to provide +/−1 cm accuracy of vertical seabed change. The intensive coverage of the beach and seabed by leveling and soundings enables mapping of erosion and deposition using surface difference modelling software. These data have been used to calibrate the longshore sediment transport rates as well as loss of nearshore sands associated with the seagrass dieback.

2.6 Discussion

The Adelaide coast provides an excellent example of the transition over the past few decades in coastal usage, development and management. Adelaide was established in 1836 and laid out on a well-planned grid and radial roadwork pattern. While much attention was paid to formal city planning the same was not the case at the coast, 10 km west of the city centre. Here development was permitted on the frontal dunes and even the active beach in places, together with six long jetties extending from the shore to deeper water. At the time there was little or no understanding of coastal processes or the impact they would have on beach-dune processes. In the nineteenth century seawalls were built at six locations, and these were expanded during the first half of the twentieth century (Fig. 2.4). Up until the 1970s the approach to coastal management was one of protection of foreshore property and infrastructure, with no attention to maintaining the beach system, understanding its processes or to public beach amenity and safety. As a consequence by 1970 the 28 km-long Adelaide coast was in a bad state. The once near continuous beach and coastal dunes were either built on, removed, mined and leveled; 50% was fronted by generally poorly engineered seawalls; groynes and training walls trapped and impeded the longshore transport; storm and waste water discharge has caused substantial seagrass dieback and consequent seabed mobilization; the coast is subsiding and beaches eroded; and beach amenity was at a low point. This was all exacerbated by a deficit in the sediment budget and rising sea level.

The transformation in the approach to coastal management commenced in the late 1960s with the publication of the Culver report in 1970 and the adoption of its

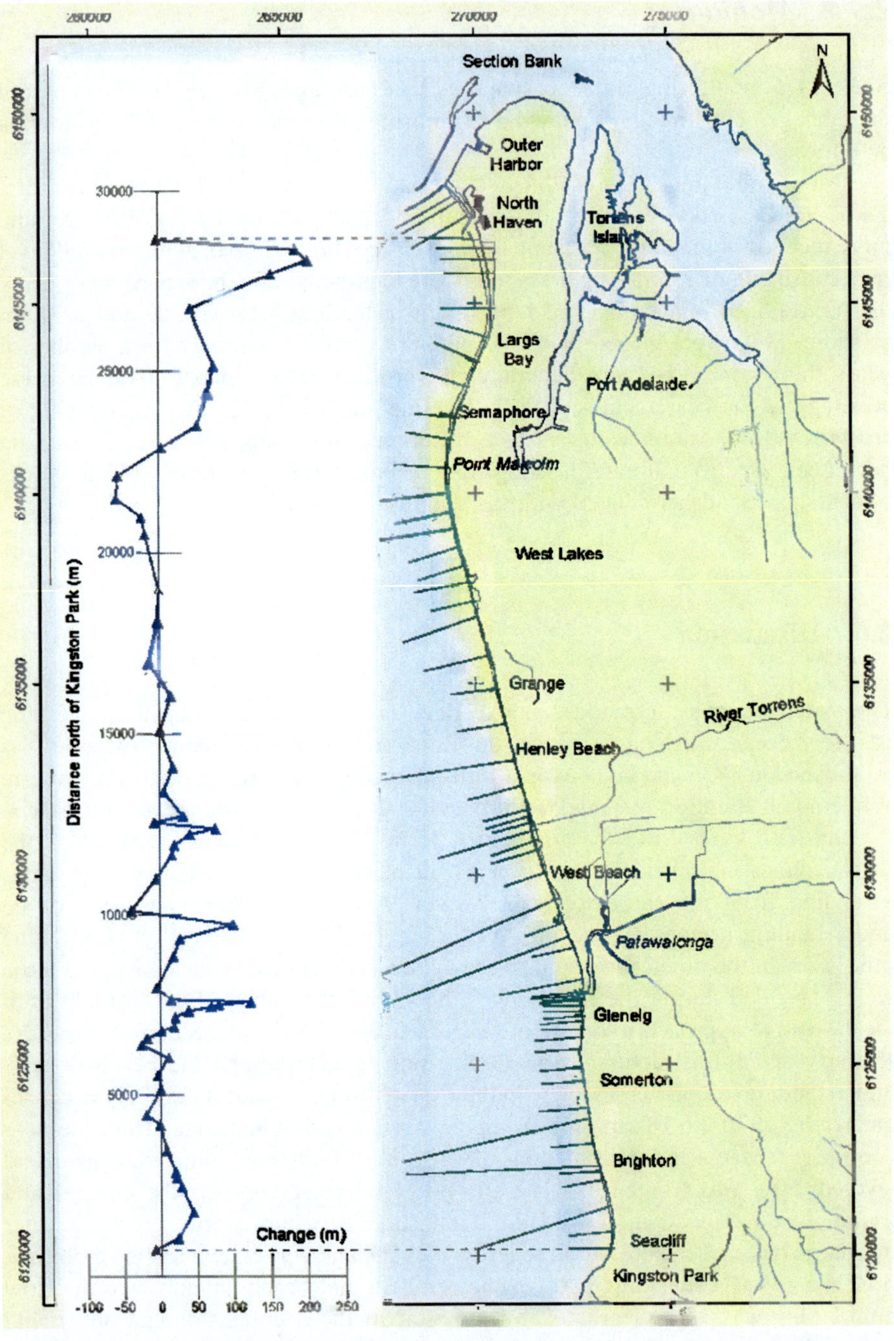

Fig. 2.11 Variation in beach width along the Adelaide coast 1975–2003. *Lines* (*right*) indicate the location and seaward extent of each of the survey lines (Source: DEH 2005)

recommendations to set up a Coast Protection Board and take stock of the coast's sand supplies. Thus began a new approach to managing the coast based on the need to minimize foreshore development; to understand the coastal processes; to increase sand supply; to maintain the dunes; and to minimize the impact of seawalls. This shift in approach, a global phenomena, was a product of the softening of coast protection works and the need to recognize and maintain public beach amenity, particularly on popular metropolitan beaches. The achievement of these two principal - softer protection and amenity, went hand in hand as beach replenishment, nourishment and sand bypassing became the principal means of beach management, coupled with dune management, more sympathetic seawalls, as well as the need to understand coastal processes and monitor the impacts of coastal works.

Today the Adelaide coast is an excellent example of best practice in contemporary coastal management. Detailed investigations have been undertaken of the processes that operate along the coast including monitoring of the beach and seabed since 1975. These data have been used to estimate rates of longshore sand transport and the impact of seagrass dieback on nearshore sand transport. The past use of seawalls has been controlled and modified and largely replaced since 1972. Beaches are replenished by recycling sand from north to south and within sand management cells created by offshore obstructions; and by beach nourishment using external sources. In addition the coastal dune systems are maintained though planting and fencing; and development controls prevent unsuitable development on the dune and beach. Finally, the entire system is monitored and the strategy reviewed on a regular basis.

2.7 Conclusions

The Adelaide metropolitan beaches are part of a 28 km-long near-continuous beach and dune system that experiences northerly sand transport at a rate of about 50 000 m^3/year. During the nineteenth century nodal development along dunes and construction of long jetties were followed by periodic storm damage and seawall construction. By the 1950s most of the dunes had been developed annihilating them from the longshore sand transport system. Net beach retreat resulted leading to the construction of more seawalls and a general degradation of the beach system.

In 1970s a comprehensive review of the Adelaide coast saw the beginning of modern coastal management and a softer approach to protecting the backing infrastructure and property, as well as maintaining the recreational amenity of the popular recreational beaches. The present strategy includes sand recycling, nourishment and trapping, dune reconstruction and more compatible structures. In 2005 a 20 year strategy for the beaches was released which will continue this program, together with monitoring of the beach and seabed, and periodic reviews of the strategy.

Acknowledgments I thank Doug Fotheringham and Jennifer Deans for their careful review of the manuscript, and Peter Johnson for supplying a number of original figures.

References

Bowman GM, Harvey N (1986) Geomorphic evolution of a Holocene beachridge complex, Lefevre Peninsula, South Australia. J Coast Res 2:345–362

Coast Protection Board (1985) Metropolitan coast protection district management plan. Department of Environment and Planning, Adelaide

Coast Protection Board (2005) Adelaide's living beaches: a strategy for 2005–2025. Technical report, Coast Protection Board, Dept Environment & Heritage, Adelaide, 206 pp

Coastal Engineering Solutions (2004) Coastal processes study of Adelaide beaches, prepared for the Department for Environment and Heritage, Adelaide

Coastal Management Branch (1984) Adelaide coast protection strategy review. Department of Environment and Planning report to the Coast Protection Board, Adelaide

Coastal Management Branch (1992) Review of alternatives for the Adelaide metropolitan beach replenishment strategy. Coastal Management Branch, Department of Environment and Natural Resources, Adelaide

Culver R (1970) Final summary report on beach erosion studies. University of Adelaide report to the Seaside Councils Committee and the State Government of South Australia, Adelaide

DEH (2005) Adelaide's living beaches: a strategy for 2005–2025. Technical report, Department of Environment and Heritage, Adelaide, 206 pp. Available online at www.environment.sa.gov.au

Department of Environment and Natural Resources (1997) Report of the review of the management of Adelaide metropolitan beaches, prepared by the Reference Group appointed by the Minister for Environment and Natural Resources, Hon. David Wotton, MP

Harvey N, Bowman GM (1987) Coastal management implications of a Holocene sediment budget: LeFevre Peninsula, South Australia. J Shoreline Manag 3:77–93

Short AD (2001) Beaches of the South Australian coast and Kangaroo Island. Sydney University Press, Sydney, 346 pp

Wynne AA, Fotheringham DG, Freeman R, Moulds B, Penney SW, Petrusevics P, Tucker R, Ellis DP (1984) Adelaide coast protection strategy review. Department of Environment and Planning, Adelaide

Chapter 3
Pitfalls of Ebb-Shoal Mining

Charles W. Finkl

Abstract Ebb-tidal deltas or ebb shoals in moderate to high-energy environments often contain well-washed sandy sediments. These clean sands attract the attention of coastal engineers because they are commonly suitable for shore protection efforts such as beach renourishment (replenishment), dune restoration, and marsh remediation. Ebb-tidal deltas are sediment sinks where sandy materials are sequestered in deposits that can build up to such significant proportions that they form large coastal salients. In addition to sediments debouching from the inlet, longshore sediments are either trapped in the shoal or are bypassed in swash bars. Accumulation of sediments in these depocenters at the mouths of inlets often appear as a ready-made point borrow source that can be accessed for shore protection by dredging and placement on adjacent eroded shores. Ebb-tidal deltas are, however, in delicate balance with inlets, longshore drift, beach-dune systems and overall coastal stability. Removal of sediment volume by dredging from deltas interrupts the sand-sharing balance between inlets, ebb-tidal shoals, coupled beach-dune systems, wetlands, and shoreline stability.

3.1 Introduction

There are many pitfalls associated with coastal engineering where sometimes the best of intentions results in unwanted environmental impacts. Rivers that transport abundant loads of sediments to the sea deposit their fluvial materials in deltaic environments, new settings where transport is reduced and sequestering begins to build deltas containing various types of sedimentary accumulations. Deltas are in delicate balance with coastal marine environments and disturbance of the deltaic regime by engineering works can wreak havoc on adjacent shorelines or even adversely impact estuarine and upriver

C.W. Finkl (✉)
Department of Geosciences, Florida Atlantic University, Boca Raton, FL 33431, USA
e-mail: cfinkl@fau.edu

J.A.G. Cooper and O.H. Pilkey (eds.), *Pitfalls of Shoreline Stabilization: Selected Case Studies*, Coastal Research Library 3, DOI 10.1007/978-94-007-4123-2_3,

wetlands (Dean and Dalrymple 2002). Dams on tributaries leading to trunk rivers hold back sediment that would naturally accumulate downstream in a river delta. River training structures, such as commonly occur along the lower Mississippi River, for example, help to jet the suspended sediment load farther downstream past the delta system instead of letting fine-grained sediments settle to the bottom and build the delta. Inlets associated with the mainstream or larger distributaries are often jettied to stabilize them for navigational purposes. Whether mainstream or distributary channels are commonly dredged to maintain sufficient depth for shipping, the spoil is disposed of in various ways depending on the size of the channel and distance to disposal sites.

Although deltas are in delicate balance with a wide range of environmental conditions and display geomorphological variation and sedimentary characteristics that are in tune with local energy conditions, several basic types of deltas are commonly recognized to simplify the complexity of reality. Interference with different types of deltas by engineering works has different impacts but the attraction of deltas occurs in the first instance because that is where sediments accumulate and they are seen as tempting sources of borrow materials for a wide range of engineering uses that include construction materials, beach renourishment (replenishment), wetland restoration, dune rebuilding, and land reclamation, among others.

Described here first is a brief rundown of the main types of deltas followed by descriptions of sediment exploitation that results in unwanted environmental impacts. The temptation to exploit deltas is great because they are basically a storehouse of materials that can be used for a variety of purposes, they are easily accessible by dredging large volumes of material, and they are economically obtainable. These factors combine to put deltas in peril by engineering use when not properly studied from a scientific standpoint to determine whether sand mining is appropriate at a particular location, when extraction of sediment could take place if appropriate, and what volume can be removed without disrupting the natural balance between river, inlet, delta, beach, barrier island, dune, and the coast in general.

There are several different types of tidal deltas (geological term) or ebb-tidal shoals (engineering term) that accumulate at the mouths of rivers and inlets where there is an abundant supply of sediment (FitzGerald 2005), *i.e*. enough to accumulate in a deposit under existing energy conditions. Different types of deltas are related to several factors that combine to produce distinct forms that are related to the amount of sediment that is supplied by downstream flow or by alongshore flow, wave climate (Maa et al. 2001), tidal regime, size of the inlet, and volume of flow *etc*. Deltas are defined as coastal accumulations, both subaqueous and subaerial, of river (fluvially)-derived sediments adjacent to, or in close proximity to, the source stream, including deposits that have been secondarily modified by waves, currents, or tides (Wright 1977). For simplicity, deltas are commonly characterized as wave-, river- or tide-dominated, but most deltas show some effects of all these processes (Davis and FitzGerald 2004). River-dominated deltas, for example, show thin sedimentary fingers that protrude into a water body where there are no strong currents or waves. A smooth coastline is the main characteristic of wave-dominated deltas. Tide-dominated deltas generally show wide lobes of land

and submerged shoals oriented perpendicular to the coast by tidal currents. All types of deltas are subject to threats from dredging, no matter how large or small.

Deltaic configuration, in plan view, is also an important identifying characteristic that can be summarized in terms of: (1) arcuate deltas, (2) digitate or bird foot deltas, (3) estuarine deltas, and (4) cuspate or tooth shaped deltas. Processional development of shapes and materials influence the morphodynamic stability of deltaic environments and dredging activities (*e.g.* channel maintenance, sand mining) can severely upset the natural equilibrium to the point in extreme cases where the delta collapses if too much sediment is removed or long jetties are erected (Kelley and Brothers 2009). Disruption of delta sedimentary regimes by dredging is the subject of this chapter which attempts to show that adverse impacts can accrue from such activities when due diligence is not followed or when proper scientific studies are not conducted prior to dredging activities (e.g. Cooper et al. 2009). With increased soil erosion but reduced sediment supply to many of the world's deltas by dams in the catchments (Syvitski et al. 2005), repair of deltaic systems damaged by dredging becomes problematic due to lack of downstream sediment supply and may contribute to a worsening of the shore erosion problem that was intended for correction.

As with most systems in Nature, morphodynamic systems are interconnected and disruption of one part of a system can magnify effects on other parts of related (interconnected) sand-sharing systems. In the coastal zone, there are multiple nexuses such as the inlet-beach system, beach-dune system, beach-bar system, delta-beach-bar system, delta-wetland system, and so on. It is recognized that a certain amount of dredging is required to keep channels open and to maintain safe navigational entrances to ports and harbors for reasons of commerce, recreation, and military use. What is under discussion here is dredging of ebb-tidal deltas for shore protection, or more explicitly the identification of cases or situations where the dredging of ebb-tidal deltas is inadvisable because of potential adverse impacts such dredging might cause to the balance of the sand-sharing system.

3.2 Disturbance of the Sand-Sharing System

Dredging of deltas is usually not a solitary activity, but is commonly associated with the construction of hard structures such as jetties to help maintain the benefits of the dredging process. Because deltas are composed of unconsolidated sediments and channels are prone to suddenly change course (avulsion), many dredged inlets are stabilized by adding jetties. The addition of jetties to a dredged inlet tends to compound the problems induced by dredging alone. Of the 565 km of Florida coastline, 16 of 19 inlets on the east coast alone are managed by dredging and/or jettying (Dean and O'Brien 1987). Inlet deepening, stabilization, and maintenance causes sand loss from downdrift beaches because they are starved of sediment, except where they are replenished by bypassing facilities at the Lake Worth, Boynton, Boca Raton, and Hillsboro inlets (Fig. 3.1), for example (Finkl 1993). Briefly put and as so well documented, removal and diversion (by dredging) of

Fig. 3.1 Dredged and stabilized inlets on Florida Atlantic and Gulf coasts. The top image shows the main inlets on both coasts of the Florida Peninsula. Shown here are 19 inlets on the east coast and 24 inlets on the west coast. All of these inlets are littoral drift blockers that interrupt the longshore drift of

nearshore sand in deltaic systems causes beach erosion (e.g. Bush et al. 2004; Bruun 1995; Cooper et al. 2009; Davis and FitzGerald 2004; Dean 1990; Dean and Walton 1973; Finkl 1993; Giese 2008; Hanson and Kraus 2001; Montague 2008; Young et al. 1996). The situation of compound engineering (dredging and jettying) is thus a double-pronged pitfall that disturbs the natural balance of the sand-sharing system. The process is briefly explained below and then followed by some examples to illustrate the point that has worldwide applicability.

3.3 The Sand Deficit Problem Caused by Dredging

With one exception, all of the marine mining activity along the Atlantic and Gulf coasts of the United States is for sand for use in beach renourishment projects. The sole exception is a commercial sand and aggregate project near New York Harbor. Sandy shores exist along about 13% of the world's coastline (Coleman and Murray 1976) and exhibit many different sizes and shapes. Thus, beach erosion, on a worldwide basis, is very much a natural phenomenon that is exacerbated by dredging of maintenance and access channels in ebb-tidal deltas. It has been calculated that over recent decades about 75% of the world's sandy shorelines are eroding (Bird 1985; Hanson and Lindh 1993). Leatherman (1988) states that for the United States the percentage is even greater than the world average and may be as much as 90% (see also the discussion in Finkl and Hobbs 2009).

Of the 1×10^9 m^3 of sediments removed from America's beaches by engineering works and anthropogenic activity in the past century (Douglas et al. 2003), about 650×10^6 m^3 have been returned to the beaches by dredging sand from the continental shelf, harbors, and ebb-tidal deltas. Thus, there is a sediment deficit that needs to be mitigated over the long term (Finkl et al. 2006), but the question remains as to where that sand should be obtained. Whatever the sediment source, Leonard et al. (1990) have discussed the durability and longevity of renourished beaches and the wisdom of taking sediment from Paul and giving it to Peter in an effort to right the wrongs of upsetting the delicate natural sediment balance in the

Fig. 3.1 (continued) sediments from north to south. At the Hillsboro Inlet (*lower left hand corner*), southeast Atlantic coast of Florida, the northern updrift jetty is a weir jetty that allows sediment to pass over a rock ledge and accumulate in a sediment trap landward of the jetty. A dredge is permanently stationed at this location to bypass sediment to the sediment starved downdrift beach. Note the landward offset of the shoreline south of the jettied inlet, which is due to the dredged and stabilized inlet functioning as a littoral drift blocker. This inlet was cut through unconsolidated sediment and into the underlying bedrock to accommodate sportfishing vessels. (Image source: Google Earth, imagery date 3/21/2011 from TerraMetrics). The dredged and stabilized entrance to Port Everglades (*lower right hand corner*) (Image source: Google Earth, imagery date 3/26/2011 from DigitalGlobe) also contains a sand trap on the north side of the channel landward of the north (updrift) side of the jetty. Note that channel dredging and inlet stabilization causes downdrift erosion and landward retreat of the shoreline at both inlets coastal sand-sharing system (see also Valverde et al. 1999). Sediment volumes and the costs involved for beach replenishment were summarized by Pilkey and Clayton (1989), who showed that efforts to correct the sand-deficit problem (manifested in the form of beach erosion) were and still are significant and probably not very well understood (Pilkey 1990).

Beach-sand deficit estimates are based on existing conditions, as described for example by Douglas et al. (2003) and Finkl et al. (2006), and do not take into account additional volumes that will be required to combat future erosional impacts of projected sea-level rise and continued dredging of inlets in deltaic environments. Finkl et al. (2006) estimate that about 650×10^6 m^3 of sediments have been returned to the coastal sand system by artificially placing sand on the shore through various types of nourishment. Most of this sediment volume has been dredged from the continental shelf, but some was dredged from ebb-tidal deltas as well (*e.g*. Boca Raton Inlet) (Fig. 3.1).

The major consideration here is the precise location of sand mined for beach nourishment. To reduce the deficit, this sand must come from outside the system. It can come from upland sources or from water deeper than the closure depth. Sand from the shallowest parts of shoals is already within the sand sharing system, as is sand from dunes, fillets at jetties, sediment traps in inlets, ebb-tidal deltas, and updrift beaches.

Montague (2008) reports that most of the sand placed in nourishment projects on the Florida Atlantic coast was derived either from channel maintenance or has been mined from nearshore bars and ebb-tidal delta shoals that are within the sand-sharing system. He further reports that on the basis of the borrow sites known to him for these projects, it is likely that less than 10% of nourishment material has been brought into the sand-sharing system from upland or offshore sources beyond the closure depth.

Elsewhere in Florida, previously maintained channels are already being allowed to fill. No provisions have been made to supply the necessary sand from outside the sand-sharing system. Beach and dune erosion should be expected as a result. For example, the channel at Pensacola Pass (on the Florida Panhandle), recently deepened for a nuclear powered aircraft carrier, is being allowed to fill to a certain degree because the carrier's homeport was reassigned. A similar action could soon apply to the aircraft carrier channel at Mayport (St. Johns Entrance) (Fig. 3.1) on the northeastern coast of Florida (Montague 2008).

Because sufficient sand is not imported into the sand-sharing system, the void will be filled from somewhere within it. Much of the sand that will fill the channels might now be on beaches and dunes, having been placed by dredging sediment from the nearshore continental shelf or ebb-tidal deltas (Finkl and Walker 2005; Finkl et al. 2005; Finkl and Hobbs 2009). The most insidious increase in capacity of the sand-sharing system now occurs on the updrift side of jetties protecting channels in many ebb-tidal deltas. It is unlikely that ebb shoals will eventually build to natural equilibrium levels at the seaward ends of jetties and once again allow sand transported in longshore drift to bypass because the littoral drift system is sediment starved and sediments are lost from the system by jet currents from stabilized inlets (e.g. Bruun 1978, 1995), storms, and lost through reef gaps to deeper water beyond the closure depth (e.g. Finkl 2004). It is only possible with increased sand supply to the littoral drift system that the sand-sharing system, as newly defined by jetties in ebb-tidal deltas, can fill to capacity. In the case of Florida, there are few exorheic rivers (rivers that rise and flow to the sea) and in any case, they do not bring large sediment loads to the coast. Additionally, sand is lost from the littoral drift system along the Florida Atlantic coast through reef gaps that act as leaky valves to deeper waters beyond the edge of the continental shelf (Finkl 2004). Before the sand-sharing system once again becomes self-maintaining, this unknown added capacity must be supplied, unless the jetties are removed (Montague 2008).

When inlets are stabilized, *i.e.* jetties installed, and channels and harbors deepened to allow safe passage of large vessels, the equilibrium of the sand-sharing system is disturbed, as previously noted. As a result of this engineering practice described in the most simplistic terms, the quantity of sand in the system decreases while its capacity for absorbing additional sand increases. This situation obtains because in the initial deepening process, machinery moved material from deltas and associated shoals and also removed deeper material (carbonate bedrock in the case of Florida inlets) that was not part of the sand-sharing system. Because dredged material not used for beach renourishment is typically deposited in deep water (because it may be contaminated or contain clast sizes that are too fine or too coarse for beaches), a diminished quantity of sand remains in the system (*e.g.* Montague 2008; Kana 2011). Removal of deeper material (as seen in borrow pits that remain unfilled for more than two decades after dredging along the Florida southeast coast) creates a greater accommodation space that expands the capacity to trap sand. In the case of many Florida inlets, channel dredging in inlets involves removal of bedrock (*e.g.* Hillsboro and Port Everglades inlets) because the sediment wedge is so thin, as described by Finkl (1993). Maintenance of the intended channel depth requires repeated dredging to remove trapped sand. Much of the sand removed in maintenance dredging is placed offshore in deep water or in massive upland mounds, depending on the travel distance for spoil disposal and availability of open space on dry land. These kinds of practices remove sand from the larger sand-sharing system (Montague 2008). Some sand dredged from channels and deltas is placed onshore if it meets environmental standards (is not polluted) and compatibility tests with native beach sand.

Along Florida Atlantic and Gulf coasts, jetties were initially constructed in the first part of the twentieth century to stabilize the location of inlet channels. Some channels were cut through deltas diverting the littoral drift of sand to deeper water (*e.g.* Bruun 1978; Dean 1990; Finkl 1993). If there is sufficient supply of sand in the longshore drift, sediment may accumulate on the updrift side of a jettied inlet to form a fillet. Once filled to capacity, sand will move around the end of the updrift jetty into the channel or be jetted offshore by ebb-tidal currents to fall into deeper water. If the inlet channel is deep, such as occurs at major navigational entrances for example at Mayport, Cape Canaveral, Fort Pierce, Port Everglades, and Miami (dredged to 15 m or deeper) on the Atlantic coast and Tampa on the Gulf coast (Fig. 3.1), sand tends to be jetted offshore by ebb-tidal currents. At smaller inlets, sand may accumulate in a deltaic environment if environmental conditions are favorable. So, to be fair, it must be admitted that jetties can trap sand (*e.g*. Dean and Walton 1975), but at the same time it must be recognized that by doing so, the deltaic balance is upset and components of coastal systems may respond in an unwanted manner as usually occurs with downdrift shore erosion (*e.g*. Dean 1990, 1993).

Because sand gets trapped at jetties, beach erosion occurs downdrift of jettied inlets with or without deltas (see following discussion about the inlet problem caused by dredging and stabilization with jetties). Sand in the southward flowing littoral drift that enters Florida from Georgia gets trapped on the updrift side of jettied inlets instead of helping to build deltas (Montague 2008). Over time as inlets were dredged and channels were cut through deltas to keep the inlets open all year, the equilibrium state of the dynamic coastal sediment transport system was upset by delta dredging and construction of hard structures, and a new equilibrium began to establish itself with less total sand (*e.g*. Dean and O'Brien 1987; Dean 1990, 1993; Finkl 1993). As a result, downdrift shorelines retreated (Fig. 3.1) as sand eroded from beaches and refilled channels or was jetted offshore and lost from the littoral drift system (Dean 1990). According to Quinn (1997), beach erosion downdrift of deepened inlet channels was apparently well known to those initially involved in harbor and channel management and to some as early as the late nineteenth century. Some beach and dune sand likely settled in deepened parts of harbors as well.

3.4 The Stabilized Inlet Problem

There are two main types of inlets, natural and artificial. Regardless of their origin, most inlets are in dynamic balance with their environment as they evolve with fluctuating conditions such as monthly and seasonal volumes of flow, sediment fluxes, decreasing hydraulic efficiency of the system (opening and closing of passes), alongshore migration with quasi-cyclic patterns (*e.g*. Giese 2008), and new inlet formation. Because this dynamic nature of inlets threatens developed shorelines, they are commonly stabilized for navigational purposes by jetties that fix the inlet in place. Although one problem is solved by an engineering fix, the solution entrains a whole series of additional problems that affect ebb-tidal deltas and adjacent shorelines. There are many geoindicators of disruption in the coastal sediment balance. One of the most obvious examples of disequilibrium of shoreline position is the landward retreat of the shoreline that is usually manifested in the form of beach erosion (Young et al. 1996) along sandy coasts. The problem is basically two-pronged with one part focusing on the impacts of dredging and the other part targeting downdrift erosion caused by dredging and jetties that interrupt the alongshore sediment transport (*e.g*. Bruun 1995).

Dredging of the channel between the jetties is commonly required to keep the inlet navigable, but can lead to collapse of the delta (Kelley and Brothers 2009).

Shoreline erosion downdrift of jetties has been discussed by numerous researchers who recognize the problem as universal (see Florida examples in Fig. 3.1). Dean (1990), for example, estimates that jetties are responsible for about 85–90% of the shoreline erosion downdrift of jetties in Florida. Discussing the problem of littoral drift barriers, Bruun (1995) recognized erosion fronts that moved at different rates downdrift from jettied inlets, an initial slow-moving erosion front that caused rapid shoreline retreat close to the entrance and a second faster-moving front that occurred after the first but which caused shoreline retreat farther downshore. The first or slow erosion front moved a relatively short distance downdrift from the stabilized inlet but the second faster-moving erosion front extended for long distances downshore, usually to the next inlet many kilometers downshore. In the case of Port Canaveral, Florida (Fig. 3.1), for example, the entrance was cut in 1951 and by 1965 the short-distance erosion front migrated about 5 km downdrift giving a rate of about 0.6–0.7 km per year. By 1965, the long-distance erosion front had reached about 18 km from the entrance giving an erosion front migration rate of about 1.5 km per year. Bodge (1993) further reports that the erosion front now extends 30–40 km downdrift giving a migration rate of 0.8–1.2 km per year. Bruun (1995) cites numerous examples of erosion fronts caused by stabilized inlets and the problem has been addressed in numerous papers that recognize chronic beach erosion adjacent to stabilized inlets.

Remediation of the problem and employing hard structures has been discussed, for example, by Dean (1993) and Hobbs (2002, 2007). Hanson and Kraus (2001) point out potential solutions with various shaped groins using Shinnecock Inlet on the south shore of Long Island, New York, as an example. They conclude that T-head and other composite groins can stabilize beaches by holding the sand located directly on either side of the structures. However, they acknowledge that at the same time there is potential for depriving downdrift beaches of sand and that there are high initial (construction) costs. Placement of these composite structures is thus restricted to situations where they are economically feasible and their influence outside the direct project area is acceptable or can be mitigated (Dean and Walton 1975). Hanson and Kraus (2001) concluded that natural sand bypassing around the ebb-tidal deltas and periodic artificial nourishment, if necessary, can maintain the beaches further downdrift of the project.

Where jetties are present, substantial quantities of sand can be trapped along the side of the jetty, as described by Bush et al. (2004) for inlets (most with ebb-tidal deltas) along the Florida Atlantic coast. This is most pronounced at the jetty on the "updrift" side of the inlet. Once the jetty reaches capacity, sand may flow around the end of the jetty in the channel or on to the ebb-tidal delta. Adjacent shorelines are deprived of the sand thus accumulated, and the implied erosion rate of the "downdrift" shoreline approximates the impoundment rate at the "updrift" jetty and within the ebb and flood deltas (*e.g.* Dean 1993). Dredging through the ebb-tidal delta in conjunction with channel construction or maintenance can create a sand deficit to the downdrift shoreline by interrupting any natural bypassing that may have developed prior to dredging. Van de Graaff (1990) describes this effect as 'structural erosion' and refers to 'gradients' (sediment losses and gains resulting in shoreline regression and progression) and in the longshore sediment transport as the cause of the erosion problem. This is a classical erosion problem caused by dredging and stabilization of delta inlets that interrupts longshore transport that deprives downdrift shores of sediment, as so well described long ago by Bruun (1978). Impacts on adjacent shorelines can be substantial if the channel is dredged through a portion of the ebb-tidal delta and associated shoals that are close to or attached to the shore. In this case, the tendency of the shoreline is to erode, to compensate for the sand removed from the shoal, and to readjust the slopes to a more natural shape in the form of an equilibrium slope.

One practice that historically has caused serious adverse impacts to shorelines adjacent to inlets is disposing of dredge spoil (sediment that is not suitable for beach renourishment) offshore in deep water (*e.g.* Khalil and Finkl 2011). This sand is permanently lost to the active "littoral system," and, as a consequence, the adjacent shorelines erode. The U.S. Army Corps of Engineers (USACE) is generally required to select the least-cost alternative for dredge disposal, and the least-cost alternative is often offshore disposal. Offshore disposal can sometimes appear to be the least-cost alternative only because the values of losses associated with erosion are not included within the economic calculations. Fortunately, because the Corps must also work within state guidelines, states may have some leverage to require that beach quality sand be placed back onto the beach. For example, Florida statutes require onshore placement of beach quality sand from inlet dredging. A water quality certification is required for a USACE inlet maintenance project and is often used as a mechanism to direct such sand management activities. In North Carolina, inlet maintenance projects are typically linked to federal beach nourishment projects, so that inlet maintenance and beach project maintenance coincide.

Proper design and maintenance can reduce inlet-influenced erosion. Evaluation of dredging and channel and shoreline adjustment history is useful in determining the inlet configuration and channel alignment that will result in the least maintenance. Price (1951) studied inlets along the Texas coast and found that it may be possible to reduce maintenance costs by realigning navigation channels such that they are more akin to, instead of in opposition to, the dominant natural forces (http://www.csc.noaa.gov/beachnourishment/html/geo/channels.htm). Bruun (1978) also points out that the unwanted erosion impacts of dredging inlets cut through ebb-tidal deltas can be partly ameliorated by 'trapping' sediment in constructed catchment basins, either on the updrift side of the jetty or immediately downdrift of the updrift jetty if it has a weir so that sand can drift into the interior trap, such as at Boca Raton, Hillsboro, and Port Everglades inlets on the southeast Florida coast (Fig. 3.1). When sediments accumulate in the traps, they must be mechanically bypassed around the inlet by pumping to the downdrift side to again join the littoral drift. It should be noted that this procedure, although locally effective, is not a panacea as such facilities are rare and most dredged inlets have no such arrangement for bypassing sediment to the sediment-starved downdrift coast.

3.5 Caveats to Ebb-Delta Mining

Trudnack (1977) studied the impacts on the inlet-beach system due to partial mining of ebb-tidal shoals through laboratory experiments. The main focus of the study, using a generic inlet, was to examine downdrift beach erosion and ebb-shoal borrow regeneration. The effect on downdrift erosion is of obvious importance as the goal of shoal mining is to use the mined sand to renourish and protect the downdrift beach. Clearly, mining the ebb-tidal shoal would not be worth the effort if the erosion rate increased dramatically. Such a case would negate the benefits of renourishment. This experimental study is important because it elucidates several important considerations that must be evaluated before mining an ebb-tidal delta for beach renourishment. The rate of regeneration (filling) of the shoal borrow area is of critical importance because it affects the post-dredging configuration and this in turn determines whether or how often the ebb shoal can be mined. Interestingly, Trudnack (1977) determined that ebb-tidal delta mining increased the downdrift erosion rate right after mining and also that the inlet channel shoaling increased due to ebb-tidal delta removal. The downdrift longshore transport volume as well as the transport rate increased due to ebb-tidal delta removal. When the laboratory ebb-tidal delta characteristics of three small- to medium-sized Florida east coast inlets (Trudnack 1977) are compared with the results of Wang et al. (1995), who also studied the evolution of ebb-tidal deltas in the laboratory, both studies showed evolutionary sequences similar to those in nature.

Sand volumes and morphologies of 17 ebb-tidal deltas off natural inlets on the New Zealand North Island coast, in both open-sea and pocket-bay settings, were investigated by Hicks and Hume (1996) in an attempt to assess permissible rates of sand mining from ebb deltas, to estimate ebb-delta sand entrapment associated with changes in the tidal prism, and to re-design the alignment of inlet channels in order to control ebb-delta sand volumes. They found that the main controls on ebb delta sand volume (V) are the tidal prism volume (P), the angle between the outflow jet and the shoreline (θ), and the wave climate. Their empirical equation $V = 1.37 \times 10^{-3} P^{1.32}(\sin \theta)^{1.33}$ accounted for 83% of the variance in sand volume in the dataset. Four basic ebb-delta forms were identified. 'Free form' deltas, typically 'bat-winged' in shape, occur on open shorelines. 'Constricted' deltas are similarly shaped but occur in shoreline angles lacking space for the free form to fully develop. 'High-angle half-deltas' are typically shore-normal or 'L-shaped' and occur in embayment corners where the ebb jet flows against the rocky headland, resulting in a significant shoal forming only on the beach side of the inlet. 'Low-angle half-deltas' are almost shore-parallel sand 'wedges' that form between the ebb jet and the beach where the ebb jet is forced by rock controls to flow at a low angle to the beach. Sand storage volumes ranged from 3.8×10^4 m^3 to 1.2×10^{10} m^3. Deltas on the high-energy west coast tended to be smaller than east coast deltas with similar tidal prisms. The supply of littoral drift also appears to influence delta volume in some cases. These results may be used to better understand the impacts of mining deltas.

Shoal bypassing is a natural form of beach renourishment. Kana et al. (1985) report, for example, that more than 10^6 m^3 may cycle back to the beach in a single event as a shoal breaks away from an inlet and migrates onshore. This important process where inlet deltas and shoreline evolution are related and which may experience systematic cycling of sediments has been known for some time (*e.g.* Dean and Walton 1975; Oertel 1977) but, as reported by Gaudiano and Kana (2001), not very much is known about the frequency and magnitude of discrete shoal-bypass events.

Mining of these ebb-tidal deltas has become more prevalent in recent years due to limited sources of beach quality sand available for beach nourishment projects. Olsen (2009) examined several ebb-tidal delta-mining projects completed since 1981 in an attempt to examine this relatively new practice of removing sand from the delta. The projects that he studied ranged in size from 170,000 m^3 removed from the ebb-tidal delta at Boca Raton Inlet, Florida, to 6,235,000 m^3 removed from the ebb-tidal delta at Great Egg Harbor Inlet, New Jersey. Completion of these projects and lack of systematic monitoring resulted in limited monitoring data to assess tidal-delta mining impacts on the coastal sand-sharing system. From his study, Olsen (2009) determined that most ebb-tidal deltas are mined on the outer "passive" portion of the shoal feature. Ebb-tidal delta sand was found to be compatible with the native beach material, indicating that the delta acts as a "sand bridge" between updrift and downdrift beaches. The rate of recovery of the mined area appears to be a function of the degree to which the system equilibrium is perturbed, sand availability (longshore transport rate), storm frequency, and the depth of the mined area. It is important to note that Olsen (2009) concluded that estimates of borrow area recovery were often overpredicted, probably due to poor longshore transport estimates. It is also important to note that further analysis is needed to better comprehend ebb-tidal delta mining impacts to navigation, inlet adjacent shorelines, ebb shoal equilibrium, and reusability of borrow area infill material (Cialone and Stauble 1998).

3.5.1 No-Impact Ebb-Tidal Shoal Mining

It must be admitted that not all ebb-tidal delta mining has negative impacts on the sand-sharing system. There are examples of successful mining efforts and one of them, briefly mentioned here in fairness to the discussion, is the entrance to Mobile Bay, Alabama, between Mobile Point on the western end of the Morgan Peninsula and Pelican Point on the eastern end of Dauphin Island. This is an extensive natural inlet that has been improved by channel dredging activities since 1904, primarily through the outer bar at the seaward extent of the ebb-tidal delta. The purpose of the Byrnes et al. (2010) study was to evaluate the potential impact of construction and maintenance dredging activities for the Federal navigation project in Mobile Outer Bar Channel on ebb-shoal changes and shoreline response along Dauphin Island, Alabama. Ebb-shoal changes and shoreline response relative to storm and normal forces, and dredging in the outer bar channel, were evaluated to determine the extent to which beach erosion along Dauphin Island could be attributed to U.S. Army Corps of Engineers (USACE) channel construction and maintenance dredging operations. Two distinct periods were evaluated: one representing conditions prior to significant construction and maintenance dredging activities to determine natural changes (1847/48–1917/20), and the other representing conditions after significant changes to the outer bar channel had been imposed (1917/20–1986/2002) to quantify changes on the ebb shoal and beach response along Dauphin Island. Overall, net sediment transport from east-to-west for the entire period of record has supplied sand quantities necessary to produce net deposition on the islands and shoals of the ebb-tidal delta, infill and nourish storm breaches and washover surge channels on Dauphin Island, and promote growth of western end of the island, even though channel dredging has been active. Based on available information, Byrnes et al. (2010) reported that there appear to be no measurable negative impacts to ebb-tidal shoals or Dauphin Island beaches associated with historical channel dredging across the Mobile Pass Outer Bar. There is, however, a long history of shore erosion along Gulf of Mexico shores as occurs along Dauphin Island (*e.g.* Douglas 1994; Morton et al. 2004, 2005; Sallenger 2009) and many factors that may make these conclusions contentious.

3.6 Conclusions

Dredging of ebb-tidal deltas is a common practice and takes place for a variety of reasons, mostly to maintain safe navigation in shipping channels and to gain access to sediments that can be used for beach renourishment. Deltaic sediments dredged for these purposes raise the question as to whether the dredged material is spoil or a resource (Khalil and Finkl 2011; Khalil et al. 2010) and consequently there are different methods of disposal, whether on land, at sea, or on the beach. Although there are exceptions, there are generally unwanted environmental consequences associated with the dredging of deltas. The practice is becoming more common because the size of vessels is increasing, especially container and cruise ships, and channels must be deepened to maintain safe navigation. Dredging of smaller deltas such as occur along Florida Atlantic and Gulf coasts but which is common for similar small deltas the world over, disrupts the natural balance of the sand-sharing system and causes downdrift shore erosion. This worldwide problem can be traced back in large part to the consequences of robbing sand from deltas by dredging.

The engineering pitfall here lies in the fact that ebb-tidal deltas contain sedimentary materials that are either a nuisance to shipping, commerce, and recreational activities or the sediments have intrinsic value for other purposes such as beach renourishment. Whether spoil or resource, ebb-tidal deltas are tempting targets that entice dredgers to do what they do, move sediment from one place to another. The pitfall lies in the fact that while engineering intentions may be good in the first instance, proper scientific studies are either not conducted at all or are not conducted or completed in a manner that provides sufficient information to know what the environmental impacts will be as a result of the dredging. All dredging is not bad and some activities are quite necessary, but it must be recognized that dredging of ebb-tidal deltas is generally not a good idea as the process carries many almost unavoidable caveats that lead to environmental degradation even under the best intentions. The temptation to dredge deltas is the pitfall into which many engineers fall because they are not cognizant of the potential for adverse or unwanted impacts to the environment, either in the delta itself or for many kilometers downdrift.

References

Bird ECF (1985) Coastline changes – a global review. Wiley, Chichester, 219p

Bodge KR (1993) Gross transport effects and sand management strategy at inlets. J Coast Res Special Issue No. 18, pp 111–124

Bruun P (1978) Stability of tidal inlets – theory and engineering. Elsevier, Amsterdam, p 510p

Bruun P (1995) The development of downdrift erosion. J Coast Res 11(4):1242–1257

Bush DM, Neal WJ, Longo NJ, Lindeman KC, Pilkey DF, Esteves LS, Congleton JD, Pilkey OH (2004) Living with Florida's Atlantic beaches. Duke University Press, Durham, 338p

Byrnes MR, Griffee SF, Osler MS (2010) Channel dredging and geomorphic response at and Adjacent to Mobile Pass, Alabama. Coastal Hydraulics Lab, Engineer Research and Development Center, Vicksburg, 330p

Cialone MA, Stauble DK (1998) Historical findings on shoal mining. J Coast Res 14(2):537–563

Coleman JM, Murray SP (1976) Coastal sciences – recent advances and future outlook. In: Science, technology, and the modern navy. Department of the Navy, Arlington

Cooper JAG, Anfuso G, Del Rio L (2009) Bad beach management: European perspectives. In: Kelley JT, Pilkey OH, Cooper JAG (eds) America's most vulnerable coastal communities, GSA special papers 406. Geological Society of America, Boulder, pp 167–179

Davis RA, FitzGerald DM (2004) Beaches and coasts. Blackwell, Malden, 419p

Dean RG (1990) Channel entrances: impacts on coastal erosion. In: Proceedings of the 53rd meeting of the coastal engineering research board, Fort Lauderdale/Dania, FL, 5–7 June 1990. CERC Final Report, Vicksburg, Mississippi, pp 51–53

Dean RG (1993) Terminal structures at ends of littoral systems. J Coast Res Special Issue No. 18, pp 195–210

Dean RG, Dalrymple RA (2002) Coastal processes. Cambridge University Press, Cambridge, p 475p

Dean RG, O'Brien MP (1987) Florida's East coast inlets – shoreline effects and recommended action. University of Florida, Coastal and Oceanographic Engineering Department, Gainesville, UFL/COEL-87/017, 65p

Dean RG, Walton TL (1973) Sediment transport processes in the vicinity of inlets with special reference to sand trapping. In: Proceedings of the second international estuarine research conference. Academic, New York

Dean RG, Walton TL (1975) Sediment transport processes in the vicinity of inlets with special reference to sand trapping. In: Cronin LE (ed) Proceedings of the estuarine research conference, vol 2. Academic, New York, pp 129–149

Douglas SL (1994) Beach erosion and deposition on Dauphin Island, Alabama, U.S.A. J Coast Res 10(2):306–328

Douglas SS, Bobe A, Chen QJ (2003) The amount of sand removed from America's beaches by engineering works. Coastal Sediments '03, CD-ROM, Clearwater, FL

Finkl CW (1993) Pre-emptive strategies for enhanced sand bypassing and beach replenishment activities: a geological perspective. In: Mehta A (ed) Conference on beach-inlet interactions, Jacksonville, FL, 1992. J Coast Res Special Issue No. 18, pp 59–89
Finkl CW (2004) Leaky valves in littoral sediment budgets: loss of nearshore sand to deep offshore zones via chutes in barrier reef systems, southeast coast of Florida, USA. J Coast Res 20 (2):605–611
Finkl CW, Hobbs CH III (2009) Mining sand on the continental shelf of the Atlantic and Gulf coasts of the U.S. Mar Georesour Geotechnol 27:230–253
Finkl CW, Walker HJ (2005) Beach nourishment. In: Schwartz M (ed) The encyclopedia of coastal science. Kluwer Academic (now Springer), Dordrecht, pp 37–54
Finkl CW, Benedet L, Campbell TJ, Walker HJ (2005) Beach nourishment experience in Florida, USA: the last half-century (1944–1997). International coastal planning forum, Kaohsiung, Taiwan, 28–29 Nov 2005, pp 23–42
Finkl CW, Benedet L, Campbell T (2006) Beach nourishment experience in the United States: status and trends in the 20th century. Shore Beach 74(2):8–16
FitzGerald DM (2005) Tidal inlets. In: Schwartz ML (ed) The encyclopedia of coastal science. Springer, Dordrecht, pp 958–965
Gaudiano DJ, Kana TW (2001) Shoal bypassing in mixed energy inlets: geomorphic variables and empirical predictions for nine South Carolina inlets. J Coast Res 17(2):280–291
Giese GS (2008) Tidal inlet migration "Cycles" and tidal hydrodynamics – a cautionary tale. Geological society of America abstracts with programs, vol 40(6), p 186
Hanson H, Kraus NC (2001) Chronic beach erosion adjacent to inlets and remediation by composite (T-head) groins, ERDC/CHL CHETN-IV-36. U.S. Army Corps of Engineers, Vicksburg, 15p
Hanson H, Lindh G (1993) Coastal erosion – an escalating environmental threat. Ambio 22 (4):188–195
Hicks D, Hume TM (1996) Morphology and size of ebb tidal deltas at natural inlets on open-sea and pocket-bay coasts, New Zealand. J Coast Res 12(1):47–63
Hobbs CH III (2002) An investigation of potential consequences of marine mining in shallow water: an example from the mid-Atlantic coast of the United States. J Coast Res 18(1):94–101
Hobbs CH III (2007) Considerations in marine sand mining and beach nourishment. Oceans 2007. MTS/IEEE annual meeting, Vancouver, BC. Presentations on CD, 10p
Kana T (2011) Coastal erosion and solutions: a primer. Coastal Science and Engineering, Columbia, 37p
Kana TW, Williams ML, Stevens D (1985) Managing shoreline changes in the presence of nearshore shoal migration and attachment. In: Proceedings of the coastal zone '85, ASCE, New York, pp 1277–1294
Kelley JT, Brothers LL (2009) Camp Ellis, Maine: a small beach community with a big problem. . .its jetty. In: Kelley JT, Pilkey OH, Cooper JAG (eds) America's most vulnerable coastal communities, GSA special papers 406. Geological Society of America, Boulder, pp 1–20
Khalil SM, Finkl CW (2011) Spoil or resource? Managing sediment for coastal restoration. In: Proceedings of the international coastal symposium (ICS), Szczecin. J Coast Res Special Issue 64 pp 1433–1437
Khalil SM, Finkl CW, Roberts HH, Raynie RC (2010) New approaches to sediment management on the inner continental shelf offshore coastal Louisiana. J Coast Res 26(4):591–604
Leatherman SP (1988) Beach response strategies to accelerated sea-level rise. In: Proceedings of the 2nd North American conference on preparing for climate change, The Climate Institute, Washington, DC, pp 353–358
Leonard L, Clayton T, Pilkey OH (1990) An analysis of replenished beach design parameters on U.S. east coast barrier islands. J Coast Res 6(1):15–36
Maa JP-Y, Hobbs CH III, Hardaway CS Jr (2001) A criterion for determining the impact on shorelines caused by altering wave transformation. J Coast Res 17:107–113

Montague CL (2008) Recovering the sand deficit from a century of dredging and jetties along Florida's Atlantic coast: a reevaluation of beach nourishment as an essential tool for ecological conservation. J Coast Res 24(4):899–916
Morton RA, Miller T, Moore L (2004) National assessment of shoreline change: part 1, historical shoreline changes and associated land loss along the U.S. Gulf of Mexico. U.S. Geological Survey, Open file report (2004–1043)
Morton RA, Miller T, Moore L (2005) Historical shoreline changes along the US Gulf of Mexico: a summary of recent shoreline comparisons and analyses. J Coast Res 21(4):704–709
Oertel GF (1977) Geomorphic cycles in ebb deltas and related patterns of shore erosion and accretion. J Sediment Petrol 47:1121–1131
Olsen EJ (2009) Shoreline preservation workshop: village of Bald Head Island. Olsen Associates, Inc, Jacksonville, FL. Special Purpose Paper, 16p
Pilkey OH (1990) A time to look back at beach replenishment. J Coast Res 6(1):iii–vii
Pilkey OH, Clayton TD (1989) Summary of beach replenishment experience on U.S. east coast barrier islands. J Coast Res 5(1):147–159
Price WA (1951) Reduction of maintenance by proper orientation of ship channels through tidal inlets. In: Proceedings of the second conference on coastal engineering, Houston, TX, pp 243–255
Quinn ML (1997) The history of the Beach Erosion Board, U.S. Army, Corps of Engineers, 1930–63. Department of the Army, Coastal Engineering Research Center, Fort Belvoir, VA, Miscellaneous Report MR 77–9, 181p
Sallenger A (2009) Island in a storm: a rising sea, a vanishing coast, and a nineteenth century. Public Affairs, New York
Syvitski JPM, Vorosmarty CJ, Kettner A, Green P (2005) Impact of humans on the flux of terrestrial sediment to the global coastal ocean. Science 308(5720):376–380
Trudnack ME (1977) Impacts on the inlet-beach system of Ebb tidal shoal mining. Masters thesis, University of Florida, Gainesville, 102p
Valverde HR, Trembanis AC, Pilkey OH (1999) Summary of beach nourishment episodes on the U.S. east coast barrier islands. J Coast Res 15(4):1100–1118
Van de Graaff J (1990) Dune and beach erosion and nourishment. In: Pilarczyk KW (ed) Coastal protection. Balkema, Rotterdam, pp 99–120
Wang H, Lin L, Wang X (1995) Laboratory mobile bed model studies on inlet. Part 2: ebb tidal shoal evolution process, UFL/COEL-95/021. University of Florida, Gainesville, 117p
Wright LD (1977) Sediment transport and deposition at river mounts: a synthesis. Geol Soc Am Bull 88:857–868
Young RS, Bush DM, Pilkey OH (1996) Evaluating shoreline change and associated risk from coastal hazards: an inexpensive qualitative approach. In: Berger AR, Iams WJ (eds) Geoindicators: assessing rapid environmental changes in earth systems. Balkema, Rotterdam, pp 193–206

Chapter 4
Documenting Beach Loss in Front of Seawalls in Puerto Rico: Pitfalls of Engineering a Small Island Nation Shore

Chester W. Jackson Jr, David M. Bush, and William J. Neal

Abstract The island of Puerto Rico is densely populated and heavily developed in some places, particularly along the shore and in coastal lowlands. Hard shoreline engineering is commonplace, even in low development density portions of the island. There are many examples of small trash revetments or cemented rock seawalls in front of individual buildings. In some cases, buildings themselves located within reach of waves and tides at the shoreline are behaving as seawalls too. Gabions have proliferated, even though they are decidedly not designed for either high-wave energy or salt-water usage.

Along the island's approximately 500-km long shoreline, and not counting major port city developments, 48 shoreline stretches were identified where segments of the shore contained a seawall and an immediately adjacent sandy stretch. A comparison between beach width in front of the walls and beach width of the adjacent sandy stretch showed that the ratios of natural (unstabilized) dry beach widths to those of beaches in front of seawalls is between 2:1 and over 4:1, respectively. These data corroborate findings of Wright (The effect of hard stabilization on the sediment transport system along the shoreline of Puerto Rico. Unpublished Master's thesis,

C.W. Jackson Jr
Department of Geology and Geography, Georgia Southern University,
68 Georgia Avenue, Building 201, Statesboro, GA 30460, USA
e-mail: cjackson@georgiasouthern.edu

D.M. Bush (✉)
Department of Geosciences, University of West Georgia,
1601 Maple Street, Carrollton, GA 30118, USA
e-mail: dbush@westga.edu

W.J. Neal
Department of Geology, Grand Valley State University,
1 Campus Drive, Allendale, MI 49401, USA
e-mail: nealw@gvsu.edu

J.A.G. Cooper and O.H. Pilkey (eds.), *Pitfalls of Shoreline Stabilization: Selected Case Studies*, Coastal Research Library 3, DOI 10.1007/978-94-007-4123-2_4,

Duke University, Durham, NC, 200 p, 1989) and lend credence to the claim that seawalls actively influence narrowing of beaches. The fate of beaches along developed stretches of the shoreline is grim if efforts continue to concentrate on hard structures to armor the coast.

4.1 Introduction

Puerto Rico provides a model of the pitfalls of shoreline engineering for small island nations, in economies that cannot afford massive engineering solutions, and where governments do not have coordinated coastal management plans dedicated to conserving coastal environments. The island faces the universal problem of multiple coastal hazards – persistent shoreline erosion, storm-surge flooding and wave attack from hurricanes (e.g., Hurricane Irene 2011), flooding from high-rainfall events, the rare but real threat of tsunamis, earthquakes, and mass wasting events. Sea-level rise and climate change now contribute to coastal retreat, and loss of critical coastal environments. Unfortunately, shore-hardening engineering has contributed to these impacts. In part, Puerto Rico's approach of adopting shore-hardening as a solution predates a good understanding of coastal processes, and has been driven in part by the economic dependency on port facilities and tourism, as well as a high-density population concentrated in the near coastal zone. The result is a shoreline with widespread and numerous types of hard stabilization structures (Fig. 4.1).

The purpose of this paper is to (1) give a general idea of how widespread shoreline hardening structures are around the island of Puerto Rico, (2) update the history of gabion use on the ocean shoreline of Puerto Rico, and document the deleterious effects of hard stabilization use, and (3) specifically document the impact of seawalls on beach width along the Puerto Rico shore.

4.2 Study Area

With a surface area of 8,896 km^2 (3,435 square miles) and a population of over 3.8 million, Puerto Rico's population density is about 430 people per square kilometer. However, the island's mountainous terrain constrains private and commercial development to sites concentrated in the coastal zone (e.g., level coastal areas, flood plains, areas of artificial fill, or on unstable hilly terrain). The high population density along with the extensive industrial, commercial, public, and private development since the 1950s in the coastal zone has placed both increased population and property at risk. Much development took place without knowledge or regard for the geologic hazards which affect the coastal zone. The predictable results are recurring disasters with increasing total property damage losses, and greater potential for loss of life. Such disasters

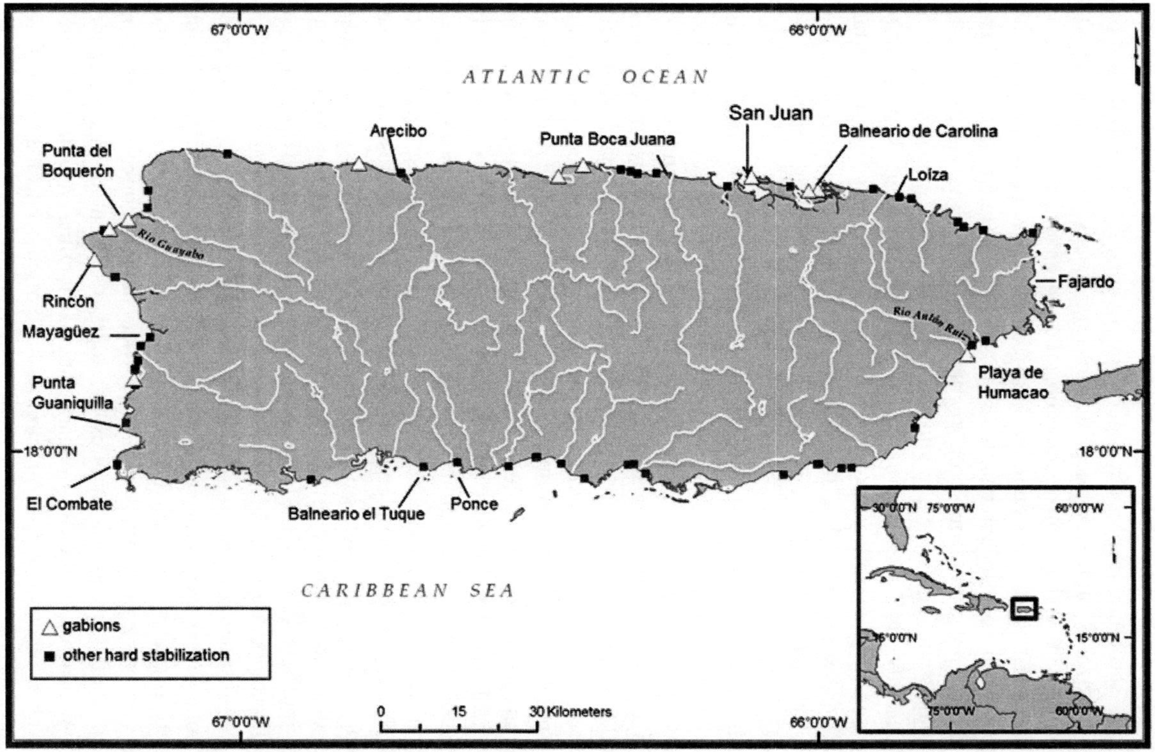

Fig. 4.1 Location of Puerto Rico's sandy shorelines engineered with seawalls, revetments, and gabions. The major port cities of San Juan, Arecibo, Mayagüez, Ponce, and Fajardo are almost completely engineered in places. The gabion locations represent the more recent engineering projects. *Inset* shows Puerto Rico's location in the Caribbean

occurred with the arrival of Hurricane Hugo in September 1989, and the passage of Hurricane Georges in October, 1998 (Thieler and Bush 1991), as well as several less intense hurricanes in the interim.

Storms are only one of many hazards affecting coastal areas. An integrated assessment of all potential coastal zone hazards is necessary for a complete understanding of shoreline response to geologic events. The multiplicity of coastal geologic hazards and their identification are discussed in Bush et al. (1995). Bush et al. (1996, 2001) presented coastal-zone hazard maps for eastern Puerto Rico (the area most impacted by Hurricane Hugo), depicting coastal geology and geomorphology, beach characteristics, offshore (inner shelf) characteristics, and hazard potential from such events as flooding, marine overwash, erosion, earthquakes, and landslides. In addition, special consideration was given to areas where shoreline engineering or dense development significantly increased the overall vulnerability (potential for property damage) of a coastal stretch. A detailed description of Puerto Rico's shoreline with information on the coastal hazards of each shoreline reach, and an extensive bibliography, can be found in Bush et al. (1995).

Puerto Rico is part of the Greater Antilles island chain (see inset on Fig. 4.1). It is approximately 160 km long (east–west) and 50–60 km wide (north–south). The surrounding insular shelf ranges in width from less than 0.5 km on the northern shelf to over 25 km on the western shelf. Including all cays and shoreline crenulations, the island has close to 1,200 km of coast line, naturally compartmentalized because of differences in hydrodynamics and geology. Adjacent compartments often have independent origins and controls (e.g., different geologic materials and structure; different orientations to marine conditions), so this compartmentalization provides a physical basis for considering compartments as separate entities. Such understanding should result in a compartmentalized approach to coastal management, rather than regarding long reaches of the coast in a uniform fashion (Jackson et al. 2009). Both environmental and economic management decisions can then be focused by compartment, and based on environmental sensitivity, geologic setting including natural hazards, and the level and type of existing development. However, an initial understanding of the vulnerability to natural hazards rests on an overview of the island's climatic, oceanographic, and geologic settings.

Shore-hardening structures are present along more than 50 coastal stretches (Fig. 4.1). It is not an overstatement to say that most of the shoreline erosion problems can be attributed to one or more of three main types of attempts to stabilize the shoreline. First, coastal communities in Puerto Rico have historically opted for various types of shore-hardening structures in order to prevent further shoreline retreat, even though such structures may enhance lateral shoreline erosion and disrupt natural beach cycles. Second, coastal communities have often been sited too close to the shoreline (and new construction largely continues to be so). And third, a history of beach, dune, and river sand mining has been a major contributor to shoreline erosion (loss of sediment supply), thus requiring perceived engineering responses.

The state of Puerto Rico's shoreline and the breadth of coastal hazards and problems facing coastal communities are discussed in several recent works. *Living With the Puerto Rico Shore* (Bush et al. 1995) discusses the coastal zone from a multi-hazards approach and gives a mile-by-mile description of coastal hazards around the entire island. *Summary of Puerto Rico's Vulnerability to Coastal Hazards: Risk, Mitigation, And Management with Examples* (Bush et al. 2009) lists seven anthropogenic actions that have contributed to overall shoreline hazard vulnerability. These are: failure to take shoreline erosion into account, constructing seawalls and revetments at the back of beaches, insufficient construction set-back from the shore, various other types of shoreline engineering structures, sand mining, inconsistent or uncontrolled shoreline stabilization, and unrealistic cost/benefit considerations. Other reports on the problems Puerto Rico faces by shoreline retreat of its sandy coastal reaches include Morelock (1978, 1984), Morelock and Barreto (2003), Thieler and Danforth (1994), and Thieler et al. (1995). Quantification of the negative impacts of shoreline structures is given by Wright (1989) who documented that beach width in front of seawalls was consistently less than beach width of sandy stretches of beach adjacent to seawalls.

4.3 Coastal Engineering in Puerto Rico

In Puerto Rico there are numerous examples of coastal engineering structures that have been destroyed already, or are in grave danger from the encroaching sea. Often this is the result of unsound shoreline management decisions, or attempting a quick fix to an acute problem. Bush et al. (2009) summarize the coastal hazard problems facing Puerto Rico.

Although the goal of this paper is not to review the literature on the negative impacts of seawalls and groins on beaches, Puerto Rico provided an early example of correlating beach narrowing and loss in front of seawalls (Pilkey and Wright 1988; Wright 1989).

Figure 4.2 shows a miscellany of the widespread impact of the seven anthropomorphic actions noted above resulting in beach loss and threatened property along the west coast of Puerto Rico. An uncounted number of similar examples exist around the entire island. Figure 4.3 shows the end point of the same dense development too close to the original shore in which buildings and their reinforced fronts become seawalls in effect, and building offsets take on the character and effect of short groins. This typical development pattern on the south coast near Ponce leads to a trashed urbanized shore.

Figure 4.4 provides a striking comparison of beach widths of different shoreline settings along the more than 45-km reach of the San Juan metropolitan area shoreline. The impact of shore hardening is apparent with natural sandy beaches being the widest, averaging over 19 m. In contrast, beaches with seawalls average around 1 m, which is about the same as natural rocky beaches.

Fig. 4.2 Examples of shore-hardening structures destroyed, damaged, or threatened by shoreline retreat on the west coast of Puerto Rico (From Bush et al. 2009). (**a**) Damaged wall and beach loss west of Rincón in 2000. (**b**) Rip-rap and building-front bulkheads offer little protection from storm waves and contribute to beach narrowing at Punta de Boquerón in 2002. (**c**) Development at the immediate back of the beach (as in all of these examples) near Punta Guaniquilla (2000) contributed to total beach loss. (**d**) Dense along-shore development at El Combate (2000) resulted in a variety of small, inefficient walls, bulkheads, short groins, and frontal reinforcement of buildings that cut off sediment supply and resulted in beach narrowing

4.4 The Gabion Pitfall on the Puerto Rico Shore

Gabions are low-cost engineering construction units for hardening structures in common use in Puerto Rico to combat coastal and riverine erosion. Individual units consist of wire cages or baskets filled with rocks and stacked on top of one another to construct shore-stabilization structures (e.g., bulkheads, revetments, seawalls, and groins). Traditionally, gabions are not used along open-ocean coasts or in high-energy coastal environments, and where they have been used in such settings they fail (USACE 1986). Gabion retainer walls and bulkheads are more commonly used to stabilize slopes, as levees for streams and rivers, and as check dams on small streams. However, the low-cost and ease of assembly on the construction site make gabions an attractive alternative to other coastal shore-hardening construction materials. As a result, coastal communities in Puerto Rico, often with

Fig. 4.3 No beach remains in front of these damaged buildings and collapsed portion of small seawalls in front of houses along Route 2 near Balneario El Tuque. The offsets of the closely spaced buildings and hodge-podge structures also have the effect of a groin field (From Bush et al. 2009)

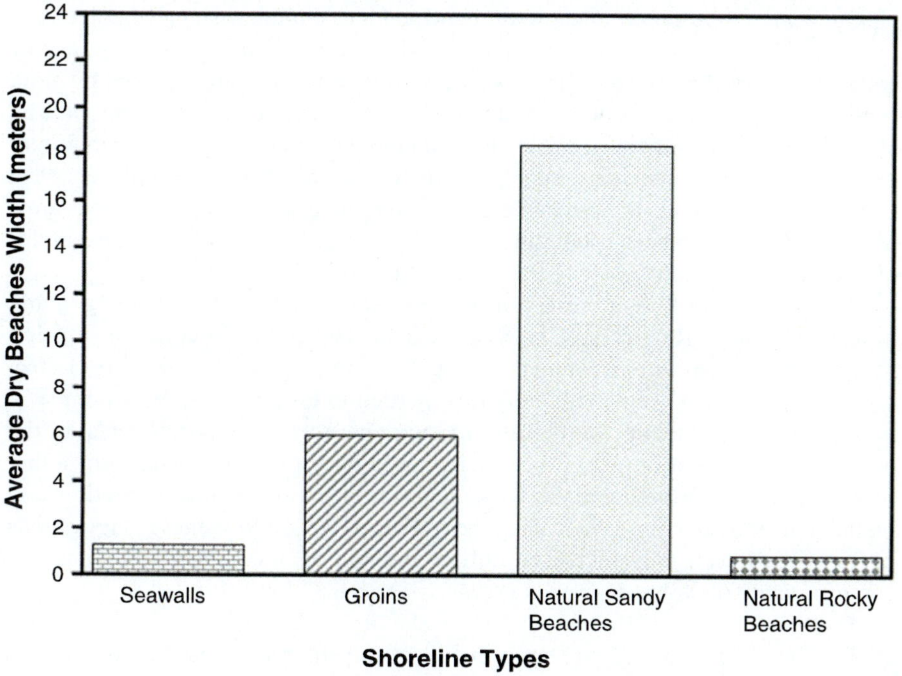

Fig. 4.4 Dry beach widths versus shoreline type along the San Juan metropolitan shoreline (From Bush et al. 2009). The approximately 45-km reach of shoreline stretches from Punta Boca Juana in the west to Punta Uvero in the east. Dry beach width is taken as an indicator of beaches having recreational value as well as providing some protection from average storm waves. Some of these data were presented in Pilkey and Wright (1988, their Fig. 4.6)

limited resources, have opted for gabions in shore-hardening construction, even if they are not ideal (Jackson et al. 2006).

In Puerto Rico, the following problems with gabions on open-ocean shorelines have been observed: (1) gradual failure by wire degradation and basket rupturing, (2) instantaneous failure during storms by undercutting, piping, subsidence, and toppling, (3) rock leakage onto the beach creating hazards for beach users, (4) protruding wire on the beach creating a recreational hazard, and (5) loss of the recreational value of the beach due to these hazards and the narrowing or loss of the beach. Failure or ineffectiveness of gabions has resulted in their replacement, reconstruction, or encasement in other materials (e.g., grouting, cement and asphalt covers), defeating the initial economy of the gabion structure and illustrating their ineffectiveness on ocean shorelines. In at least two cases in Puerto Rico (Rincón and Carolina) the gabion structures were unnecessarily built as walls on public swimming beaches where no development was at risk. In Rincón (on the west coast) the wall was built on the back of a wide sandy beach. The gabion eventually failed and was replaced with a concrete wall. In Carolina (on the north coast just east of San Juan) the wall was built right on the shoreline. Several generations of gabions finally gave way to a revetment wall (Fig. 4.5).

Four stages of gabion life have been identified: (1) emplacement—newly emplaced gabions can cause active, passive, or placement loss of the beach, similar to seawalls and revetments (Pilkey and Wright 1988), (2) initial weakening—piping, wire deterioration and rupture, and gabion slumping signal the eventual failure of the gabion structure, (3) failure—the gabions leak rock content and/or collapse onto the beach, and (4) replacement—failed gabions are replaced by more substantial hard structures or covered in place with concrete or other material. The alleged low-cost of gabions is misleading and more than offset by their high failure rate, negative environmental impact, and threat to the safety of beach users. The Puerto Rico experience indicates that even with regular maintenance, gabions are a poor choice for open-ocean shoreline protection. Figures 4.6 and 4.7 give examples of gabion use and failures along the Puerto Rico shoreline.

Gabions, as well as more traditional seawalls, can be found along many beaches in Puerto Rico (Fig. 4.1). Several important points to consider in using gabions in the coastal environment are spelled out in detail by USACE (1986); most of which have been completely disregarded in Puerto Rico. For example, it is suggested that gabions not be used in high energy wave environments, in the active surf zone, nor on public beaches where injury to bathers from protruding wire is possible. Moreover, the final suggestion given is that inspection and maintenance plans must be established. Failure to follow these suggestions has resulted in not only increased dollar cost, but also a loss of aesthetics and compromising of safety for beach users.

A better management alternative is to identify high erosion risk zones (Bush et al. 1996, 2001; Thieler and Danforth 1994), and enforce set-back regulations as well as to develop retreat/relocation strategies. If shore-hardening structures are warranted as a last resort, communities should opt for the best design and strongest

Fig. 4.5 Balneario de Carolina (previously known as Balneario de Isla Verde). Balnearios are public swimming beaches with amenities and maintained by a local government. Major erosion caused by swell from a January, 1988, winter storm, and the passage of Hurricane Hugo in September, 1989, resulted in the emplacement of gabions. Continued erosion along with deterioration of the gabions has led to several generations of modifications and repairs to the gabions. Ultimately the gabions have failed, spilling rock and wire debris onto the recreational beach (From Bush et al. 2009). (**a**) Immediately after the January, 1988, winter storm swell erosion. View to the east. The offshore boat ramp marks the shoreline previous to the mid-1970's. (**b**) Similar view as 2-a, but in 1994 after gabion has been built and begun to deteriorate. Note filter cloth behind gabion. (**c**) A wooden observation deck was added and threatened by erosion by 2001. View to the west. The gabion continued to be undermined and to collapse. (**d**) By 2006, the gabion wall had failed so completely that it had been replaced by a rip-rap wall. Beach narrowing continued and the "Removable Porous Groin System" had been emplaced, but this alternative device also failed. It was promised by the contractor that this porous groin system would restore the beach and prevent further erosion. No such beach restoration was noticed and the structures were removed by 2009

materials. The latter is not likely to fall into the "low cost" category and will result in loss of the recreational beach (Jackson et al. 2001).

Why gabions came into such favor as the structure of choice in Puerto Rico is not clear, or how these structures met permitting and construction requirements, but the perception of a relatively low initial cost played a part. Perhaps an enterprising contractor convinced the decision makers that gabions were the best choice for combating shoreline erosion in Puerto Rico, although they clearly failed the test.

Fig. 4.6 Gabion wall near Aguada's Parque de Colon, built in the late 1980s (From Jackson et al. 2006). (**a**) and (**b**) By 1992 the wall had already deteriorated and been "improved" with a concrete covering, which in turn had been broken by storm wave action. Fill rock from the failed gabion was observed to be washed several tens of meters up into the small drainageway visible in upper right of photo a. Sharp, angular cobbles of leaked fill stone and protruding ends of wire strands made the narrowed beach hazardous to walk on. (**c**) Just to the south, the entire bluff was covered with a sloping gabion revetment/wall, visible in this 1992 photograph. Note that the wire mesh had trapped driftwood and flotsam. The steep wall cuts off the underlying sand scarp sediment source, and increases the erosional effect of waves. (**d**) By summer of 2000, the gabion in photograph (c) was severely compromised, ruptured and devoid of rock fill over much of its length, and partially slumped and toppled. The scarp was again eroding, providing some new sand to the beach

4.5 Seawall Impact on Beach Width

Coastal geologists generally agree that shoreline armoring (e.g., various types of groins, breakwaters, and seawalls) contributes to beach erosion. Groins and breakwaters interrupt longshore sediment transport and trap sand, resulting in the cutoff of sediment supply to down-drift beaches. The effects of seawalls (used here for all shore-parallel structures designed to hold the land in place and dissipate wave energy) are more controversial, and disagreement as to their erosional effect exists between some geologists and coastal engineers (see Kraus and Pilkey 1988, for introduction to the early literature and Tait and Griggs 1990).

One of the early studies of seawalls' impact on beaches was that of Wright (1989) who compared beach profiles between beaches in front of seawalls to those

Fig. 4.7 After erosion caused by Hurricane David in 1979, a gabion was emplaced along a 100-m stretch of the Playa de Humacao shoreline just south of Rio Anton Ruíz (From Jackson et al. 2006). (**a**) Only a few years old in 1988, the wall was already beginning to show signs of deterioration (ruptured wire, rock leakage). (**b**) Hurricane Hugo (1989) did not cause major damage to the wall, but narrowed the beach for a short time. Some rock fill was spilling onto the beach. (**c**) By 1991 the wall had failed in several places. (**d**) By 2002 there were no signs of maintenance or upkeep of the wall. The gabions were overgrown with vegetation. The 20-year history of this wall suggests that it was an unnecessary response to the extreme hurricane event, at the expense of the beach that had been there previously. Failure to maintain the wall would have contributed to its loss if shoreline erosion had been a serious on-going threat at this locality

of adjacent beaches on unwalled reaches of the shoreline around Puerto Rico. In particular the study compared dry-beach width between the beaches as a general indicator of the health of the beach. Wright concluded that beaches in front of seawalls were consistently narrower or absent in comparison to the adjacent unwalled beaches.

Beach degradation and loss in front of seawalls is attributed to three mechanisms (Pilkey and Wright 1988): passive loss, placement loss, and active loss. *Passive loss* results when a fixed barrier is built adjacent to an eroding shoreline. The beach is retreating and will eventually reach the position of the seawall. The process causing the retreat (e.g., storm wave run-up; sea-level rise) will not be affected by the seawall. When the beach can no longer move landward, it will begin to narrow, and erosion will continue until the beach is gone. Often one can see where the beach position would be if there were no seawall by looking at the adjacent beach.

Placement loss occurs when a seawall is built seaward of the high-tide line, removing part of the beach at the time of construction. *Active loss* is the result of erosion generated by the interaction of the seawall with waves and currents, particularly during storms. Active loss along and adjacent to seawalls might vary in spatial extent due to one or more processes enhancing erosion, but possible contributing factors include:

- wave reflection off of the seawall causing erosion in front of the wall
- wave refraction and energy concentrated at the end of the wall
- cut-off of sand supply to the beach by eliminating back-beach erosion
- blocking beach retreat (profile adjustment) during storms (passive).

Wright (1989) took approximately 230 profiles at 48 locations around Puerto Rico (Fig. 4.8) and grouped his results regionally (Fig. 4.9).

Since 1999, an ongoing project has been to reoccupy as many of Eric Wright's profile locations as possible and to re-profile these sections. The data are treated regionally, and a longer-term goal, not discussed here, is to determine what may be controlling these trends (e.g., natural setting vs. human interference with shoreline processes).

The 48 areas originally chosen for study reflected those parts of the Puerto Rico shoreline most heavily armored, or stabilized by seawalls and groins (Wright 1989). The first comparison of stabilized vs. unstabilized beach widths in Puerto Rico was in the San Juan area on the north coast. Pilkey and Wright (1988) reported that dry beach widths for unstabilized pocket beaches between Punta Uvero and Punta Boca Juana averaged over 18 m (Fig. 4.4) while stabilized beaches in the area averaged only 1 m wide. This marked difference reflected, in part, the nearly continuous stabilized shore of the San Juan metropolitan area where the beaches have been degrading for some time.

The heavy reliance on seawalls as a shoreline erosion management tool and their suspected contribution to additional erosion was apparent by the mid-1980s. Wright's (1989) general conclusion is shown in Fig. 4.10: in all areas of Puerto Rico, dry beach width is wider for natural (unstabilized) beaches than for beaches in front of seawalls by a factor of 2:1 to over 4:1.

In attempting to expand the study by Eric Wright (1989), it became clear that not all of the Wright profile locations could be reoccupied because of seawall failure and replacement or enlargement of the structures. A regional study in 2000 represents a partial sampling of 37 of Wright's 48 localities, matching 106 out of 230 (45%) of his beach profile locations (Table 4.1). This sampling density gives a good approximation of the four coasts with the exception of the West coast that was densely sampled by Wright. Subsequent data collection continues in order to continue to fill out the data set.

The 2000 profile results are summarized in Fig. 4.11, and compared to the Wright (1989) summary in Fig. 4.12 and Table 4.1. Overall, dry beach widths are narrower in front of seawalls than for unstabilized beaches in 97% of the locations. No dry beach exists in front of 55% of the shoreline stabilization structures, and many lack any beach at all. Exceptions do occur where accretion is resulting in spite

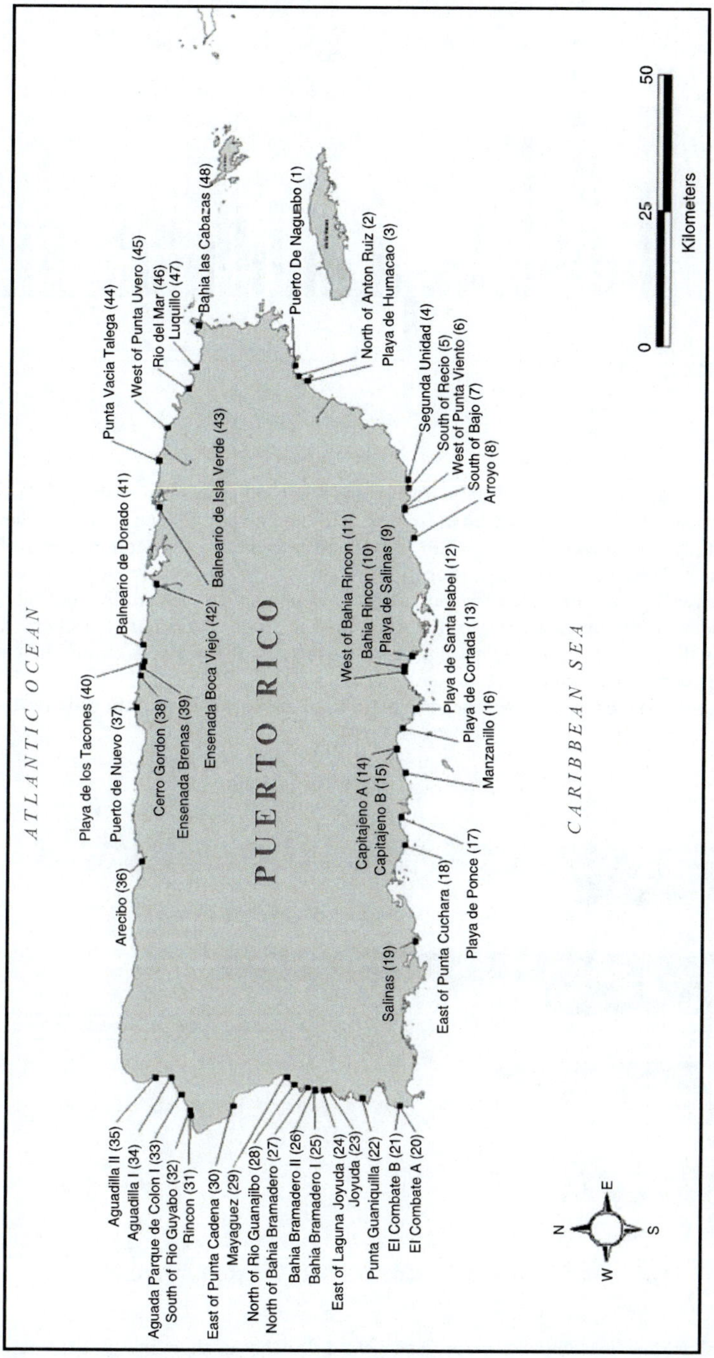

Fig. 4.8 Location of the 48 beach profile sites included in the Wright (1989) study. Most of these sites were reoccupied for the current study. Only beaches that had walled and adjacent unwalled beaches were selected for the study. A total of 230 individual profiles were taken in front of walls and adjacent unwalled beaches

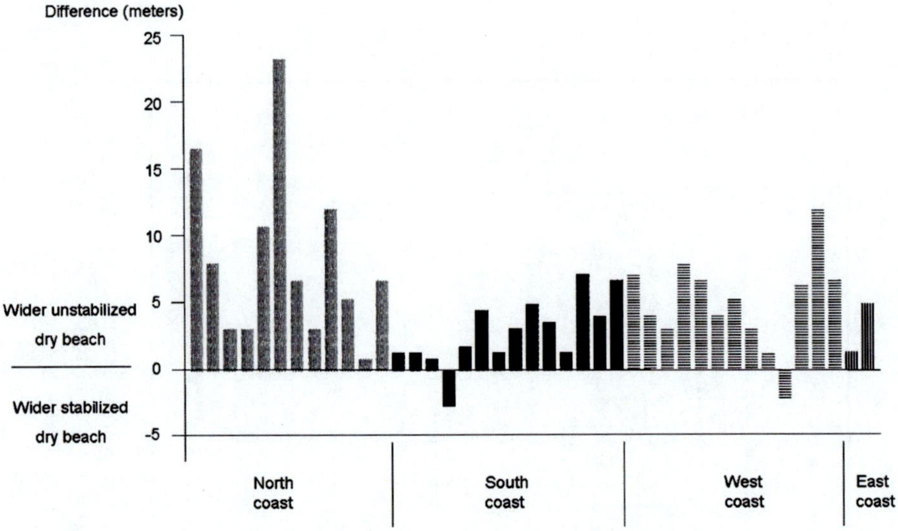

Fig. 4.9 Results of the Wright (1989) study. Each vertical bar represents one of the profile sites. At each site dry beach width was measured in front of walls and on adjacent unwalled beaches. In particular, the study compared dry-beach width between the beaches as a general indicator of the health of the beach. The data are arranged geographically around the coast showing the difference in width between walled and unwalled beaches. Bars extending above the zero line indicate that the beach is wider on the unwalled stretch of beach than in front of the walled sections. Bars extending below the zero line indicate that the beach is wider on the walled section. The causal relationship between beach width and presence of seawalls is still argued in some circles. However, almost without exception, beaches in Puerto Rico are narrower in front of seawalls

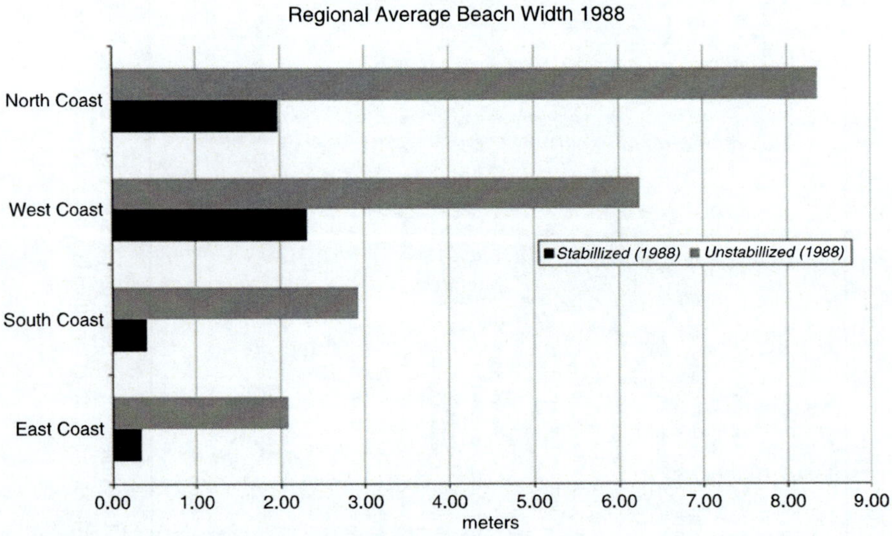

Fig. 4.10 Beach widths on stabilized versus unstabilized stretches of shorelines in 1988 based on the Wright (1989) data. Wright's general conclusion is that in all areas of Puerto Rico, dry beach width is greater for natural (unstabilized) beaches than for beaches in front of seawalls by a factor of 2:1 to over 4:1

Table 4.1 Summary of 2000 beach width data and comparison to 1988 (Wright 1989) data

Total number of sites	48
Number of beach profiles measured by Wright (1989)	230
Number of beach profiles reoccupied and measured in 2000	106 (46%)
Number of sites with profiles that were lost due to increased erosion/stabilization	35 (73%)
Number of sites with increased erosion from 1988 to 2000	43 (90%)
Number of sites with increased beach width 1988 to 2000	5 (10%)

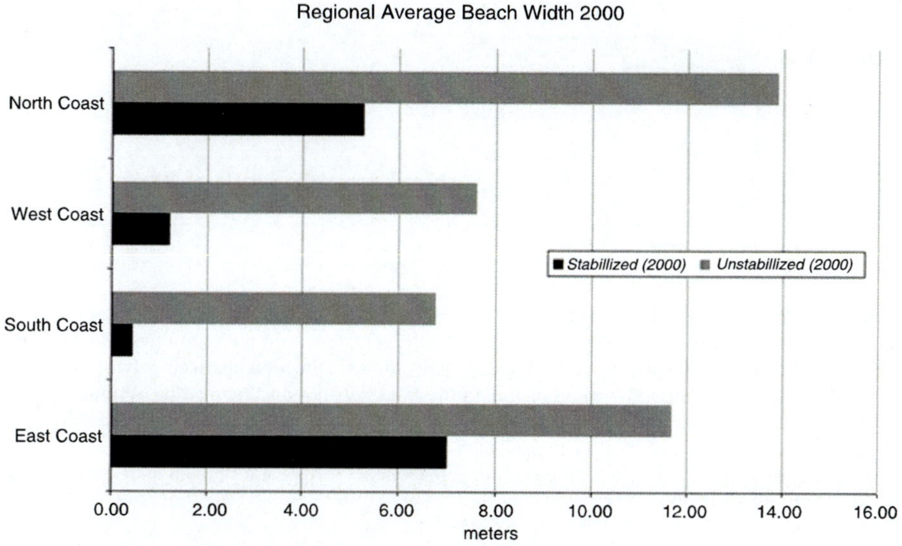

Fig. 4.11 Beach widths on stabilized versus unstabilized stretches of shorelines in 2000. Overall, dry beach widths in 2000 were narrower in front of seawalls than for unstabilized beaches in 97% of the locations. No dry beach exists in front of 55% of the shoreline stabilization structures

of the seawalls, such as near stream mouths or the down-drift terminus of littoral cells where sand supplies are high. Beach widths along these sites, when averaged together with remaining localities within the region, can help explain the appearance of an increase in beach widths from 1988 to 2000. Playa de Humacao (East coast), North of Rio Anton Ruíz (East coast), Cerro Gordo (North coast), Ensenada Brenas (North coast), and Luquillo (North coast) had unstablized and stabilized beach widths that ranged from 15 to 32 m in 2000, increases more than double what was recorded in the previous study along those localities. However, these locations only account for 10% of the study sites, while net erosion was observed along the remaining 43 sites (Table 4.1). Ignoring sites with net accretion, average unstabilized and stabilized beach widths were approximately 6 m and 1 m or less respectively.

Another problem encountered in the 2000 study that accounts for the appearance of net increase of beach widths is that in both data. The work of sets the profiles are not tied to a fixed-point of known elevation or feature. Wright (1989)

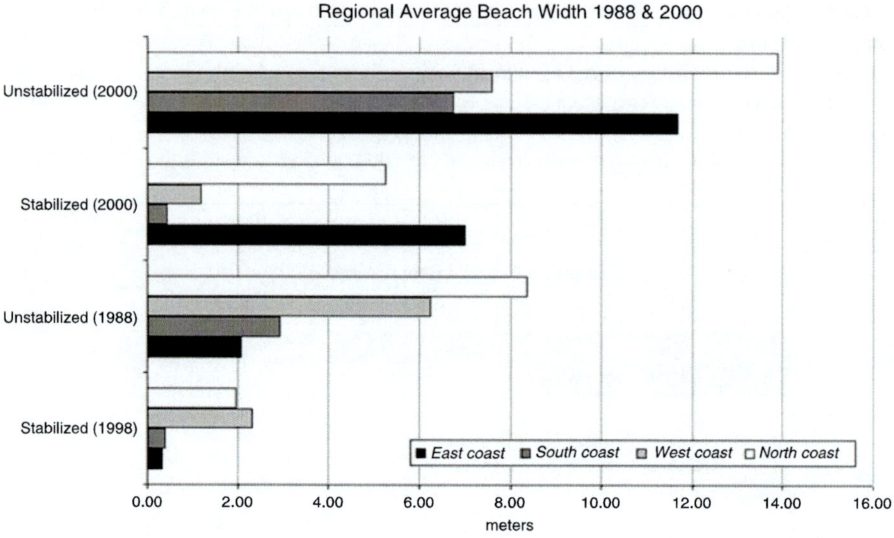

Fig. 4.12 Comparison of beach widths from the Wright (1989) study and the 2000 study. There appears to be an increase in beach width from 1988 to 2000, however, this is not the case. Overall trends within each region are about the same in both studies. Direct comparison between the two data sets is limited because the origin of the profile lines was not documented by Wright (1989). In addition, a small handful of profile sites are near stream mouths or the down-drift terminus of littoral cells where sand supplies are high. Beach widths along these sites, when averaged together with remaining localities within the region, can help explain the appearance of an increase in beach widths from 1988 to 2000

was vague in its data presentation as to the definition of the zero line, and the current study had the same problem in how to define a common point for comparing profiles. Also, in the 2000 study the maximum landward extent of the beach sometimes included the terminus of overwash deposits, toe of incipient dunes, and other active sandy environments of the backshore. It is unclear if the widths of these features ending at the wet/dry line (shoreline) were measured as the dry beach width by Wright (1989) or were even present in matching profiles between the two studies. In other words, a beach might appear wider based on a sandy overwash deposit extending landward well beyond the edge of where dunes and/or stable vegetation once existed.

Ultimately, the comparison of the averaged beach widths for the two studies (Fig. 4.12) is misleading in that the year 2000 data give the impression that most beaches in Puerto Rico have accreted over the last 10 years, but this is not the case. Overall trends within each region are about the same in both studies. In general, the narrowest beaches, both natural and in front of seawalls, are along the south coast. The data for the east coast show the widest natural beaches, but the average is for a small number of beaches. The relative ratios of natural (unstabilized) dry beach widths to those of beaches in front of seawalls are similar between both data sets, 2:1 to over 4:1.

This relationship needs to be conveyed to coastal managers and city officials so that better alternatives are sought to mitigate shoreline erosion. Beach nourishment, construction setbacks, and relocation of buildings are a few of the alternatives available to reduce property loss from erosion, and which also reduce the impacts of other coastal hazards.

General conclusions that can be drawn from the profile data are as follows:

1. Dry beach widths in front of seawalls were consistently narrower than adjacent natural beaches. Beaches are completely absent in front of many seawalls whereas natural beaches exist in positions landward of the ends of the seawalls. Some of these settings are definitely due to passive loss, and some may now be contributing to active beach loss, but the original conditions at the time of construction are not known, so placement loss cannot be evaluated. Newer walls that caused placement type loss are known in Puerto Rico, but were not included in this study because they were not in the original data set (Wright 1989).
2. The determined regional and overall patterns of narrower beaches in front of seawalls replicates Wright's 1989 results and his conclusions (Figs. 4.9, 4.10, 4.11, and 4.12).
3. Although the reference methods for determining the dry beach width are not consistent between the two studies, similar orders of magnitude were obtained for the differences between unstabilized and stabilized beach widths within regions (2:1 to over 4:1) with closest agreement on the north and east shores.
4. The greatest difference was for the narrow beaches on the south coast, and it is plausible to conclude that beach degradation in front of seawalls along this coast has worsened over the last decade.
5. Profiles between the two studies are not tied to a common reference point so direct measurements of decadal beach shape change cannot be determined, but slopes can be compared. Steepening has occurred in front of some walls, and new walls have been placed in front of some of the structures that were included in Wright's study. Comparison with old photos for a few locations provides qualitative evidence of continued loss of beach in front of seawalls.

4.6 The Future of Puerto Rico's Shoreline

Coastal managers and communities continue to look for panaceas to solve their coastal erosion problems. The "low cost" tag on a device or approach with a catchy copyrighted name and the claim that it solved the same kind of problem in a far-away place is an attractor to such planners and officials. See McQuarrie and Pilkey (1998) for a description of "non-traditional" shoreline stabilization devices. Puerto Rico has fewer examples than the Atlantic Coast of the U.S.A., however, one example is sufficient to illustrate the problem. Balneario Carolina (formerly Balneario Isla Verde) is a popular recreational beach in the San Juan metropolitan

Fig. 4.13 Installation, failure, and removal of a "Removable Porous Groin System" along Balneario de Carolina from 2006 to 2011 (see text for further discussion). The recreational beach in 2006 along the rock seawall is no longer present in 2008 and there is severe deterioration of the groin structure. Note the removal of abandoned buildings and light posts between 2006 and 2008 that were once part of the public park behind the chronically eroding beach. Between 2008 and 2011, additional rocks were added to the seawall. A stated advantage of the groin system is, indeed, that it is removable. The system is purported to "regenerate" beaches. Obviously that was not the result in Puerto Rico

area. A persistent erosion problem was exacerbated by the cut-off of the sand supply by the jetty at Boca de Cangrejos, and earlier removal of sand from the beach system when the channel for the entrance to the Boca de Cangrejos marina was dredged. At the eastern end of this reach, rock revetments have been in place for many years to protect the road. The response was to build a gabion wall which failed and was replaced by a revetment (Fig. 4.5). When erosion continued, the local government paid over $4 million for a "Removable Porous Groin System" in 2006 (Fig. 4.5, 2006 photo). It was promised that this groin system would restore the beach and prevent further erosion, however, it too failed leaving the beach littered with structural debris. Figure 4.13 shows the history of the porous groin at Balneario Carolina, including the groin's removal and its replacement with a revetment. The shoreline segment was left with ultimately less dry beach. This proves the old adage that you get what you pay for, and the coastal protection corollary that if you opt for shore hardening you will need the most expensive structure, and you must be willing to give up your beach.

History suggests that it is likely that Puerto Rico will continue with crisis-based responses to coastal management. That is, reacting to storms or sea-level rise, instead of formulating a coordinated long-term approach. Scarcity of funds often forces that type of approach. Puerto Rico's compartmentalized coast, however, provides a setting whereby more foresight can be utilized at least along short stretches of the shoreline (Jackson et al. 2009).

Acknowledgments Many agencies and individuals have provided support for our various Puerto Rico projects over the past three decades. We express our gratitude to FEMA, NSF, USGS, the University of Puerto Rico Sea Grant College Program, University of West Georgia Faculty Research Grants and Student Research Assistantship Program, the Michigan Space Grant Consortium, and the University of Puerto Rico-Mayagüez Marine Science Station. In addition, special thanks go to Charlie González and Elisabeth Hyman for hosting students and collaborators over the years, and to Rob Young and Orrin Pilkey for insight and recommendations. To any others we have neglected to thank, we apologize.

References

Bush DM, Webb RMT, Hyman L, González LJ, Neal WJ (1995) Living with the Puerto Rico shore. Duke University Press, Durham, 193 p

Bush DM, Richmond BR, Neal WJ (1996) Coastal zone hazards maps of Puerto Rico: Hurricane Hugo impacted portion of the shoreline, Cibuco (Punta Garaza) to Punta Viento. U.S. Geological Survey open-file report, pp 96–506, 12 p, 12 sheets

Bush DM, Richmond BR, Neal WJ (2001) Coastal zone hazard maps: Eastern Puerto Rico. Environ Geosci 8:38–60

Bush DM, Neal WJ, Jackson CW (2009) Summary of Puerto Rico's vulnerability to coastal hazards: risk, mitigation, and management with examples. In: Kelley JT, Pilkey OH, Cooper JAG (eds) America's most vulnerable coastal communities, Geological Society of America special paper 460. Geological Society of America, Boulder, Colorado, pp 149–165

Jackson CW, Adams A, Bush DM, Wright EE, Neal WJ (2001) Beach width and profile comparison between natural beaches and beaches fronting seawalls: abstracts with programs, vol 33, no. 2, Geological Society of America Southeastern Section Meeting, Raleigh, NC, p 71

Jackson CW, Bush DM, Neal WJ (2006) Gabions, a poor design choice for shore hardening: the Puerto Rico experience. In: Proceedings of the 8th international coastal symposium, Itajai. J Coast Res SI 39:852–857

Jackson CW Jr, Bush DM, Neal WJ (2009) The coastal compartment management plan: using Puerto Rico as a model. Southeast Geol 46(2):69–84

Kraus NC, Pilkey OH (eds) (1988) The effects of seawalls on beaches. J Coast Res Special Issue no. 4, CERF, Charlottesville, VA, 146p

McQuarrie ME, Pilkey OH (1998) Evaluation of alternative or non-traditional shoreline stabilization devices. In: Finkl CW, Bruun P (eds) Proceedings of the international coastal symposium (ICS98), West Palm Beach. J Coast Res Special Issue 26:269–272

Morelock J (1978) The shoreline of Puerto Rico. Department of Natural and Environmental Resources, San Juan, 45 p

Morelock J (1984) Coastal erosion in Puerto Rico. Shore Beach 52:18–27

Morelock J, Barreto M (2003) An update on coastal erosion in Puerto Rico. Shore Beach 71:7–11

Pilkey OH, Wright HL III (1988) Seawalls versus beaches. In: Kraus NC, Pilkey OH (eds) The effects of seawalls on beaches. J Coast Res Special Issue 4:41–64

Tait J, Griggs G (1990) Beach response to the presence of a seawall. Shore Beach 58:11–28

Thieler ER, Bush DM (1991) Gilbert and Hugo: hurricanes with powerful messages for coastal development. J Geol Educ 39:291–299

Thieler ER, Danforth WW (1994) Historical shoreline mapping (II): application of the digital shoreline mapping and analysis systems (DSMS/DSAS) to shoreline change mapping in Puerto Rico. J Coast Res 10:600–620

Thieler ER, Rodriguez RW, Carlo M (1995) Beach erosion and coastal development at Rincón, Puerto Rico. Shore Beach 63:18–28

USACE (1986) Use of gabions in the coastal environment. U.S. Army Corps of Engineers, Coastal Engineering Technical Note, CETN-III-31, 12/86, 5 p

Wright EE (1989) The effect of hard stabilization on the sediment transport system along the shoreline of Puerto Rico. Unpublished Master's thesis, Duke University, Durham, NC, 200 p

Chapter 5
Narratives of Shoreline Erosion and Protection at Shishmaref, Alaska: The Anecdotal and the Analytical

Owen K. Mason, James W. Jordan, Leanne Lestak, and William F. Manley

Abstract Mitigating, or adapting to, the impacts of environmental change on coastal landscapes, from both social and engineering perspectives, requires accurate baseline data that must be related to geomorphic processes. However, the inherent social and environmental dynamics of the coastal zone set up a contentious situation for decision makers and researchers because of the real, perceived, and stochastic nature of catastrophic threats to human life and property loss. Anecdotal accounts and first person observations generally propel the media and influence governmental policy far more effectively than scientific data. However, claims of extreme erosion rates are more adequately addressed through photogrammetric studies of erosion. Contrary to anecdotal accounts from Shishmaref, Alaska, sequential aerial photographs from 1950 to 2007 reveal that erosion has increased on the south-facing shores of the Chukchi Sea, while prior to 1977, erosion was higher on the north-facing shores such as Shishmaref. In addition, comparisons of property records indicate that high rates of erosion prevailed prior to 1950. Several engineering solutions were attempted in Shishmaref between 1983 and 2003, including gabions and cinder block/boulder/cobble revetments, leading to increased end-around erosion downdrift and an erosion rate twice that of undeveloped, unarmored shorelines. To adapt to heightened erosion rates, societies should either retreat from the shore or confront ever-increasing engineering costs.

O.K. Mason (✉) • L. Lestak • W.F. Manley
Institute of Arctic and Alpine Research, University of Colorado,
Campus Box 450, Boulder, CO 80309-450, USA
e-mail: Owen.Mason@colorado.edu; geoarch@ptialaska.net; Leanne.Lestak@colorado.edu; William.Manley@colorado.edu

J.W. Jordan
Department of Environmental Studies Program, Antioch University New England,
40 Avon St., Keene, NH 03431, USA
e-mail: jjordan@antioch.edu

J.A.G. Cooper and O.H. Pilkey (eds.), *Pitfalls of Shoreline Stabilization: Selected Case Studies*, Coastal Research Library 3, DOI 10.1007/978-94-007-4123-2_5,

5.1 Introduction

Shishmaref's erosion crisis has generated several metaphors: it is the proverbial "panting" global change "canary" (Murphy 2001), or the "poster child of the Arctic global change and sea level rise problem" (Pilkey and Young 2009:8)—a public relations coup for a small, impoverished Native community (pop. 625) coping with both the reality and unquestioned assumption of anthropogenically-forced environmental change (Marino and Schweitzer 2009). Though still living on the island, Shishmaref residents are characterized as "climate refugees" (Argos Collective 2008),[1] visually equated on the web with Maldive Islanders or herders in the parched Sahel. The prevailing narrative from Shishmaref represents it as "the front line" of climate change, the "one place most identified with global warming and climate change" (Lord 2011:141). Shishmaref *does* face a dual threat, both from coastal erosion and from the thinning and disappearance of sea ice that may cripple its subsistence economy; the latter issue is outside the scope of this paper (Wisniewski 2005, 2010; Lord 2011). The threat to the community's survival was considered imminent even in the early 1970s as the first engineering firm assessed its coastal erosion problem; DOWL (1975:2–3) anticipated that "one third to half" the village would be in the sea "in 1–10 years." A generation later, in 2003, the same imminent threat is routinely stated on the web (Argos Collective 2008). Missing from the media and community conversation is that the 1 km-long bluff on which the modern village is concentrated is a developed coastal reach that has been subject to nearly 75 years of erosion control efforts and that its erosion history differs significantly from that of adjacent undeveloped coasts on the Seward Peninsula. In terms of historic erosion processes, Shishmaref more resembles some areas of the New Jersey shore and is better understood as a battle in the on-going "war" between the U.S. Army Corps of Engineers (ACE) and the shore (Pilkey and Dixon 1996). How can anectodal and media accounts be reconciled with geological reality?

Field observations and photogrammetric analyses can authenticate and validate any changes in landform and vegetation cover and in anticipating future changes. Shifts in bluff and beach erosion can be readily documented by sequential measurements on geographically-referenced images (Manley et al. 2007). In this paper, we report on the first geo-referenced analyses of sequential aerial photography that documents 53 years of shoreline change at Shishmaref. We compare these analyses with anecdotal accounts, as well as earlier erosion studies completed in 2006 by the Army Corps of Engineers and by geo-engineering consultants from the 1970s.

---

[1] A photo caption states: "Shishmaref, Sarichef island, Alaska. Climate change is melting the permafrost which the village is built on. The erosion of their home gives the Inuits up until 15 years to find another home (*Guy-Pierre Chomette*)." Argos Collective (2008).

5.1.1 Environmental Problems in Coastal Alaska

Numerous coastal villages in northwest Alaska are situated on sand dunes and barrier islands and face the loss of property from an apparent intensification of erosion, linked by climatic models to an anthropogenically-accelerated loss of seasonal ice cover and shifting storm tracks (Lynch et al. 2008; Serreze et al. 2007). The Army Corps of Engineers estimates that over 60 villages along coasts and rivers in Alaska face imminent threat from erosion; engineering mitigation costs for the entire state are calculated in the hundreds of billions of dollars (US ACE 2006a). Communities built on sand are particularly vulnerable to storm waves and sea level rise; in fact the greatest natural disaster to hit the United States struck the barrier island city of Galveston in 1900, killing 8,000 (Hebert et al. 1996). Barrier islands in particular are dynamic land forms that can be modified in hours to the force of storm surges and hurricanes – events that can readily destroy structures, cut new channels across islands, and redistribute beach sand inland atop back-barrier marshes (Hayes 1967).

5.2 Geomorphic Constraints

The coasts of northwest Alaska are beyond the effects of late Pleistocene continental glaciation and lie within a zone with minimal seismicity; earthquakes and tsunami are rare and insignificant (Fujita et al. 1990). Ice-covered annually for between 8 and 9 months, the southern Chukchi Sea is microtidal (<1 m), but is wave-dominated and susceptible to fall storms that can develop across several 100 km, generating surges of up to 4 m (Wise et al. 1981). Sand and gravel barrier island chains front a substantial portion of the coasts of the Chukchi Sea, as a consequence of a shelf reservoir of mobile sediments in the littoral zone, transported by onshore waves (Hameedi and Naidu 1988; Shepard and Wanless 1971). The sandy Shishmaref barrier island chain fronts 90 km of the northern Seward Peninsula and encloses one major and several minor shallow lagoons (Fig. 5.1). Sand derived from the shallow continental shelf has migrated landward with the Holocene rise in sea level (Mason and Jordan 2002), much as it did during the last interglacial high stand of sea level (Isotope Stage 5e; Brigham-Grette and Hopkins 1995), when the coastal dunes of a barrier island chain were deposited as a thin veneer on the present mainland (Sainsbury 1967).

5.3 Shishmaref's Geographic Setting

Shishmaref is one of several communities in Alaska situated on sandy barrier islands or spits, at least two of which, Pt. Lay and Pt. Hope, have re-located since 1960 (Mason et al. 1997; Mason 2006). Drumstick-shaped Sarichef Island,

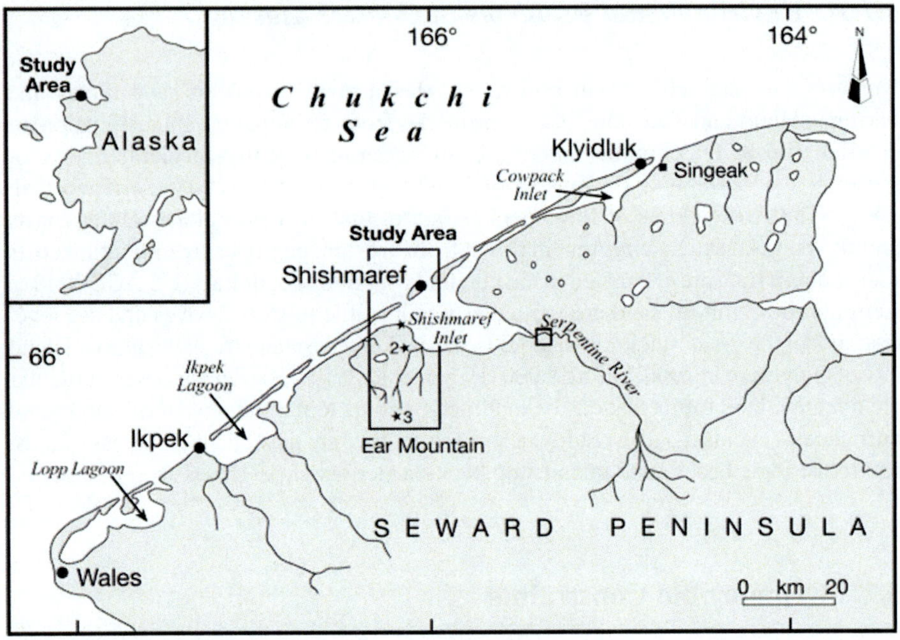

Fig. 5.1 Study area showing the location of Shishmaref and proposed relocation sites

5 km long, lies within the central portion of the sandy Shishmaref barrier island chain and is defined by two 1 km wide, tidal inlets at both north and south (Figs. 5.1 and 5.2). The Shishmaref barrier islands form the northwest shore of Seward Peninsula, on the southeast margin of the Chukchi Sea. During maximum open water, the coast can be subject to >500 km of fetch. Sand dunes up to 6 m ASL are superimposed atop Sarichef Island; dunes have built landward to the east, along the tidal sand flats along the two inlets. Sarichef Island consists of four litho- and chronostratigraphic depositional units, as inferred from aerial photography, stratigraphic exposures and several ^{14}C ages on peat and driftwood (Sainsbury 1967; Mason 1996). Cross-cutting relationships between the superimposed and truncated dune sets indicate the past configuration of the island during the last millennia of barrier island retreat. Sarichef Island has eroded from the southwest where a portion of its former configuration is preserved. This sediment package is a series of recurved dunes. The margin of an infilled tidal inlet, oriented north to south, has a limiting age between AD 1000 and 1200, based on basal peat (Mason 1996).

The present village of Shishmaref lies on the highest dunes on the island, along its western shore, at the nexus of greatest wave attack. The dunes in this area are broad and have formed parallel to the coast and perpendicular to the wind. The active layer migrates through the upper 2–3 m of the dunes but permafrost persists below this elevation if not exposed to wave action. At depth, ca. 1 m

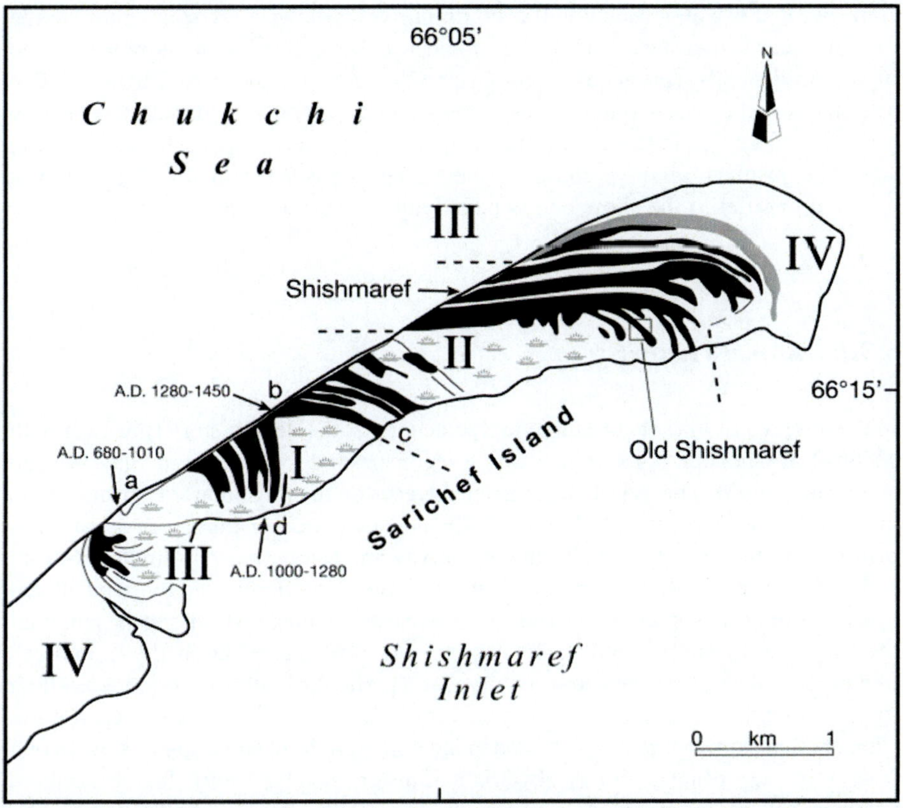

Fig. 5.2 Map of Sarichef Island showing geological radiocarbon ages and location of old village (cf. Mason 1996 for data on the dates)

above sea level, the dunes are frozen, a factor that complicates erosion, as evident in the 1997 storm surge that produced thermoniche formation and large scale block collapse. Subsequent colluviation serves to inhibit erosion, if only for a brief period. The role of melting permfrost in accelerating erosion remains an open question since less thermoniche formation occurs with melting, but melting is believed to be considerable by community residents. The abandoned ancestral village of "Old" Shishmaref lies inland on one of the oldest (and lowest) dunes along the northern tidal inlet. While no archaeological research has occurred at the site, informal collections by local residents are useful in estimating its age at AD 1400–1500 (Mason 2011, observations on National Park Service records). The site is no more than 3 m above sea level and shows no signs of recent erosion—it is >1 km from the eroding bluff. Several other cultural modifications are apparent on Sarichef Island; a landfill is maintained below on the southern tidal flat and the northern quarter of the island is considered a snow collection zone, with the water used for the community washeteria. A small part of the unoccupied northeastern part of the island was designated as a snow watershed in the mid-1980s (Farmwald and Crum 1986). Two airstrips further define settlement – the oldest one crosscuts the island and consists of a thin veneer of asphalt constructed in the 1940s; the strip is inactive and serves as a new housing site. The primary airstrip, south of the village, was built in the 1960s and is oriented parallel to the shore and is paved and gravel-bedded.

5.3.1 Storm History

In the absence of historic or ethnographic accounts, the deep history (pre-twentieth century) of Chukchi Sea storms must be reconstructed from geologic data (Mason and Jordan 1993). The oldest backbarrier surfaces on the Shishmaref Barriers are no more than 2,000 years old (Jordan 1990) and evidence of human occupation is restricted to the last 500 years, although survey and testing are very limited (Schaaf 1988). The last several centuries, the Little Ice Age, were dominated by frequent and catastrophic storm surges that forced open up to a dozen tidal inlets along the length of the Shishmaref Barrier island chain (Jordan 1990; Jordan and Mason 1999). In much of northern Alaska the nineteenth century was a particularly stormy period, according to the eyewitness accounts of Lt. Kashevarov (Van Stone 1977). A powerful north Chukchi Sea storm surge in 1893 flooded the 4 m high Point Hope spit to a depth of a least 1 m, according to its Presbyterian minister (Kindle 1909). In the northern Chukchi Sea, major storms occurred in the late 1940s, 1954, and 1963 (McCarthy 1953; Hume and Schalk 1967); the 1963 surge was a catastrophic disruption in Barrow and the events and community responses are well described by Brunner et al. (2004). In the adjacent northern Bering Sea, storms have decreased in frequency and intensity since 1920 (Mason et al. 1996), likely in response to shifts in the El Nino Southern Oscillation and other climatic controls.

At Shishmaref, four or five storms are recounted by local informants (SERC 2011a) as major events: the 1973, 1974 (less so), 1997 and 2001, 2003; most are linked to substantial erosion. A powerful storm in early November 2011 was associated with extremely high southerly winds but had minimal effects on Sarichef Island (Hopkins et al. 2011a, b; Burke and DeMarban 2011). The major storms of the early 1970s also had variable effects due to differing ice cover and shifting storm trajectories: the October 1973 storms generated waves >4 m above MLW (Wise et al. 1981:I–11) and covered all the low-lying areas of Sarichef Island, except for the airstrip (DOWL 1975:6). In 1973, open water waves directly hit the Shishmaref bluff from the northwest and produced a wave-cut niche 3 m into the bluff (DOWL 1975:6). The situation the next year was considerably different; the community escaped substantial harm during the great storm of 1974

(Fathauer 1975), and, for a large extent, during the massive Bering Sea storm of November 2011 (Hopkins et al. 2011a, b; Burke and DeMarban 2011).

5.4 The Uses and Abuses of Anecdotal Accounts

Anecdotal accounts have served in constructing the Shishmaref narrative, as clarified from examining a decade of reporting by the *Anchorage Daily News*; the *Nome Nugget* and the *Bering Strait Record*, summarized in Table 5.1. The powerful 1997 storm has assumed an iconic role in the media (Stanton 1997a, b; Demer 1997; Hunter 1997). Following the storm surge, Shishmaref residents told the *Anchorage Daily News* (Stanton 1997a) that Sarichef Island had contracted by "one quarter mile" [400 m] between the early 1980s and 1997—a rate of 30 m per year (cf. Table 5.1). Storms in the early 2000s were nearly as fierce (O'Harra 2001, 2002). Anecdotal accounts are commonly reported without qualification in scientific journals and testimony. In 2006 a news-feature in *Science* presented a dramatic photograph of an off-kilter structure on the Shishmaref bluff, accompanied with the statement that "[v]illage officials say that since 2001, the island has lost an average of nearly 23 ft [7 m] of shoreline per year" (Lempinen 2006:609). Even higher rates, "15 m over night in a single storm," are reported without citation in the *Arctic Climate Impact Assessment*, commissioned by the Arctic Council, an intergovernmental forum funded by NSF and NOAA (ACIA 2004:80).

The role of public relations and political action by the Shishmaref community has placed data collection in a subsidiary role; precisely measuring erosion rates is not a high priority for agencies tasked with engineering solutions. The web site of the Shishmaref Erosion and Relocation Committee (SERC) serves as a clearing house for the erosion narrative, including a summary of erosion rates and of several storms, including one of a 2003 storm, possibly the worst since 1974 (Weyiouanna 2003a, b).

> On November 21-23, 2003 the community of Shishmaref, Alaska was hit by a storm with northwest winds coming from the Chukchi Sea blowing 45 mph [72.4 km/h] and gusting up to 61 mph [98.2 km/h] at times. Seas were as high as 14' [4.26 m] along with high tides.

The 2003 storm purportedly led to more erosion than the 1974 storm (DOWL 1975:6), up to 10 m in places. The drama and near-tragedy associated with each storm enabled the Shishmaref Relocation Committee to garner media attention (Roosevelt 2004), by repeatedly offering congressional testimony (US Senate 2004), to secure funds for several revetments (TetraTech 2004; US ACE 2006a). The plight of Shishmaref has reached global dimensions; in 2005 alone, over a dozen media outlets visited the island, including film crews from France, Sweden, Germany, Switzerland, *TIME* magazine, *National Geographic* and others (Shishmaref Relocation Newsletter, 30 August 2005, p. 6, cf. SERC 2011b). Shishmaref has attracted independent film producers, environmental bloggers and environmental essayists. Dutch film-makers Jan Louter and Welle van Essen (2011)

Table 5.1 Shishmaref news accounts, 1997–2003

The need for accurate measurements

"During a storm last year (1996) villagers fighting erosion tossed any old machinery they could find over the embankment, hoping to build a barrier between land and sea. But this year village emergency service told people to hold on to their old generators, dinged up trucks and snow machines because when thrown onto the ocean, the items actually contribute to erosion by forcing more water around the machinery…"

Stanton 1997a (*Anchorage Daily News*, 5 Oct 1997, p. A-12)

"A raging sea whipped by 55 mph gusts ate away at Shishmaref's sea wall Saturday [Oct 4]…this weekend's torrent is the worst to hit the village in about 15 years"
Many of the houses were built about 20 years when the ocean bluffs were about 50–60 ft away

Stanton 1997a (*Anchorage Daily News*, 5 Oct 1997, p. A-1)

"Sarichef Island was a mile wide 10 or 15 years ago, erosion has taken more than a quarter mile [1,300 ft] of its width…embankment eroded by up to 30 ft in places"

Stanton 1997a (*Anchorage Daily News* Oct 5, 1997, p. A-1 and A-12)

Extrapolated erosion rate: 88 ft per year…
Waves of up to 30 ft crashed against and eroded the embankment of up to 30 ft in some spots.

Stanton 1997a (*Anchorage Daily News*, 5 Oct. 1997, p. A-12)

"…the livable portion of the island is just 2 miles long and 3/4ths of a mile wide, with the recent storm gobbling up another 30 ft…"
Latest attention comes after a storm pounded Shishmaref with 55 mph winds Friday and Saturday [Oct 4 and 5], whipping up waves of up to 30 ft

Demer 1997 (*Anchorage Daily News*, Tues. 7 Oct 1997, p. B-1)

"…the barrier has lost 30 ft of its sandy bluffs."

Stanton 1997b (*Anchorage Daily News* 9 Oct 1997, p. A-1)

northwest winds 55 mph
"the latest in generations of fall storms thrashed this slender barrier island last weekend, chewing through 20–30 ft of sandy shorefront, …and undercutting the bluff by as much as 12 ft"……
"Recent studies say Sarichef Island. is a mile wide, it's maybe half that where Shishmaref is"

Hunter 1997 (*Anchorage Daily News*, 12 Oct 1997, p. A-1)

"….we lost 25 ft in 1 day," Percy N. said

Hunter 1997 (*Anchorage Daily News*, 12 Oct 97, p. A-12)

"Eleven families evacuated after the Chukchi Sea, driven by 50 mph winds undercut the shoreline by up to 16 ft. Damage to the bank was accelerated by unusually warm waters which at high tide caused exposed permafrost to melt and destabilize"

Braem 1997 (*Bering Strait Record*, 15 Oct 1997, p. 1)

"…we lost 20 ft and we have 40 ft left"
Nayokpuk trading company estimate,

Braem 1997 (*Bering Strait Record*, 15 Oct 1997, p. 13)

"12 foot waves during high tide," "By the time the storm ceased Monday [Oct 8], 25–foot to 50–foot swaths of the north beach had been consumed…"
"… recent years have brought more intense and damaging storms, some times made worse by Global Warming, and the late arrival of the protection afforded by sea ice"

O'Harra 2001 (*Anchorage Daily News*, 10 Oct 2001, p. B-9)

A powerful storm slammed the island with 12 foot waves in early October, threatening to undercut houses…

O'Harra 2003 (*Anchorage Daily News*, 30 Oct 2001, p. B-3)

(continued)

Table 5.1 (continued)

"In Shishmaref 45 mph winds drove 14 foot seas into the beach during a high tide and demolished drying racks...threatened to undercut a dozen homes Another casualty was the gabion sea wall built with $110,000 of state disaster funding ...1 year ago." O'Harra 2002 (*Anchorage Daily News*, 18 Oct 2002, p. B-3)
"Strong winds and big waves off the Chukchi Sea last Saturday [Nov. 8] mauled the beach at Shishmaref, eroding as much as 30 ft of shoreline in place" *Anchorage Daily News*, 12 Nov 2003a, p. B-3
Picture caption (only): "Strong winds and big waves off the Chukchi Sea last weekend [Nov. 8–9] eroding as much as 30 ft of shoreline in places. ...gabions and boulders disappeared along most of the beach." Photograph by Tony Weyiouanna. *Anchorage Daily News*, 13 Nov 2003b, p. B-3
Second storm pounds Shishmaref. ...winds steady at 35 mph, gusted to 45 mph, accompanied by 8 foot waves. ...winds shifting to NW. ...large chunks of bluff fell all today... " 'Worst storm since 1997, the water is almost as high as the big storm in 1974'..." *Anchorage Daily News*, 22 Nov 2003c, p. B-3

produced a docudrama, followed by a book, both entitled *The Last Days of Shishmaref: An Inupiaq Community Swallowed by the Sea*. Perhaps attracted by the parallel to the Netherlands, this film implies that the government refuses to move the village or to improve its infrastructure. The erosion problem is presented by the inclusion of a CNN film clip broadcast on a television in the home of a watching resident; the clip offers the standard talking points about "global warming" and erosion. Erosion estimates from a journalistic account by Kolbert (2006) are linked into the wikipedia entry on Shishmaref. Nancy Lord (2011), devotes a chapter to Shishmaref in her book, *Early Warming: Crisis and Response in the Climate-Changed North*. While sensitive to the complexities of erosion control, Lord (2011:142) repeats the assertion that the 1997 storm took out "125 ft" of shoreline. Ironically, media discourse on "climate change" itself may contribute to a certain bias in discussing the community's real concerns (Marino and Schweitzer 2009).

5.5 Photographic Data Sources for Sarichef Island

The earliest aerial photographic imagery for the Chukchi Sea coast derives from the Defense mapping projects in 1950, and is supplemented at intervals from 1972 to 1984, until a 2003 run. *In toto*, six photographic images document the last 60 years of geomorphic change on Sarichef Island: 1950, 1972, 1974, 1980, 1984 and 2003. Several sets of aerial photographs are archived by Aeromap, Inc. (Anchorage), the GeoData center at UAF, the USGS (Sioux City) and the National Park Service (Alaska office). A variety of ancillary photographic imagery are also available, including on the ground and oblique aerial views from low-flying, fixed-wing air craft. Two of the authors (OKM, JWJ) have visited Sarichef Island at several intervals from 1988 to 2007 and maintain photographic negatives and slides, as well as digital images. Data also can be extracted from scattered newspaper accounts that contain photographs from the early 1980s and the 1990s. In addition, community members have provided video and digital images to National Park Service personnel from storms in 2002–2005 and have posted images, videos and anecdotal accounts on the internet (SERC 2011a).

5.6 Studies of Erosion Rates on Sarichef Island

The bluff erosion of Sarichef Island is documented, with variable accuracy, from several sources: (a) longitudinal comparisons of property records and maps between 1937 and 1973 (Peratrovich and Nottingham 1982:2, Fig. 2.2); (b) anecdotal accounts from Shishmaref residents; and (c) sequential comparisons of 1950 and 1972–2003 aerial photographs (US ACE 2006a, this paper); (d) 20 years of repeated field observations by Mason and Jordan and (e) on-the-ground measurements by community members for 2 years 2001–2003 (Section 117, p. 6). The period prior to 1982 represents the baseline, prior to the first revetment built along the bluff. The first study by Peratrovich and Nottingham (1982:2) established four "banklines" for 1962, 1972 and 1974 and concluded that "the rate of erosion was not constant. Far greater rates of erosion occur as a result of storm events than under normal conditions." Major storms in the late 1940s produced 27.5 m of erosion while from 1963 to 1972 few if any major storms affected the coast, leading to 3–6 m of erosion, while, along shore, some areas experienced no loss. Erosion rates were an order of magnitude higher during two 1973 storms which produced up to 13 m of erosion. From these data, one might conclude that erosion was highest in the 1940s and declined until the early 1970s.

The Army Corps of Engineers undertook two erosion studies in the early 2000s; one involved sequential aerial photography while the second employed on-the-ground measurements by community members (Table 5.2). The methodology employed by the ACE is briefly described, although an error analysis is unavailable. The ACE study scanned four aerial photos (1972, 1980, 1984 and 2003) and used buildings and the airstrip as stations to measure distance from autocad measuring tools to the top of the bluff line. The ACE study (Table 5.2) infers that the Shishmaref bluff eroded *annually* between 80 cm and 2.71 m from 1972 to 2003—for an unspecified reason the 1950 imagery was not used by ACE. Although the ACE erosion rates were presented as annualized rates, between 25.5 and 84 m eroded in the shore-perpendicular direction from 1972 to 2003, as extrapolated from multiplication. The extent of temporal variability is evident in the years 2001–2003 where on-the-ground measurements record a considerably higher rate, from 4 to 7 m (US ACE 2006a). Because the field measurement (2001–2003) is incorporated into the longer record, and assuming the accuracy of the on-the-ground measurements, roughly 10% of the total erosion occurred in just those 2 years.

Table 5.2 Erosion rates calculated by the Army Corps of Engineers (ACE) and community volunteers. Measurements converted to metric units

Erosion rates	Source of data	Measurement year	Low range (annual)	High range (annual)
Rate #1	Aerial photograph	1972, 1980, 1984, 2003	0.82	2.71
Rate #2	Community measurements	2001, 2002, 2003	3.96	6.89

An erosion rate calculation undertaken for this project by University of Colorado (CU) researchers[2] shows broadly similar results to the previous analyses, especially for the decades of the mid-twentieth century. According to the CU calculations (Fig. 5.3), between 35 and 48 m of bluff retreat occurred along the developed Shishmaref shorelines, with considerably more to the northeast, downdrift from the revetments. Examining its history, erosion on Sarichef Island was far greater in the earlier decades, from 1950 to 1980, ca. 27–0.9 m annually, and has slowed considerably since 1980 to <13 m, 0.57 m on annual basis (Fig. 5.3, cf. Table 5.1). This is contrary to anecdotal accounts. However, the period after 1980 does differ considerably from the 30 year period before 1980: erosion has slowed considerably, falling by nearly half. In total, only about 50 m of erosion has occurred since 1950, according to the CU Study. The reliability of the CU study shows a strong signal with little noise, as calculated in an error analysis.[3] The Shishmaref record can also be compared to

[2] CU Methodology: We used the DSAS-software and methodology outlined in U.S. Geological Survey Open-File Report 2005–1304 (http://woodshole.er.usgs..gov/project-pages/dsas/version3/index.html) to calculate shoreline change. Once all shorelines were digitized a baseline was created that mirrored the general shape of the shoreline and its offset approximately 150–250 m inland from the blufftop. According to the DSAS manual, Thieler et al. (2005): "the DSAS extension generates transects that are cast perpendicular to the baseline at a user specific spacing alongshore. The transect/shoreline intersections are used by the program to calculate the rate of change statistics." Transects were cast perpendicular to the base line and spaced 10 m apart for its entire length, each transect was linked to tables containing a series of shoreline change statistics and distance measurements. The same transects were recast three times to capture coastal changes for different time periods. In each case, the change between just two time periods was calculated. For this analysis, the end-point rate (EPR) statistic was used. The EPR statistic is the measured distance between two shorelines divided by the time elapsed between the two shore line dates. The final EPR value is the yearly rate of change, positive or negative, for a given time period. For our study, all EPR units are meters per year.

[3] Measurement errors are related to the geocorrection of aerial photographs and onscreen digitization, a simple calculation of the shoreline feature. Calculating shoreline position errors: A single yearly shoreline position error can be calculated by taking the square root of the sum of squares (Morton et al. 2004 and Fletcher et al. 2003) of geocorrection error (g) and bluff top digitization error (Bt) for each year. Thus the position error for a given year for the bluff top is shown in Equation 1: $E_{hear} = \pm\sqrt{g^2} + bt^2$. Then errors are calculated for each time period (early, late or long term) by taking the square root of the sum of the squares of the two bracketing yearly error values and annualizing over the time period of interest. The "early" time period calculation is Equation 2: $E_{year} = \pm — {E_{1980s}}^2$. This error value can be applied to any given transect. The annualized error value for the "Early" period (ca. 1950-ca. 1980) is ± 0.20 m, "Late" period (ca. 1980–2003) is ± 0.23 m and for the "Long Term" period (ca. 1950–2003) is ± 0.09 m. The signal to noise ratio is very good: the error values ("noise") are substantially less than the annualized shoreline change rates ("signal"), particularly for the "Long Term" rates.

Fig. 5.3 Comparative photographs of 1980s concrete block revetment, *upper* as failed in 1992, *lower*, replaced in 1993

broader trends across the entire north shore of the Seward Peninsula. Baseline data are available for a 20 km section of the Shishmaref Barriers from measured sections and photogrammetric comparisons from 1950 to 1987 (Jordan 1990), extended by additional photographs and a broader regional coverage (Manley et al. 2007). From this comparison, Sarichef Island exhibits an erosion rate that is nearly twice that of non-developed shorelines (Manley et al. 2007).

The three erosion studies, those of PNB (1937–1973), ACE (1972–2003) and CU (1950–2003), employ different methods and overlap for the last 30 years. Nonetheless, allowing for the methodological and temporal peculiarities, all three studies may be linked to establish a single long term record covering the period from 1937 to 2003. However, the ACE study produced an erosion total for the last 31 years that is 57% higher than CU rate, although the annualized rates are similar. The PNB study indicated erosion was higher prior to 1950, a circumstance not evident in other studies or in anecdotal accounts. Combining the PNB and CU efforts, the entire period of 76 years witnessed about 77 m of erosion; with nearly one third of that occurring prior to 1950 during a cycle of massive storms in the late 1940s. Erosion was highest during two other periods, in the early 1970s that produced ca. 13 m of land loss and in the early 2000s when up to 8 m were eroded.

5.7 Shoreline Stabilization and Protection at Shishmaref

Over the last 65 years, engineering consultants have proposed a wide variety of responses to cope with heightened bluff erosion (DOWL 1975; Peratrovich and Nottingham 1982; US ACE 2006a; TetraTech 2004). Some of these were actualized, with mixed results on erosion rates, as discussed above. The most common remedies sought to mitigate bluff erosion include retreat from the bluff margin, the emplacement of gabions or concrete block and rock revetments and free-standing sea walls. In addition, several beach or marine remedies were proposed, including beach nourishment, groins and an offshore shore-parallel breakwater. In cost, the various methods have ranged from the inexpensive, e.g. gabions, to the extravagant, life-time cost of $260 million for groins, although beach nourishment costs exceed $81 million. Relocation off the island is typically placed at ca. $180 million, probably a low estimate. By contrast, the ACE "preferred alternative," a 1,000 m revetment has an estimated cost of $13 million and would involve considerably less up-keep costs (US ACE 2006b:12).

The first effort in shoreline protection at Shishmaref was in the late 1940s and employed a series of 55 gal drums placed at the north side of a landing strip (Peratrovich and Nottingham 1982). The 1973 storm apparently removed these drums and any other protective measures (DOWL 1975:6). The major storm of 1973 led to a massive sand-bagging effort (>50,000 bags) in the following year, with the effect of mitigating erosion by a considerable amount during a subsequent storm in 1974, until sea ice gouged the bags (DOWL 1975:2). Following the 1970s storms, a cement block revetment was emplaced in 1982 seaward of a BIA school for about 100 m of the Shishmaref bluffs (Nome Nugget 1984). While the engineers recommended a revetment height above the dunes—to prevent wave overtopping, this design was not implemented. This concrete block revetment failed within a year but it was resuscitated by the community several times into the early 1990s (Mason et al. 1997), supplemented by a series of gabions on its south margin. End-around erosion was a consequence of the revetment; especially on the north end. In 1995 residents employed a more problematic strategy to mitigate erosion: "numerous volunteer[s] pushed old truck bodies, etc. over the edge to try to slow the erosion (City of Shishmaref 1995). At the cost of $100,000 gabions were placed along 150 m of the dune in 2001 (O'Harra 2001) but "most [gabions] have sunk into the sand," according to Tony Weyioanna (O'Harra 2002).

The massive storms of the early 2000s led to more elaborate constructions, several independently designed stone-faced revetments. The first effort was in 2004 when the Bureau of Indian Affairs (BIA) and Kawearak Native corporation installed a boulder revetment at the western end of Shishmaref covering 135 m of dune margin (US ACE 2006a; Inklebarger 2005). This structure extends 2.75 m above MHW and is graded to the contour of the bluff but stands no higher than the original bluff (Mason, unpublished field notes August 2006). In September 2005, two further revetments were constructed by ACE, one protects the old BIA school for 75 m and another 180 m reveted by the City of Shishmaref, (DOWL 1975) funded by the State of Alaska (Anchorage Daily News 2005). The three projects used rip rap and small boulders (dia. 1 m) to a height of between 2.75 and 3 m above MHW (Mason unpublished field notes, August 2006). The three revetments are idiosyncratic and non-linear: the jagged, eroded bluff contour is preserved. As a consequence of lithology, clast size, the rock revetments along the Shishmaref bluff resemble a cobble beach more than the standard large boulder sea wall, as at Nome. This may be partly due to a difference in the source of materials; the Shishmaref revetment contains rock from the Shumagin Islands (Lord 2011). Larger clasts, large boulders, 1–2 m dia., are distributed on its upper part, similar to the storm beach while the lower portion is composed of clasts in the cobble range, 10–30 cm dia. (Mason, unpublished field notes, 2006).

After 75 years of repeated storms and a range of engineering responses (gabions, barrels, revetments), what is the state of the Shishmaref shoreline and what are the prospects for the community? Despite the warm, limited sea ice of the past several years, Shishmaref has not reported heightened erosion, even during the anomalously high southerly winds of the "Great Storm of 2011" (Hopkins et al. 2011a, b); (Burke and DeMarban 2011). The three revetments have perhaps played a role in mitigating any effects of high water but this palliative effect remains uncertain and edge-around erosion is apparent (Fig. 5.4). Although the community voted to move inland about 5 years ago, it is possible that the newly placed shore line protection may have lessened the sense of urgency so apparent in the community (SERC 2011a, b). One should recall that Shishmaref has defied the first dire warnings of DOWL engineers and may yet dodge the bullet of storm surge-driven erosion (Figs. 5.5 and 5.6).

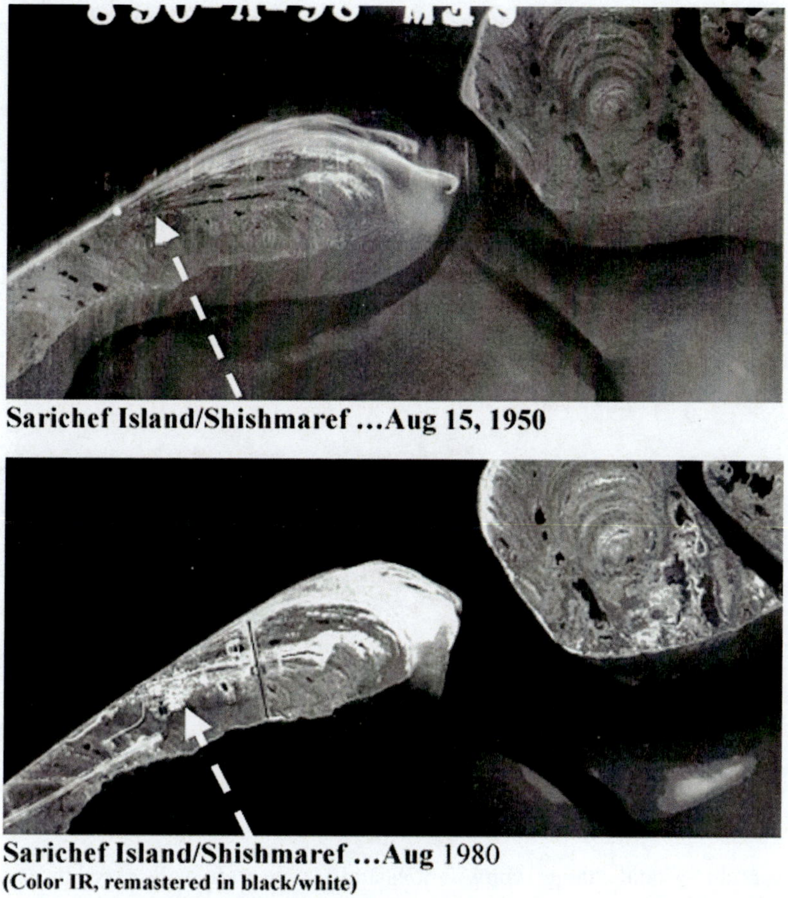

Fig. 5.4 Aerial photographs of Sarichef Island, 1950 and 1980

Fig. 5.5 Inferred beach margins, as calculated by CU study, this paper. *Upper*: 1948, *Middle, dashed*, 1980, *lower solid*, 2003. All three lines are plotted on the 2003 image. The lines may differ from the estimates of erosion based on bluff margin

Fig. 5.6 Serial revetments in place along Shishmaref

5.8 Conclusions

To respond to global change, communities and planners are well served by accurate data that can be related to geomorphic processes and to short-term changes in weather and longer-term changes in climate. However, the shoreline can provide a contentious platform from which the emotions of catastrophic threats to human life and property loss are projected. Because anecdotal accounts are not referenced, such comments take on a life of their own and can propel media and influence governmental policy more effectively than scientific data. Contrary to anecdotal accounts from Shishmaref, sequential aerial photographs from 1950 to 2007 reveal that erosion has increased on the south-facing shores of the Chukchi Sea, while prior to 1977, erosion was higher on the north-facing shores. In addition, comparisons of property records indicate high rates of erosion prevailed prior to 1950. Several engineering solutions were attempted in Shishmaref between 1983 and 2003, including gabions, as well as cinder block and boulder/cobble revetments, leading to increased end-around erosion down drift and an erosion rate twice that of undeveloped shorelines. To adapt to heightened erosion rates, societies either retreat from the shore or confront ever-increasing engineering costs.

A call to collective action is an emotional process, requiring the engagement of an audience to the plight of a victimized group—in this case, the tiny village of

Shishmaref facing down the huge and seemingly increasing waves of the globally warmed Chukchi Sea.[4] The immediacy of the testimony of Shishmaref residents is often painful and profound, as people face the imminent loss of their lives, culture and property (U.S. Senate 2004, SERC Newsletters 2004, 2005, Lord 2011). By contrast, scientific or engineering studies are generally tepid and dispassionate, filled with sterile economic calculations (TetraTech 2004; US ACE 2006a). For this reason, anecdotal testimony often gains wide currency and generates sympathy and may lead to action. However, such personal testimony can also preclude effective policy decisions and be employed uncritically by the media and policy makers—persisting in the literature, on the web and in policy discourse for decades. It is not the intent of this article to belittle or question the plight of Shishmaref people; however, scientific methods should offer data-based perspectives and will assist in a solution to the problem.

References

ACIA (2004) Impacts of a warming world: arctic climate impact assessment. Cambridge University Press, Cambridge

Anchorage Daily News (2003a) Violent wind, wave action work to undermine community's foundation. Anch Daily News LVIX, no 316, p B-3, Nov 12

Anchorage Daily News (2003b) Shoreside Retreat. Anch Daily News LVIX, no 317, p B-3, Nov 13

Anchorage Daily News (2003c) Second storm pounds Shishmaref. Anch Daily News LVIX, no 326, p B-3, Nov 22

Anchorage Daily News (2005) Boulders protect Shishmaref Anch Daily News LXI, no 339, p B-1, 5 Dec 2005

Argos Collective (2008) Photos: climate refugees from New Orleans to Germany. Posted on October 28th on Cafébabel.com: the European magazine. http://www.cafebabel.co.uk/article/26974/argos-collective-climate-refugees-photos-worldwide.html. Accessed 11 June 2011

Braem NM (1997) Can Shishmaref be saved from the sea? Bering Sea Rec 2(18):1

Brigham-Grette J, Hopkins DM (1995) Emergent marine record and paleoclimate of the last interglaciation along the northwestern Alaskan Coast. Quat Res 43:159–173

Brunner RD, Lynch AH, Pardikes J, Cassano EN, Lestak L, Vogel J (2004) An arctic disaster and its policy implications. Arctic 57:336–346

Burke J, DeMarban A (2011) Shishmaref residents escaped Bering Sea storm's wrath. Alaska dispatch, posted November 11. http://www.alaskadispatch.com/article/shishmaref-residents-escaped-bering-sea-storms-wrath?page=full

City of Shishmaref (1995) Unpublished notes accompanying a half hour VHS videotape of the high water experienced on Sarichef Island on October 30, 1995. In the possession of the authors

Demer L (1997) Villagers fight to save island home. Anch Daily News LII, p B-1, 7 Oct 1997

DOWL (Dickinson, Oswald, Walch and Lee) Engineers (1975) Shishmaref erosion protection alternatives feasibility and cost study. Unpublished report to Department of Community Affairs, State of Alaska, Contract CPA-AK-10-19-1010

---

[4] Mark Lynas (2003) in the New Statesman wrote: "Shishmaref lives in perpetual fear. The cliffs on which the 600-strong community sits are thawing, and during the last big storm 50 ft of ground was lost overnight. People battled 90 mph winds to save their houses from the crashing waves."

Farmwald JA, Crum JA (1986) Developing a community water system for Shishmaref, Alaska. In: Ryan WR (ed) Proceedings of the 4th international conference cold regions engineering, Anchorage, American Society of Civil Engineers, New York

Fathauer TF (1975) The great Bering Sea storms of 9-12 Nov 1974. Weatherwise 28:76–83

Fletcher C, Rooney J, Barbee M, Lim SC, Richmond B (2003) Mapping shoreline change using digital orthophotogrammetry on Maui, Hawaii. J Coast Res Spec Issue 38:106–124

Fujita K, Cook DB, Hasegawa H, Forsyth D, Wemiller R (1990) Seismicity and focal mechanisms of the arctic region and the North American plate boundary in Asia. In: Grantz A, Johnson L, Sweeney JF (eds) The arctic ocean region, Geology of North America vol L. Geol Soc Am, Boulder

Hameedi MJ, Naidu AS (1988) The environment and resources of the southeastern Chukchi Sea. NOAA OCS Study MMS, Anchorage, pp 87–0113

Hayes MO (1967) Hurricanes as geological agents: case studies of Hurricanes Carla, 1961, and Cindy, 1963. Rep Investgns 61, Bur Econ Geo, U Tex, Austin

Hebert PJ, Jarrell JD, Mayfield M (1996) The deadliest, costliest and most intense United States hurricanes of this century. Tech Memorand NWS TPC-1 NOAA, Nat Hurricane Cent, Miami

Hopkins K, Grove C, Dunham M (2011a) Epic storm bears down: coastal communities prepare for fury of 100 mph winds and blinding snow. Anch Daily News, XLV, no 313, 9 Nov 2011, p A-1, A-12

Hopkins K, Grove C, Dunham M (2011b) Tempest scours coast: water still rising in some communities as storm isn't done yet. Anch Daily News, XLV, no 314, 10 Nov 2011, p A-1, A-12

Hume JD, Schalk M (1967) Shoreline processes near Barrow, Alaska: a comparison of the normal and the catastrophic. Arctic 20:86–103

Hunter D (1997) Crumbling Shishmaref ponders future. Anch Daily News, LII, no 288, p A-1, 12 Oct 1997

Inklebarger T (2005) Shishmaref to extend rock wall to slow erosion. Anch Daily News LXI, no 189, p B-2, 8 July 2005

Jordan JW (1990) Late Holocene evolution of barrier islands in the southern Chukchi Sea, Alaska. Master's thesis, Quaternary Studies, University of Alaska Fairbanks, Fairbanks

Jordan JW, Mason OK (1999) A 5000 year record of intertidal peat stratigraphy and sea-level change from northwest Alaska. Quat Int 60:37–47

Kindle EM (1909) Notes on the Point Hope Spit, Alaska. J Geol 17:178–189

Kolbert E (2006) Field notes from a catastrophe: man, nature, and climate change. Bloomsbury, New York

Lempinen EW (2006) In arctic Alaska: climate warming threatens a village and its culture. Science 314:609

Lord N (2011) Early warming: crisis and response in the climate changed north. Counterpoint Press, Berkeley

Louter J, Van Essen M (2011) The last days of Shishmaref: an Inupiaq Community swallowed by the Sea. http://www.thelastdaysofshishmaref.com/shishmaref3/cms/cms_module/index.php. Accessed 23 June

Lynas M (2003) It's Later Than You Think. New Statesman 132:16–17

Lynch AH, Lestak LR, Uotila P, Cassano EN, Xie L (2008) A factorial analysis of storm surge flooding in Barrow, Alaska. Mon Weather Rev 136:898–912

Manley WF, Jordan JW, Lestak LR, Mason OK, Parrish EG, Sanzone DM (2007) Coastal erosion since 1950 along the southeast Chukchi Sea, Alaska, based on both GIS and field measurements. In: Proceedings of the 38th Arctic workshop, Univ. Col Boulder, Boulder

Marino E, Schweitzer P (2009) Talking and not talking about climate change in northwestern Alaska. In: Crate S, Nuttall M (eds) Anthropology and climate change: from encounters to action. Leftcoast Press, Walnut Creek

Mason OK (1996) Surficial map of part of the Shishmaref A-3, B-3 and Teller D-3 quadrangles. Pub Data File Rept 96/7, Alaska Div Geol Geophys Surv Fairbanks. Incl. text: Geological and Anthropological Aspects of Relocating Shishmaref

Mason OK (2006) Living with the coast of Alaska revisited. In: Smith O (ed) Coastal erosion responses for Alaska. Alaska Sea Grant Program, Fairbanks

Mason OK, Jordan JW (1993) Heightened storminess and synchronous truncation of northwest Alaska beach ridges. Quat Res 40:55–69

Mason OK, Jordan JW (2002) Minimal late Holocene sea level rise in the Chukchi Sea: Arctic insensitivity to global change? Glob Planet Change 32(1):13–23

Mason OK, Salmon DK, Ludwig SL (1996) The periodicity of storm surges in the Bering Sea from 1898 to 1993, based on newspaper accounts. Clim Change 34:109–123

Mason OK, O'Neal WJ, Pilkey OH (1997) Living with the coast of Alaska. Duke University Press, Durham

McCarthy GR (1953) Recent changes in the shoreline near Point Barrow, Alaska. Arctic 6:195–197

Morton RA, Miller TL, Moore LJ (2004) National assessment of shoreline change: part 1: historical shoreline changes and associated coastal land loss along the U.S. Gulf of Mexico. US Geol Surv Open-file Rep 2004–1043

Murphy K (2001). Front row exposure to global warming: engineers say Alaskan village could be lost as sea encroaches. LA Times, 8 July 2001

Nome Nugget (1984) Seawall races against time. Nome Nugget 83, no 3, p 1, 3, 1 Aug 1984

O'Harra D (2001) Storm erosion of bluff puts 10 homes in danger on Shishmaref. Anch Daily News LVII, no 283, p B-1, 10 Oct 2001

O'Harra D (2002) State assesses northwest coast storm damage. Anch Daily News LVIII, no 291, p B-3, 18 Oct 2002

Peratrovich and Nottingham, Inc. (1982) Shishmaref erosion control engineering studies. Report Department of Transportation Public Facilities, State of Alaska, Anchorage

Pilkey OH, Dixon KL (1996) The corps and the shore. Island Press, Washington, DC

Pilkey OH, Young R (2009) The rising sea. Island Press, Washington, DC

Roosevelt M (2004) Vanishing Alaska. Time, 4 Oct. http://www.time.com/time/magazine/article/0,9171,995264,00.html

Sainsbury DL (1967) Upper Pleistocene features in the Bering Strait area. Geol Research 1967 US Geol Surv Prof Pap 575-D

Schaaf J (1988) The Bering land bridge: an archaeological survey, vol 14, Res Managmt Rep. Nat Park Serv, Anchorage

SERC Shishmaref Erosion and Relation Center (2011a) Web site. http://www.shishmarefrelocation.com/. Accessed 23 June

SERC Shishmaref Erosion and Relation Center (2011b) Web site. http://www.shishmarefrelocation.com/newsletters/08_2005.pdf

Serreze MC, Holland MM, Stroeve J (2007) Perspectives on the Arctic's shrinking sea-ice cover. Science 315:1533–1536

Shepard FP, Wanless HR (1971) Our changing coastlines. McGraw-Hill, New York

Stanton D (1997a) Storm batters village: families forced out in Shishmaref. Anch Daily News LII, no 278, p A-1, 5 Oct 1997

Stanton D (1997b) Villagers fight sea with sand bags. Anch Daily News, LII. no 285. p A-1. 9 Oct 1997

TetraTech (2004) Shishmaref partnership: Shishmaref relocation and collocation study, Shishmaref, Alaska. Rep to U.S. Army Corps of Engineers, Alaska District, Anchorage

Thieler ER, Himmelstoss EA, Zichichi JL, Miller TL (2005) Digital shoreline analysis system (DSAS) version 3.0; An ArcGIS© extension for calculating shoreline change. US Geol Sur Open-File Rep 2005–1304

U.S. Army Corps of Engineers (2006a) An examination of erosion issues in the communities of Bethel, Dillingham, Kaktovik, Kivalina, Newtok, Shishmaref and Unalakleet. Alaska Village erosion technical assistance program. U.S. Army Corps of Engineers, Alaska District, Anchorage

U.S. Army Corps of Engineers (2006b) Shishmaref, Alaska Section 117 Project Fact Sheet. Posted on http://nome.colorado.edu/HARC/members/Scanned_docs/ArmyCorps_2006_ShishText.pdf

U.S. Senate, Committee on Appropriations (2004) Hearings and testimony, committee field hearing on Alaska native villages affected by flooding and erosion. Lucy Eningowuk, Chairperson, Shishmaref erosion and relocation coalition. http://appropriations/senate.gov/hearmarkups/record/cfm?id=223530

Van Stone JW (ed) (1977) A.F. Kashevarov's coastal explorations in northwest Alaska, 1838, vol 69, Fieldiana Anth. Field Museum of Natural History, Chicago

Weyiouanna T (2003a) Strong winds and big waves off the Chukchi Sea. Photograph and caption, Anch Daily News LVIII, no 317, 13 Nov 2003

Weyiouanna T (2003b) Shishmaref battered by waves. Arct Sound 17(50):1, Dec 4

Wise JL, Commiskey AL, Becker R (1981) Storm surge climatology and forecasting in Alaska. Arctic Environmental Information and Data Center, Anchorage

Wisniewski J (2005) "We're always going back and forth": Kigiqtaamiut subsistence land use and occupancy for the community of Shishmaref. Rep to US AC Engr, Alaska Dist, Anchorage

Wisniewski J (2010) Knowings about Sigu: Kigiqtaamiut hunting as an experiential pedagogy. In: Krupnik I, Aporta C, Gearhard S et al (eds) Siku: knowing our ice, documenting Inuit sea-ice knowledge and use. Springer, New York

Chapter 6
Portballintrae Bay, Northern Ireland: 116 Years of Misplaced Management

Derek W.T. Jackson

Abstract Portballintrae has had a protracted history of human interference ranging from small-scale sand removal to hard coastal engineering. A small, horse shoe embayment and a once popular seaside destination on the north coast of Northern Ireland, it has suffered from progressive sediment loss over the last 116 years. From a once sediment-rich system, with a wide sandy beach, it now contains only a limited amount of sand draped over bedrock and/or gravel substrate and a relatively narrow beach. Installation of a pier in its western section is thought to have interrupted the natural hydrodynamics and set in motion a progressive longshore transport and removal of sand into deeper water. Successive hard engineering 'solutions' prompted through public pressure and engineers keen to do business, have been largely ineffective as they failed to address the root causes of erosion in this sediment-starved beach system.

6.1 Introduction

The small town of Portballintrae is located on the County Antrim coast of Northern Ireland, U.K. around 70 km north of Belfast. The original town developed as a popular seaside resort, particularly during the Nineteenth century due to its close proximity to the nearby Giant's Causeway and the fact that it was a relatively sheltered and safe bathing beach.

Its formation along a north to south fault through geology dominated by Tertiary basalt (Wilson 1972) was driven by glaciation and marine erosion. It currently

D.W.T. Jackson (✉)
School of Environmental Science, University of Ulster, Cromore Road, Coleraine BT52 1SA, Northern Ireland, UK
e-mail: d.jackson@ulster.ac.uk

J.A.G. Cooper and O.H. Pilkey (eds.), *Pitfalls of Shoreline Stabilization: Selected Case Studies*, Coastal Research Library 3, DOI 10.1007/978-94-007-4123-2_6,

forms a small north-facing horseshoe-shaped bay, approximately 350 × 400 m that is bounded by two rocky headlands (Carter et al. 1983). The bay is located within a high energy wave regime where significant swell wave events from the Atlantic with fetch distances in excess of 2,000 km, approach predominantly from the west and northwest but undergo refraction as they encounter the northwest coast of Ireland. Refracting around the Inishowen peninsula in Co. Donegal, they still retain enough energy to bring significant waves to this shoreline (Jackson et al. 2005; Jackson and Cooper 2009; Backstrom et al. 2009). Average annual wave heights are 1.3 m, with a mean period of 8.5 s (UK Met Office, 2000–2005); however, offshore storm waves of 10–12 m and periods of 14–16 s are not uncommon in this area (Carter 1982). The tidal regime increases from east (microtidal) to west (mesotidal) and the area is subject to a mean tidal range of about 1.9 m.

Regionally, many of the beaches around the north coast of Ireland are currently operating within a finite sediment regime comprising reworked glacial sands (Cooper et al. 2002; Jackson et al. 2005). Beaches exist in embayment settings bounded by rocky headlands (Cooper 2006; O'Connor et al. 2007, 2011). Many stretches of sandy coastlines owe their origins to the glacial history of the area with the location of former ice margins dictating the sculpting of bedrock and acting as the main source of sediments (Jackson et al. 2005). Contemporary sea level can also dictate the behaviour of beaches, adjusting the way sediments are reworked and how beach morphology is sculpted. Historical sea level in Northern Ireland has been stable as local isostatic uplift has kept pace with eustatic sea level rise (Orford et al. 2006).

6.2 Historical Changes

The sandy beach at Portballintrae has had a long history of change since the late 1800s, involving a succession of beach alterations largely instigated by human activities. An oblique aerial photograph of the site in 1938 shows a relatively extensive beach, over 150 m wide that is in stark contrast to the modern gravelly beach (Fig. 6.1). A ground view of the beach (Fig. 6.2) sometime around 1920 also shows the abundant sandy beach volume at the site. However, a study by Carter et al. (1983) of vertical aerial photos and maps between 1935 and 1980 showed a progressive loss of intertidal beach sand area at a rate of 1,000 m^2 $year^{-1}$, to finally form a much steeper (and narrower), sand-limited, boulder-strewn beach by 1980.

Carter et al. (1983) calculated an overall volumetric loss of around 60,000 m^3 over a 75 year period. Field observations and aerial photography show that the beach has operated in a sediment-starved state since the 1980s and has shown little shoreline retreat since then. This suggests it has now adjusted to the prevailing forcing factors (mainly wave energy) and the limited amount of sand now available

Fig. 6.1 A 1938 oblique (*top image*) aerial photo (Portballintrae, 8th April 1938, © Belfast Telegraph, Collection Ulster Folk & Transport Museum) showing a wide sandy beach system and in 1999 (*bottom photo*) the beach made up of only a thin veneer of sand on the lower beach face and a large cobble/boulder back beach, particularly along the western section (*right* of picture)

in the system. Examination of the contemporary beach sediments (cobbles to sand size) currently located along the beach also shows that waves have graded sediments alongshore according to size. Larger material is located on the western (higher wave energy) side of the bay and the smallest is found toward the eastern section (Fig. 6.3).

Fig. 6.2 (*Top photo*) Ground view looking east along the shore. Photo is dated before 1920's showing a substantial swash-aligned sandy beach backed by a supra-tidal gravel beach (General view, Portballintrae, © Belfast Telegraph Collection Ulster Museum). (*Bottom* two pictures) Views of the same stretch of coast in 2011, *left* image looking northwest towards Leslie's Pier, *right* image looking southeast. These show a sand starved beach, engineering groynes and palisade barrier on the back beach. The supratidal back beach (gravel) as well as the sandy lower beach present in the older photograph has now disappeared and been replaced by a series of ineffective groynes

6.3 Where Has All the Sand Gone?

This question is frequently posed by local residents and tourist visitors to the site. The change from a relatively sand-abundant beach at the turn of the twentieth century to a sand-depleted one has not been matched on any nearby beach system. For the answer we must examine the history of human activities at Portballintrae (Fig. 6.4).

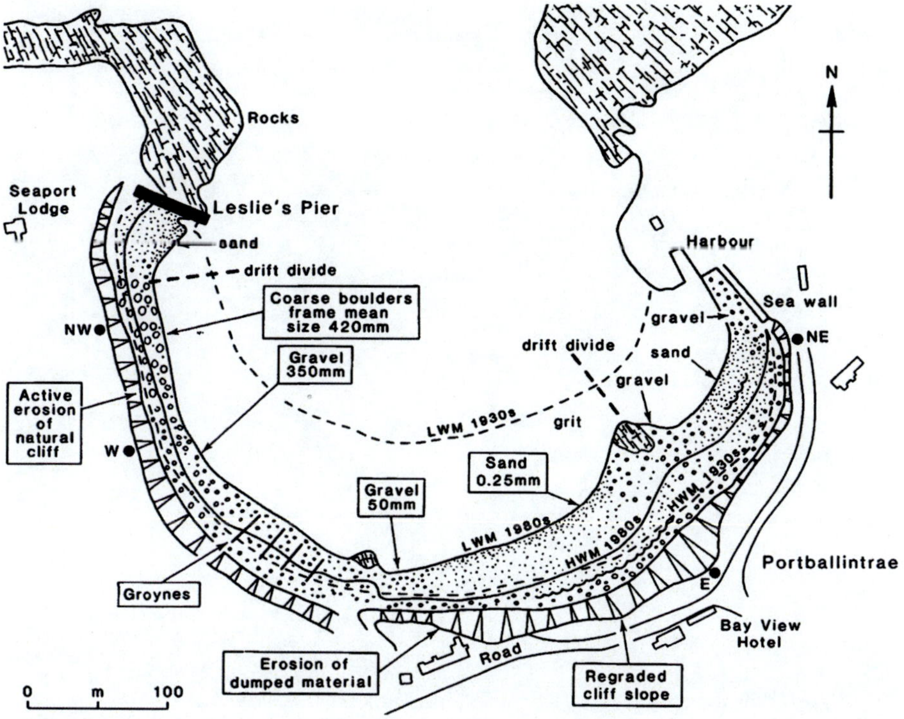

Fig. 6.3 Schematic of the changes that have taken place in Portballintrae Bay, showing the beach grading and shifts in shoreline position (low and high water marks) (Figure from Carter 1991)

Carter et al. (1983) attributed the onset of beach sand loss to the extension of the small fishing pier, known locally as Leslie's Pier, in the northwest section of the bay. Built around 1760, its original construction was an open-work wooden or wood and stone design, relatively small in dimensions, and useful for boats only at high tide. Its destruction in 1830 by a large storm saw it being replaced in 1895 by a significantly larger pier of solid stone that extended into deeper water on the same orientation as the original. At this point the problems for Portballintrae's beach are deemed to have started. Before the extension of the pier, waves entering the bay, even under high energy storm conditions, were refracted in a radial fashion from the two headlands. Wave energy would have been rapidly depleted through shoaling processes over the low angle dissipative shoreface, resulting in low energy waves eventually reaching the shoreline parallel with the beach and contributing to a swash-aligned equilibrium. This situation was associated with little or no longshore transport of sediment.

Introduction of the pier provoked an immediate interruption of this natural pattern of wave energy. During high energy wave events and at high stages of the tide, the structure captures, reflects and also diffracts incoming swell waves. The zone of diffraction down-wave of the pier is estimated to extend up to some 200 m. Beyond this wave energy 'shadow', high energy reconnects to the shoreline, setting up an easterly artificial energy gradient along the beach. This in turn helps promote

Fig. 6.4 Aerial views of Portballintrae in 1949 (*top*) and in 2004 (*bottom*) showing significant changes in beach width and sand volumes

strong longshore currents (and transport of sand) that flow to the centre of the bay. As a result of this new current regime the sand on the western part of the beach was eroded revealing the underlying gravel and boulders. According to Carter et al. (1983), initial impacts would have been restricted to local regions near the pier but later spread eastward to the rest of the shoreline.

The progressive transport of sediment from the beach into an alongshore current quickly reduced the limited amount of beach sand and created an offshore current flow (effectively a large mega rip), in the central section of the bay, similar to those reported elsewhere (Smith et al. 2010; Loureiro et al. 2011). Sediments were rapidly transported into deeper waters beyond wave base impeding natural replenishment of the bay.

6.4 Engineering Cavalry to the Rescue...

Coincident with the loss of sand and the deepening of the bay, back beach erosion and shoreline retreat occurred during this period (Kirk McClure and Morton 2003). Carter and Bartlett (1990) showed these rates of retreat were on the order of 0.03 m $year^{-1}$ during 1833–1904, 0.15 m $year^{-1}$ (1904–1966) and 0.25 m $year^{-1}$ (1949–1975). The then common practice of sand extraction from beaches by farmers for livestock bedding likely added to the beach lowering and shoreline retreat pattern that was underway (Carter et al. 1992).

With the obvious paucity of sand at the beach, its loss of amenity value as well as the relatively close proximity of housing right along the bay's shoreline, the first response of the local authority (Coleraine Borough Council and its predecessors) charged with managing the site was to employ engineers to help counteract the beach loss. In 1904, 14 years after the extension of Leslie's Pier, a series of wooden groynes was emplaced in the western sector of the bay downdrift of the diffraction zone from the distal end of the pier (Fig. 6.5).

After these fell into disrepair, further upgrading and additional groynes were installed to make up a set of six 25 m-long structures. Over the years up until the 1970s, these were maintained in an attempt to stabilize the shoreline and promote sand accumulation. Shoreline retreat over this period was particularly prevalent along the western section of coastline coincident with the pier's diffraction zone, and rapid erosion of a bluff of late glacial and Holocene deposits occurred.

During the period 1992–2003, further engineering works were completed (Kirk McClure and Morton 1992, 2003). In 1997, six 26 m-long replacement groynes were installed along with 1,200 m^3 (2,100 tonnes) of sand nourishment placed in the inter-groyne spaces to an approximate depth of 0.5 m. The re-nourishment sand was of unknown origin but was likely from an inland quarry source. In addition to the groynes, an extensive 150 m-long palisade barrier was also installed along the base of the western cliffs that were experiencing the most erosion (Fig. 6.6).

A recent report by Kirk McClure and Morton (2003) showed obvious beach surface elevation changes in the 6 years since the works carried out in 1997. Further lowering had taken place in the diffraction shadow zone down-drift of the

Fig. 6.5 Early emplacement of groynes at Portballintrae originally in 1904 shown here in a later photo, probably around the 1920's in a dilapidated state. Note that the beach had steepened and only minimal sand volume covers the lower beach face

Fig. 6.6 Present (2011) ground view of current (recently upgraded) groynes and palisade barrier in place at Portballintrae's western shoreline, just down-wave from Leslie's Pier

Fig. 6.7 Oblique aerial image of the site showing the latest groyne emplacements now lying within the wave diffraction zone down wave of Leslie's Pier. Sediment re-nourishment to a depth of 0.5 m in between the new groynes has now lowered itself to a thin veneer of sand locked inside glorified sand pits between each groyne structure (Photo courtesy of Northern Ireland Water)

pier, up to and including the northern sectors of the 1997 groynes. This was enough evidence for the engineers to recommend yet more engineering for the site.

These additional recommendations to the local authority involved the further seaward extension (by up to 26 m in some cases) of all of the 1997 groynes, as well as emplacement of additional groynes within the diffraction zone of the Leslie's pier. Further recommendations included the construction of three offshore breakwaters in the centre of the bay accompanied by a large scale re-nourishment programme.

To date, extensions of the groynes have taken place as proposed in 2003, as well as the installation of the additional (three) groynes to give a total of nine shore-normal groynes in the western section of the bay (see Fig. 6.7). The offshore breakwaters were not installed.

To date, costs for engineering works at Portballintrae are estimated to be in the region of £250,000 for schemes carried out between 1970 and mid 1990s, up to £590,000 in 1997 and, if recommendations of the latest scheme proceed in their entirety, then an additional £1.1 million (2003 price estimates) is envisaged.

6.5 Pitfalls of Shoreline Stabilisation

The successive interventions involving hard coastal engineering at Portballintrae along with the ignorance of a local authority being under immense local pressure to do something, has resulted in a series of damaging human impacts on natural

Fig. 6.8 The now rarely used Leslie's Pier, constructed in 1895, is most likely the main culprit in the removal of Portballintrae's sand. Natural in-bay wave refraction patterns were interrupted by the structure, in turn setting up artificial longshore currents in the western sector of the bay to ultimately invoke sand depletion of the beach system. Its removal is now imperative for future management of the site

coastal processes at Portballintrae. Since its construction, the modern concrete form of Leslie's Pier (Fig. 6.8) has instigated a coastal system that has created a sediment deficiency by inducing longshore sand transport and ultimate loss from the bay. At significant expense, a number of engineering solutions have been implemented to try and arrest the sediment loss.

Unfortunately, these have been installed in an environment that now operates where sediment volume is depleted and no fresh supply is available in any significant quantities. Small scale sediment re-nourishment of the inter-groyne cells was attempted (in 1997) but this formed only a thin veneer of sand that appears periodically to shift up and down within the steepened beach faces of each groyne cell.

Throughout the last 116 years of human interference, Leslie's Pier, central in the instability of the bay's hydrodynamics, has consistently been ignored. A game of 'engineering hopscotch' appears to be playing out at Portballintrae, whereby previous engineering seems to set the basis for the next set of engineering proposals and so on. Little attention has been paid to academic advice offered to help in the understanding of the natural dynamics of the bay. Instead, the local authority has continuously buckled under public pressure to put in place schemes for 'protection' against erosion. As with many coastal engineering schemes in beach systems, engineering options have been taken as they provide tangible evidence that something has been done.

6.6 The Remedy

The solution to Portballintrae's long running problem of sediment loss is quite simple but would require a new approach by the custodians of the site. The answer is to simply remove the main cause of perturbation of the sedimentary system i.e. Leslie's Pier. The natural processes of wave refraction and energy dispersion in a natural horseshoe bay create relatively sheltered conditions, even within a high energy stretch of coastline such as this. However, introduction of the pier close to one of the natural headlands resulted in focusing of high wave energy in the bay, forcing an artificial longshore current and ultimately leading to sediment loss. Removal of the pier would resolve this instability. Today the pier is not heavily used and serves little purpose for the local community.

Of course this action alone does not then immediately solve Portballintrae's beach loss problem. The next step would be to remove all the groynes and the palisade barrier. After the removal of the pier these would no longer be required and would in fact cause subsequent problems for natural wave processes approaching the shore if left in place. Localised scouring, refraction and reflection of wave energy around these structures would likely be detrimental to any restoration programme at the site.

Since the bay has now been artificially denuded of most of its natural sand and the local hydrodynamics preclude any natural reintroduction of sand, for beach build-up to occur, only one option remains – re-nourishment. A significant volume of sand would be required -probably close to the 80,000 m^3 that has been lost- to restore the bay to its original pre-Leslie's Pier state. In the absence of pier-induced megarips, it is likely that such a nourished beach would remain in place without repeated re-nourishment.

Whether or not the local council can wean itself off its dependency on coastal engineering remains to be seen. However, a heavily engineered bay is quite incongruous on this scenically beautiful coastline adjacent to the Giant's Causeway World Heritage Site. Further investment in coastal engineering at the site seems to be far from the public interest and seems difficult to justify on cost or environmental grounds. The recommended mitigation, in this case total removal of non-functioning engineered structures from the site, is the only logical option for long-term management.

Acknowledgments This work is a contribution to the IMCORE (Innovative Management for Europe's Changing Coastal Resource) project that has received European Regional Development Funding through INTERREG IVB. The author is grateful to Coleraine Borough Council for access to recent engineering reports on Portballintrae. Gratitude is also expressed to Northern Ireland Water and the Ulster Museum for the use of a number of archival photos.

References

Backstrom JT, Jackson DWT, Cooper JAG (2009) Shoreface morphodynamics of a high-energy, steep and geologically constrained shoreline segment in Northern Ireland. Mar Geol 257:94–106

Carter RWG (1982) Sea-level changes in Northern Ireland. In: Proceedings of the geological association, vol 93. London, pp 7–23

Carter RWG, Lowry P, Shaw J (1983) An eighty year history of shoreline erosion in a small Irish bay. Shore Beach 51:34–37

Carter RWG (1991) Shifting sands: a study of the coast of Northern Ireland from Magilligan to Larne, Countryside and wildlife research series no.2. HMSO, Belfast

Carter RWG, Bartlett DJ (1990) Coastal erosion in Northeast Ireland – Part I: sand beaches, dunes and river mouths. Ir Geogr 23:1–16

Carter RWG, Eastwood DA, Bradshaw P (1992) Small scale sediment removal from beaches in Northern Ireland: environmental impact, community perception and conservation management. Aquat Conserv Mar Freshw Ecosyst 2:95–113

Cooper JAG (2006) Geomorphology of Irish estuaries: inherited and dynamic controls. J Coast Res SI 39:176–180

Cooper JAG, Kelley JT, Belknap DF, Quinn R, McKenna J (2002) Inner shelf seismic stratigraphy off the north coast of Northern Ireland: new data on the depth of the Holocene lowstand. Mar Geol 186:369–387

Jackson DWT, Cooper JAG (2009) Application of the equilibrium planform concept to natural beaches in Northern Ireland. Coast Eng 57:112–123

Jackson DWT, Cooper JAG, del Rio L (2005) Geological control of beach morphodynamic state. Mar Geol 216:297–314

Kirk McClure and Morton (1992) Summary report on the coastal erosion of Portballintrae Bay. Internal consultancy report to Coleraine Borough Council, 42 pp

Kirk McClure and Morton (2003) Portballintrae beach study. Internal consultancy report to Coleraine Borough Council, 34 pp

Loureiro C, Ferreira Ó, Cooper JAG (2011) Extreme erosion on high-energy embayed beaches: influence of megarips and storm grouping. Geomorphology. doi:10.1016/j.geomorph.2011.10.013

O'Connor M, Cooper JAG, Jackson DWT (2007) Inlet-associated beach dynamics, northwest Ireland. J Coast Res SI50:626–630

O'Connor M, Cooper JAG, Jackson DWT (2011) Decadal behavior of tidal inlet-associated beach systems, northwest Ireland, in relation to climate forcing. J Sediment Res 81:38–51

Orford JD, Murdy J, Freel R (2006) Developing constraints on the relative sea-level curve for the northeast of Ireland from the mid-Holocene to the present day. Philos Trans R Soc A 364:857–866

Smith A, Mather A, Guastella L, Cooper JAG, Ramsay PJ, Theron A (2010) Contrasting styles of swell-driven coastal erosion: examples from KwaZulu-Natal, South Africa. Geol Mag 147:940–953

Wilson HE (1972) Regional geology of Northern Ireland. HMSO, Belfast, 115 pp. ISBN 337 06012 6

Chapter 7
Beach Nourishment in the United States

Andrew S. Coburn

Abstract Beach nourishment is a "soft" coastal erosion control strategy that involves the importation and emplacement of sand along dynamic shorelines in an attempt to stabilize and artificially maintain a minimum subaerial beach width. In the United States beaches are typically nourished to protect human economic development vulnerable to shoreline erosion/migration and coastal storms. While preferable to hard erosion control structures such as seawalls and groins, fiscal and economic issues have cast doubt on the efficacy of beach nourishment as an affordable, equitable and sustainable erosion response measure. Of particular concern are the delineation of costs and benefits, how costs and benefits are distributed and the possibility that storm damage reduction benefits – the economic force that drives nearly all beach nourishment projects – may never materialize. An accurate accounting of potential future costs must be combined with a pragmatic expectation of potential benefits before beach nourishment can be compared, let alone considered preferable, to non-traditional coastal erosion response measures such as retreat.

7.1 Introduction

The main approach to beach erosion and storm damage in the United States prior to World War II was the use of fixed structures such as groins, seawalls and jetties. These met with varying degrees of success and by the 1920s and 1930s, the use of structures had proliferated along certain sections of the nation's coastline to such an

---

A.S. Coburn (✉)
Program for the Study of Developed Shorelines, Western Carolina University, 294 Belk, Cullowhee, NC 28723, USA
e-mail: acoburn@wcu.edu

J.A.G. Cooper and O.H. Pilkey (eds.), *Pitfalls of Shoreline Stabilization: Selected Case Studies*, Coastal Research Library 3, DOI 10.1007/978-94-007-4123-2_7,

extent that they impeded the recreational use of the beaches (National Oceanic and Atmospheric Administration 2011).

By the late 1940s and early 1950s, the coastal science and engineering communities realized that the protective characteristics of natural beach and dune systems were more cost-effective and functionally successful than traditional, fixed coastal erosion control structures (Hillyer 2003). This concept emphasized the use of beaches and dunes as methods of dissipating wave energy, while also considering the aesthetic and recreational benefits of artificially created beaches.

Today, coastal communities have three basic strategies for managing beach erosion: (1) do nothing, (2) strategic retreat and (3) stabilization. "Do nothing" allows the beach to naturally migrate/erode while nothing is done to protect or relocate buildings and infrastructure. "Strategic retreat" also allows the shoreline to naturally migrate, but buildings and infrastructure are relocated/removed when necessary as the shoreline retreats.

Most coastal property owners, local government officials and beach-related businesses do not support the "do nothing" or the "strategic retreat" solutions. Instead, they favor beach nourishment, an engineered shoreline stabilization alternative that involves the importation and emplacement of sediment above the mean high water line in order to widen and artificially stabilize eroding or migrating shorelines.

Nourishment moves the shoreline seaward from property, creating a sacrificial buffer for property that dissipates wave energy (Fig. 7.1). The primary functions of beach nourishment are to:

- Provide hurricane and storm damage protection to upland property
- Increase recreational space along the shore
- Replicate natural coastal processes by augmenting coastal sand budgets

After an eroding beach has been nourished, subsequent beach nourishment *episodes* are typically referred to as beach *re*nourishment and a series of beach renourishment episodes comprises a beach nourishment *project*. Although a beach nourishment project may span several decades, individual beach renourishment episodes typically only last between 3 and 5 years before a beach reverts back to its pre-nourishment dimensions (Leonard et al. 1990). Therefore, a beach nourishment project with a 50-year life might consist of up to 20 individual beach renourishment episodes.

Every coastal state in the US has completed at least one beach nourishment episode for the purpose of erosion prevention, storm protection or ecosystem enhancement (Fig. 7.2). Since the first widely-recognized beach nourishment episode at Coney Island, NY in 1922, over 1,900 beach nourishment episodes have emplaced over 1.3 billion cubic yards of sediment along nearly 1,500 miles of US shoreline at a total cost of $5 billion in 2010 dollars (PSDS 2011).

Although the costs of beach nourishment are relatively high – the average cost along the US East and Gulf coast being about $2 million per mile – and

Fig. 7.1 A beach nourishment episode in Nags Head, NC

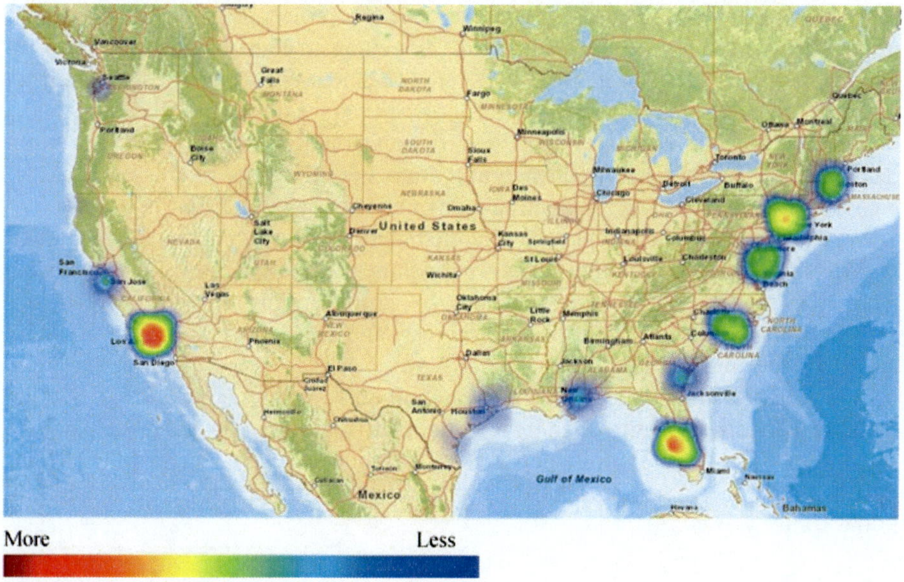

Fig. 7.2 A heat map showing the general location and intensity of beach nourishment in the United States

serious environmental issues must be resolved, there has been a clear shift from shoreline armoring to beach nourishment for most oceanfront shorelines over the past two decades (Hillyer 2003).

However, because beach nourishment provides a sacrificial, rather than fixed, barrier, any property damage mitigation, as well as recreational or ecological benefits it affords, are ephemeral. As a result, project/episode performance and environmental impacts are being increasingly scrutinized by the coastal science and management community, and the efficacy of continually replacing lost beach sand is being questioned.

As sea level rise and apprehension over the national and state economies continues, controversies over the technical merits of beach nourishment are being eclipsed by concerns regarding the identification and distribution of project benefits.

7.2 Costs of Beach Nourishment

The costs associated with beach nourishment are fairly straightforward and include the expected costs of initial construction, the present value of periodic renourishment and any costs associated with environmental mitigation. Social costs/issues such as scarcity, opportunity costs and negative externalities are not fully considered in decisions about whether to nourish, or in determining which of several nourishment project designs and implementation strategies is optimal.

Analytical models used to determine the amount of sand required to nourish a beach are used to estimate the construction cost of a project. Erosion rates and past weather patterns provide an estimate of how long a nourishment episode will provide a sufficient beach width before renourishment is needed. Costs related to sand transport, labor and other materials are estimated using market prices.

Because models estimate future events, uncertainties create problems with predicting the severity and frequency of storms, the variability in erosion rates for a given storm climate and the continuing availability of sand sources.

While estimating the future cost of nourishment involves a considerable amount of uncertainty, an examination of past nourishment episodes shows a pronounced increase in cost over the past four decades. In North Carolina, for example, the unit cost (in dollars per cubic yard) of nourishment sand has risen from around \$0.92 in the 1970s to \$11.94 in 2009 (Fig. 7.3).

It is believed that the 26.7% increase in unit cost from \$8.96 in 2004 to \$11.35 in 2005 is a direct result of increased demand for dredging and beach nourishment in Florida following the active 2004 hurricane season in which four hurricanes (Frances, Jeanne, Charley and Ivan) and a tropical storm made landfall in the state.

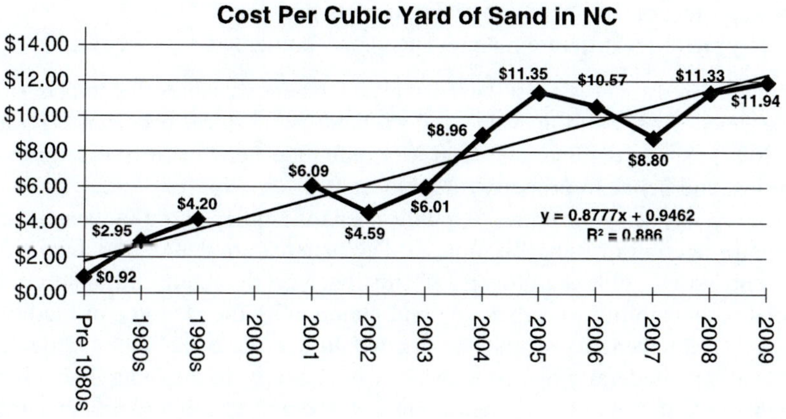

Fig. 7.3 Change in the cost of nourishment sand in North Carolina from the 1970s to 2009

7.3 Benefits of Beach Nourishment

Compared to project costs, the benefits of beach nourishment are more ambiguous. In general, however, the primary benefits of beach nourishment include:

1. **Hurricane and Storm Damage Reduction (HSDR) benefits**: HSDR benefits represent the protection against storm damage to upland property and infrastructure. They accrue to the owners of beachfront property and are distributed according to the residence patterns of affected property owners.
2. **Recreational benefits**: Recreational benefits accrue to beach visitors and are distributed by the residence patterns of the beach users.
3. **Other benefits**: Other benefits include the impact on business, tax revenues and property valuations. This category may also include environmental, aesthetic and/or unique site considerations.

7.3.1 Hurricane and Storm Damage Reduction (HSDR) Benefits

A wide beach provides measurable economic benefits to owners of upland property, both public and private, by protecting property from losses or destruction due to storms. The total HSDR benefits attributed to a particular property are typically determined by subtracting the expected value of damages with a beach nourishment project in place from the expected value of damages without that project in place. Damages considered include anticipated costs to protect property and the dollar-value of land and buildings lost to erosion. Specific categories of HSDR benefits include (US Army Corps of Engineers 2000):

- Land loss prevention
- Upland structure damage/loss prevention

- Shore protection structure construction costs prevented
- Shore protection structure maintenance costs prevented

Damage assessment is a risk-based analysis, where variables that are considered include land value, building value and probability of storm occurrence (Gravens et al. 2007). Storm damage and land loss reduction benefits accrue to individual properties, and hence to property owners.

Where a property provides a public benefit, either through beach access, resource protection or transportation, HSDR benefits are deemed to accrue to the public entity that holds the property in trust because that entity ultimately has the financial responsibility for protecting and maintaining the resource or facility. For example, HSDR benefits at a state park would accrue to the state, while HSDR benefits along a federally owned highway would accrue to the federal government.

Each of these benefits depends upon site-specific characteristics that include local erosion rates, local sand transport characteristics, the assessed value of land and upland developments, the existence and type of shoreline armoring and the permissibility of using shoreline armoring to protect endangered properties. In locations that prohibit shoreline hardening, HSDR benefits often include avoidance of future costs for structural relocation.

The land loss prevention benefit is the dollar value of land expected to be lost to erosion. For example, if the estimated average annual erosion rate of a 100-ft wide beachfront property is 2 ft per year, and land is valued at $10 per square foot, the annual land loss benefit is computed to be $2,000. For a 30-year nourishment project, the dollar value of land lost in years 2 through 30 is discounted to present day dollars using an appropriate discount rate in order to determine the total land loss benefit over the life of the project.

The upland structure damage and/or loss prevention benefit is determined from estimates of anticipated damages to buildings, including flood damage, in the absence of a project. Project benefits associated with storm damage reduction and erosion losses are generated for oceanfront properties. Properties at greatest risk of suffering structural damage include those with little or no protective beach and/or with relatively low elevation (National Oceanic and Atmospheric Administration 2011).

The actual dollar value of the expected loss depends upon the value of the property at risk and the percent of total loss that is estimated before the property owner takes action that is in his/her best economic interest. In locations that permit shoreline armoring, remedial action may include construction of a seawall – if such costs are economically justified – and the HSDR benefits of erosion control structure construction and maintenance costs.

For locales that prohibit shoreline hardening, a property owner may be faced with the expense of relocating the structure or, if relocation is not physically or economically feasible, total building loss. If erosion modeling of the beach in the absence of beach nourishment demonstrates that an unarmored property will be undermined at some point within the design life of a project, it can be assumed that the property owner will act to protect the property provided it is economically justifiable to do so. In this case, benefits include present day costs for initial construction in the year in which armoring is needed, in addition to annual maintenance costs (National Oceanic and Atmospheric Administration 2011).

It should be noted that the benefits associated with land loss prevention, upland structure damage/loss prevention, shore protection structure construction costs prevented and shore protection structure maintenance costs prevented accrue to individual property owners.

7.4 Calculating HSDR Benefits

The models of beach behavior used to evaluate project benefits attempt to describe physical beach processes and attempt to estimate the rates of beach erosion, how frequently a beach must be maintained or renourished and the amount of time before coastal structures are impacted by erosion or storms.

A segment of shoreline is studied and erosion rates over a period of time are calculated. It is usually inferred that past long-term erosions rate will apply in the future. Then, given an initial starting point or baseline for the beach, the future position of the beach and its proximity to upland structures is estimated for any year by applying the annual erosion rate over a specified time period. From this, an estimate of the land value and structure losses is calculated (Gravens et al. 2007).

While there are usually no market prices for storm damage reduction, there are several methods to assign monetary values to projected storm damage reduction benefits: (a) the contingent valuation method, (b) the averting behavior approach and (c) the hedonic approach.

The contingent valuation method involves directly asking someone how much they would be willing to pay for a specific service and is called "contingent" valuation because people state their willingness to pay, contingent on a specific hypothetical scenario and description of a service. Because the contingent valuation method is based on what people say they would do, as opposed to what people are observed to do, conceptual, empirical and practical problems associated with developing dollar estimates of economic value on the basis of how people respond to hypothetical questions about hypothetical market situations are debated constantly in the economics literature (Vincent et al. 1986).

The averting behavior or opportunity cost approach involves estimating storm damage reduction benefit from the cost that property owners incur to avert storm damages if the nourishment project were not undertaken (National Oceanic and Atmospheric Administration 2011).

Hedonic pricing uses statistical analysis to estimate benefits from related goods that are sold in a market. For example, "beach valuation" for a house may be estimated using housing prices, housing characteristics, and proximity to the beach. One of the components of the value could be beach width, which would provide insight into the significance of storm protection (and recreation) for beachfront property (Gopalakrishnan et al. 2010).

7.4.1 *Recreational Benefits*

Recreational benefits resulting from beach nourishment projects depend upon how the demand for beach visitation, and the value of a beach experience, change. While many recreational benefits have a wider geographic distribution than HSDR benefits, it is a common belief that anyone – local residents, non-local residents of the state in which the beach is located, non-state residents or residents of foreign countries – who visits a nourished beach benefits.

Two methods commonly used to estimate the recreational benefits of beach nourishment are the contingent valuation method and travel cost method. The contingent valuation method relies on survey techniques to elicit an individual's willingness to pay for recreational benefits. For example, visitors to a beach are asked a series of questions such as why they visited, how often they visit, their expenditure patterns while there and their willingness to pay for beach use (Vincent et al. 1986).

The travel cost method uses travel costs as a proxy for market prices and is used to estimate demand curves for semi-public goods. The travel cost method combines the total number and frequency of beach visits with data and information on distance from the beach to estimate the recreational value of the beach (Ecosystem Valuation.org 2011).

Prior to 1986, beach nourishment projects were primarily justified for federal participation based on their recreational benefits. The 1986 Water Resources Development Act (WRDA) changed the weighting of benefits recognizing HSDR as the primary purpose of federal participation in beach nourishment projects (National Research Council 1999).

7.5 The Economics of Beach Nourishment

Given the institutional structure in the United States and the relatively high cost of beach nourishment, most episodes/projects are financed with public funds. The use of tax dollars, however, raises a variety of economic/fiscal questions and issues including whether governmentally-funded beach nourishment projects are "worth it."

In capital budgeting, in the private market, this question is answered affirmatively if a producer's total revenue for providing a good or service is greater than or equal to the total cost of doing so. In the public sector, a government entity providing financing for beach nourishment should be most interested in knowing how much of its investment will be returned in the form of increased taxes, and other revenue, as a result of a project (Seitz and Ellison 1999).

However, when it comes to publicly-funded beach nourishment projects, fiscal issues are usually trumped by concerns over the value of property damage avoided.

With property damage reduction the engine that drives beach nourishment, three methods are available to compare storm damage reduction benefits to the costs of beach nourishment (National Oceanic and Atmospheric Administration 2011):

1. Net present value (NPV) measures the difference between the project benefits and the project costs (NPV = $ Benefits – $ Costs).
2. Benefit cost ratio (BCR) is a ratio of dollar benefits to dollar costs (BCR = $ Benefits/$ costs).
3. Internal rate of return (IRR) is the implicit interest rate returned by the project.

A positive NPV, a BCR greater than one or an IRR greater than the benchmark rate imply that the benefits of a project are greater than the costs and a project should be undertaken. A negative NPV, a BCR less than one or an IRR less than the benchmark rate implies that project benefits are less than the costs and a project should not be undertaken.

The model commonly used to ascertain whether or not a beach nourishment project is "worth it" – the benefit cost analysis – is similar to that used in traditional capital investment problems in which the present value of project benefits is compared to the present value of project costs.

In general, for a project to be economically justifiable, it must have a benefit-to-cost ratio greater than 1.0. That is, the dollar value of the benefits received from all categories must exceed projected project costs. In theory, however, project planners should seek to maximize the benefit cost ratio, thereby providing the greatest benefit for the least cost (National Oceanic and Atmospheric Administration 2011).

7.6 Cost Benefit Analysis Issues

A cost benefit analysis reduces the final evaluation of a complex beach nourishment project down to one number that is highly sensitive to the assumptions made while performing the analysis. Since the feasibility of a beach nourishment project rests almost entirely upon the ability of project proponents to demonstrate that benefits sufficiently exceed costs to justify the expenditures required over the design life of the project, questions have arisen regarding the suitability of the cost benefit analysis.

The types of benefits and costs evaluated may also depend on who is analyzing the project, as some of the benefits and costs are spatial. For example, a local entity may include local economic development as a benefit if the beach is nourished. However, it may be that without the beach nourishment, development will take place anyway, or this development will just take place somewhere else. Therefore, from a national perspective, an increase in development cannot be considered a resulting benefit of beach nourishment.

7.6.1 Temporal Issues

The temporal nature of a beach nourishment project's effects poses special problems for valuation, including extrapolating future costs and benefits, evaluating the effects of uncertainty associated with random future events, estimating future behavioral responses to the project and discounting future costs and benefits.

Since a dollar spent or received now is not the same as a dollar received or spent in the future, future values must be discounted back to the present. Construction costs are incurred at the beginning of the project, while periodic maintenance costs are incurred every few years. A project with a design life of 50 years generates an estimated dollar benefit each year for 50 years. However, because a $1 benefit received next year, 5 years from now or 50 years from now is not the same as a dollar received today, it is necessary to discount these future flows back to the present.

Present value factors and rates, which can be obtained from a present value table or computed on a financial calculator, are used in the Present Value (PV) calculation. In general, the pretax rate of return to private capital or the federal government's borrowing rate is used. Treasury borrowing rates are published at http://www.slgs.gov/opd/opdirtb.htm for various maturities.

The selection of rates is important because higher interest rates result in lower present values of future "expected" benefits for a project such that higher interest rates effectively lower the present value of benefits. If the present value of benefits falls below the present value of costs, a project should be deemed economically unjustifiable (National Research Council 1995).

7.6.2 Issues of Equity

Federal beach nourishment projects are financed through federal, state and local governments, and tax money must be raised to finance project costs. This makes the determination of project benefits and project beneficiaries essential in order to make legally defensible decisions about the allocation of project costs, as well as to develop accurate and appropriate funding mechanisms.

To address concerns over equity within the context of beach nourishment, several questions need to be asked including:

- Who are the beneficiaries?
- What specific benefits will each beneficiary receive from a proposed project?
- What is the dollar value of the benefits received by each beneficiary?

In theory, a benefit-cost analysis estimates numbers for qualitative benefits such as storm damage prevention, recreational income, changes in amenity value for local property owners and weighs them against quantitative costs including the direct costs of construction, negative physical or ecological effects and increased infrastructural burdens caused by increased usage. What the benefit-cost analysis fails to explain, however, is who benefits and who pays. In addition, little regard is paid to issues such as the temporal and spatial distribution of costs and benefits (Cooper and McKenna 2008).

According to the NOAA Coastal Service Center, HSDR benefits to oceanfront property owners comprise 60–90% of all beach nourishment benefits. Furthermore, an analysis of property records in 5 North Carolina coastal counties and communities revealed that 97% of all oceanfront properties do not serve as a primary residence (National Oceanic and Atmospheric Administration 2011; PSDS 2011). As a result, 58–87% of all beach nourishment benefits accrue to property owners whose permanent residences are not beachfront.

Because many oceanfront property owners reside in non-coastal states, the Corps considers this distribution of benefits national in scope and uses this logic to justify the expenditure of federal tax dollars for beach nourishment.

Regardless of the geographic distribution of beachfront property owners, the fact that the majority of beach nourishment benefits accrue to a very small fraction of the population brings up issues of equity regarding who pays, and who should pay. While there are many definitions of equity, the concepts of absolute equity, ability to pay and the benefit principle can all be used to help answer this question (Stiglitz 1999).

Absolute equity implies that if a beach nourishment project's cost is $1,000,000 and there are 100 people who benefit from the project, each person should pay $10,000. This is a simple concept and works well with small groups, such as when a group of friends go out for dinner and split the check. It does not generally work well with large diverse groups, as in funding a beach nourishment project.

The ability to pay principle allocates costs according to an individual's ability to pay. Generally, the higher the individual's income or the larger the individual's asset base, the more that individual is asked to pay. Federal income taxes work this way.

The benefit principle espouses that it is fairer to tax individuals on the basis of the benefits they receive, and implies that fees be charged for publicly-provided services. Explicit examples include highway and bridge tolls and parking fees at publicly-owned beach parking lots.

The benefit principle of equity is most appropriate when it comes to determining who should pay for beach nourishment since benefits accrue almost exclusively to beachfront property owners. Subsidizing beachfront property owners with federal tax revenues violates both the benefits principle as well as ability to pay principle.

According to the benefit principle, because HSDR benefits accrue to individual properties, the proportional costs for beach nourishment should appropriately be assessed to each property owner. Recreational benefits that accrue to people who use a nourished beach are more difficult to quantify. However, recreational beneficiary categories can be linked to potential funding sources such that allocation and use of funds is appropriate and defensible.

7.6.3 *What if You Build It... and It Doesn't Come?*

As previously discussed, HSDR benefits are evaluated by subtracting the expected value of storm damage with nourishment from the expected value of storm damage without nourishment. But what if a beach is nourished and a storm never hits?

Between 2001 and 2004, 4.46 million cubic yards of sand were emplaced along 149,000 linear feet of beach on Bogue Banks in Carteret County, North Carolina at a cost of $32 million (PSDS 2011). The reason given for these efforts was to protect property from hurricanes and erosion even though the island hadn't experienced a land-falling hurricane since Ginger in 1971 and the average long-term erosion rate along most of the island was about 2 ft/year (Carteret County 2011; NCDCM 2011).

In the period between 2001 and 2004, Bogue Banks experienced no land-falling hurricanes. The island was, however, impacted by Hurricane Ophelia, a category 1 storm in 2005 that remained 40 miles to the east of the island. According to the National Weather Service, moderate beach erosion occurred as a result of a 3 to 5 foot storm surge along with high waves on the beaches and sounds south of Frisco, NC (NOAA NWS 2011). In the absence of nourishment, property damage inflicted by Ophelia along Bogue Banks would likely have been a fraction of the $31.6 million spent on nourishment prior to the storm. Since Ophelia, an additional 2.9 million cubic yards of sand have been emplaced along 102,000 linear feet of Bogue Banks beaches at a cost of $35 million (PSDS 2011). To date, Ginger remains the last hurricane to make landfall on Bogue Banks.

The situation on Bogue Banks raises a profound question: If a beach is nourished and no storms hit, what are the HSDR benefits of that nourishment? Putting it another way, did the communities on Bogue Banks spend $67 million and get almost nothing in return?

7.7 Estimating the Future Cost of Beach Nourishment

A number of studies evaluating and comparing the costs of beach nourishment to other forms of erosion response, including retreat, have been undertaken by coastal economists.

A study by Wakefield and Parsons, for example, compared the costs of nourishment and retreat in Delaware over a 50 year span and concluded that the cost of retreat exceeded the cost of nourishment by more than a factor of three (Wakefield and Parsons 2003). The study determined the 50-year social cost of retreating from Delaware's Atlantic coast to be between $196 million and $375 million (in 2001 dollars), while the cost of nourishing Delaware's beaches over 50 years was estimated to be in the range of $48–$60 million (in 2001 dollars). The authors employed coastal simulation models to predict the volume of sand required to stabilize the coast via beach nourishment and based future costs on historic data such as the volume of sand used and the unit cost of nourishment.

According to data compiled by the Program for the Study of Developed Shorelines at Western Carolina University, the annual cost of beach nourishment in Delaware between 1957 and 2001 was about $1.5 million. Between 2002 and 2011, however, nearly $72 million, or $8 million/year, had already been spent nourishing Delaware's beaches (PSDS 2011).

A number of factors likely contributed to Wakefield and Parsons' failure to accurately estimate future nourishment costs in Delaware including the inability to predict the future demand for nourishment, the availability of funds, storm activity and sea level rise. Also missing from the study are the environmental costs of nourishment. Wakefield and Parsons contend that their analysis supports continuation of Delaware's nourishment policy. However, since the actual cost of nourishing Delaware's beaches over a 9-year span already has exceeded the upper limit of the study's estimated 50-year cost by $12 million, experience dictates that beach nourishment in Delaware is, and likely will continue to be, significantly more expensive than predicted and may not be the best erosion response strategy for the state of Delaware.

While Delaware's experience may not be representative of every coastal state, it is clear that computer models and historic nourishment costs are not an accurate predictor of future nourishment costs.

7.8 Conclusion

Beach nourishment is a coastal hazard mitigation strategy that involves the importation and emplacement of sand along an eroding or migrating shoreline in an attempt to maintain some pre-defined beach width in order to protect vulnerable human economic development located behind that shoreline. As such, the primary objective of beach nourishment is to reduce hurricane storm damage.

While preferable to the use of hard structures, fiscal and economic issues have cast doubt on the efficacy of beach nourishment as an affordable, equitable and sustainable erosion response measure in many coastal locations.

A cost benefit analysis, traditionally used to justify the cost of constructing an artificial beach through nourishment, reduces the final evaluation of a complex activity into a single number that is highly sensitive to the assumptions made while performing the analysis. Therefore, since the feasibility of beach nourishment rests almost entirely upon the ability of proponents to demonstrate that benefits sufficiently exceed costs over a defined period of time, questions regarding the suitability and objectivity of the cost benefit analysis have arisen.

Of particular interest are issues associated with the delineation of costs and benefits, how costs and benefits are distributed and the possibility that, in the absence of a post-nourishment hurricane, storm damage reduction benefits – the economic force that drives most nourishment projects – may never materialize.

If nourishment is to remain the socially-preferable response to the problems facing the developed shorelines of the United States, as it appears it will, the methodology used to identify, delineate and quantify nourishment costs and benefits should be revised to capture the many inherent uncertainties associated with nourishment.

At a minimum, the basis for evaluating the potential economic impacts of beach nourishment should:

- Reassess the costs and benefits that merit inclusion in episode/project evaluation,
- Incorporate uncertainty regarding demand for nourishment, storm activity and the availability of funds when assessing future costs and benefits (both with and without a project),
- Investigate behavioral responses stimulated by beach nourishment and associated policy issues,
- Consider the coupling of nourishment with local growth and land-use plans to increase net benefits and
- Investigate incentive-based financing schemes.

Pre-nourishment analyses of HSDR benefits consider the expected storm environment based on historical weather and storm data. There are few, if any, after-the-fact studies that evaluate the actual benefits a beach nourishment episode/project affords. To improve the technical basis for assessing future costs and benefits, a post-construction economic/fiscal assessment that monitors, identifies and measures a wide range of costs and benefits should be completed.

As the Delaware experience suggests, an accurate accounting of potential future costs must be combined with a pragmatic expectation of potential benefits before beach nourishment can be compared, let alone considered preferable, to nontraditional coastal erosion response measures, such as retreat.

References

Carteret County Shore Protection Office (2011) http://www.protectthebeach.com/

Cooper JAG, McKenna J (2008) Social Justice and coastal erosion management: the temporal and spatial dimensions. Geoforum 39:294–306

Ecosystem Valuation.org (2011) web site: http://www.ecosystemevaluation.org/travel_costs.htm

Gopalakrishnan S, Smith MD, Slott JM, MurrayAB (2010) Value of disappearing beaches: a hedonic pricing model with endogenous beach width, working paper EE 10-04, Nicholas School of the Environment, Duke University, Sept 2010

Gravens MB, Males RM, Moser DA (2007) Beach-fx: Monte Carlo life-cycle simulation model for estimating shore protection project evolution and cost benefit analyses. Shore Beach 75 (1):12–19

Hillyer T (2003) The Corps of Engineers and Shore Protection: History, Projects, Costs. U.S. Army Corps of Engineers Institute for Water Resources, Alexandria, VA

Leonard LA, Dixon KL, Pilkey OH (1990) A comparison of beach replenishment on the U.S. Atlantic, Pacific, and Gulf coasts. J Coast Res SI 6:127–140

National Oceanic and Atmospheric Administration, Coastal Services Center (2011) http://www.csc.noaa.gov/beachnourishment/html/human/socio/index.htm

National Research Council (US) (1995) Beach nourishment and protection. National Academy Press, Washington, DC

National Research Council (US) (1999) New directions in water resources planning for the US Army Corps of Engineers. National Academy Press, Washington, DC

National Weather Service, National Oceanic Atmospheric Administration (2011) Newport/ Morehead City, NC http://www.weather.gov/storms/ophelia/PSHMHX.txt

North Carolina Division of Coastal Management, Long-Term Average Annual Erosion Study (2011) http://dcm2.enr.state.nc.us/Maps/carteret_rate.htm

PSDS (2011) Beach Nourishment Database, Program for the Study of Developed Shorelines Western Carolina University. http://psds.wcu.edu/1038.asp

Seitz N, Ellison M (1999) Capital budgeting and long term financing decisions, 3rd edn. Dryden Press, Fort Worth

Stiglitz JE (1999) Economics of the public sector, 3rd edn. W. W. Norton & Company, New York

US Army Corps of Engineers (2000) Engineering regulation 1105-2-100, 22. Headquarters, Washington, DC

Vincent MK, Moser DA, Hanson WJ (1986) National economic development procedures manual – recreation, vol 1 recreation use and benefit estimation techniques. IWR Report 86-R-4. US Army Corps of Engineers

Wakefield J, Parsons G (2003) A comparison of nourishment and retreat costs on Delaware's Ocean Beaches. Shore Beach 71:23–26

Chapter 8
Failed Coastal Stabilization: Examples from the KwaZulu-Natal Coast, South Africa

Alan M. Smith, Simon C. Bundy, and Andrew A. Mather

Abstract Coastal stabilization, mostly involving coastal dune cordon destruction, has been practised on the KwaZulu-Natal coast for the last two decades. During this time no major swells occurred. The last Lunar Nodal Cycle (LNC) tidal peak was associated with a series of unusually high swells. The temporal coincidence of exceptionally high tide and large swell led to catastrophic coastal erosion and underlined the folly of coastal stabilization, especially at reversing erosion hotspots, many of which are urbanized and defended. These locations remain at risk and will experience future erosion.

8.1 Introduction

The period 2006–2007 was one of marked erosion on the KwaZulu-Natal (KZN) coastline during which the pitfalls of previous efforts in coastal stabilization were demonstrated. The centrepiece of this erosion episode was a high swell ($H_s = 8.5$ m; $H_{max} = 14$ m; $T = 16$s; Direction SE) which struck the KZN coast on the 2007 March equinox, just 6 months after the 18.6 year Lunar Nodal Cycle (LNC) peak maximum which fell on 7th October, 2006 (Smith et al. 2010). Estimates of the March 2007 large swell recurrence interval range from 20 years

A.M. Smith (✉)
School of Geological Sciences, University of KwaZulu-Natal, Durban 4001, South Africa
e-mail: asconsulting@telkomsa.net

S.C. Bundy
Sustainable Development Projects CC, P.O. Box 1016, Ballito 4420, South Africa
e-mail: simon@ecocoast.co.za

A.A. Mather
Ethekwini Municipality, P.O. Box 680, Durban 4000, South Africa
e-mail: mathera@durban.gov.za

J.A.G. Cooper and O.H. Pilkey (eds.), *Pitfalls of Shoreline Stabilization: Selected Case Studies*, Coastal Research Library 3, DOI 10.1007/978-94-007-4123-2_8,

(based on the historic record: Guastella et al. 2008) to 35 years (based on wave height): (Mather 2008). Extensive erosion resulting in the loss of infrastructure and property was exacerbated by a number of factors, including previous attempts to stabilize and defend portions of coastline. We give three examples of the pitfalls of stabilization and forewarning of the effects that a continued trend towards coastal zone stabilization may have in the future. There are indications that wave heights are on the increase although the frequency of storms remains constant (Guastella and Rossouw 2009; Mavume et al. 2009) and consequently coastal storms may wreak further destruction, assisted and abetted by the practice of decoupling of dune and beach by shoreline stabilization.

8.2 KZN Coastline

The KZN coastline (Fig. 8.1) is generally eroding in the long-term at rates between −10 and −50 cm per year (Cooper 1991a, b, 1994). Coastal erosion is attributed to various global and regional factors such as sea level rise (Mather 2007; Mather et al. 2009), coastal and terrestrial land use changes (Smith et al. 2007), river sand mining and dam construction (Garland and Moleko 2000). In many places ribbon coastal development has been allowed on the primary dunes closing off this natural sand reservoir. The KZN coast is a high-energy coastline dominated by coarse-grained, reflective to intermediate beaches (Cooper 1991a; Cooper and Flores 1991).

The average H_s for Durban is 1.8 m, diminishing northward to 1.5 m for Richards Bay (Moes and Rossouw 2008) (Fig. 8.1). Durban and Richards Bay have average spring tidal ranges of 1.8 m and 1.84 m, respectively, with Highest Astronomical Tides (HAT) of 2.3 and 2.47 m, respectively. The KZN coast comprises a number of headland-bound bays containing sandy beaches (Cooper 1991a, b, 1994), some of which form barriers across river mouths (Cooper 2001). In the Durban area, a net sand volume of 360 000–620 000 m^3 is moved northward annually by the littoral drift (Schoonees 2000). Seasonal swell-driven beach rotation is a feature of these headland-bound bays (Smith et al. 2010). A continuous sand beach in the form of a 20 km-long zeta bay extends through the Durban Bight (Fig. 8.1). Steep coastal dunes back most beaches, although some are backed by rocky outcrops. Coastal dunes commonly reach heights of 30–60 m, and are often compound features, varying in age from Late Pleistocene to Holocene (Tinley 1971). Coastal development, especially over the last 20 years, has covered large parts of the coastal dune cordon, leaving them dysfunctional, and any bare sand patches have been planted and stabilized, often with alien vegetation.

Prior to 2007, the previous comparable high swell event on the KZN coast was in February 1984 driven by tropical storm Imboa (H_s of 9 m: Guastella and Rossouw 2009), but there are records of similar sized swell events stretching back to 1904. The period following tropical storm Imboa was characterized by a 'quiet' marine environment with swells rarely exceeding 3 m. These conditions

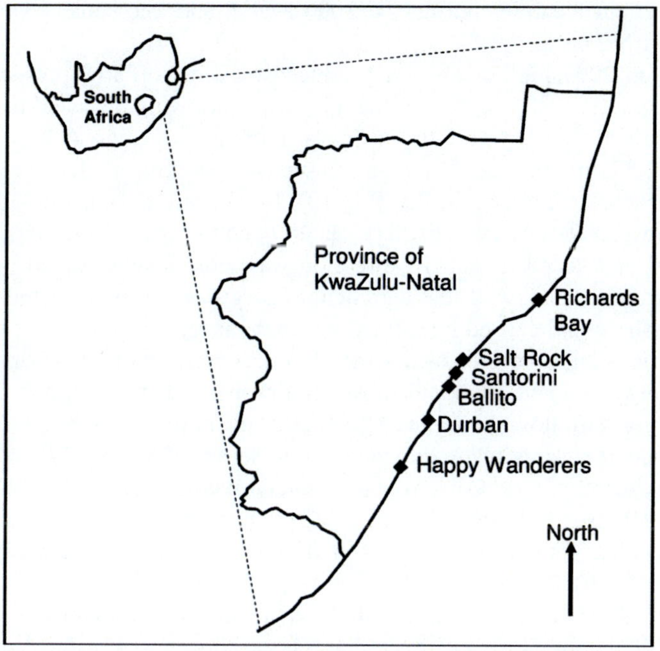

Fig. 8.1 Location of KwaZulu-Natal coastline in regional context

persisted during the 1990s and early twenty first century. During this time a massive coastal development boom occurred, coincident with widespread coastal stabilization, including coastal dune fragmentation (or complete excavation and removal of dune), the construction of seawalls, together with land reclamation and site raising. Development on the coast rose from 28% to around 50% in the period 1994 to 2007 (Cilliers, personal communication 2011). The extended period between high swellevents gave an impression of coastal stability and hence coastal infrastructure development took little account of the possibility of wave erosion.

8.3 2006–2007 Erosion Event

During 2006–2007 the KwaZulu-Natal coast of South Africa was exposed to several high swell events (H_{so} >3 m), which struck near the peak of the 18.6 year lunar nodal cycle (LNC), causing shoreline recession (Mather 2008; Smith et al. 2007, 2010). The largest swell (H_s = 8.5 m) struck the coast on the March equinox (18–20th) and generated a strong northward littoral drift and an offshore-directed storm-return flow (Smith et al. 2010). This high swell cannot be viewed in isolation since minor (and in some cases catastrophic) erosion, connected with the LNC peak was a feature at certain localities during the 2006 March and September equinoxes and the October LNC peak.

The March 2007 high swell was produced by a cut-off low pressure system, stopped in its west-east progress by a ridging high, allowing the swell to grow. The ensuing southeast swell (more-or-less coast-normal), together with an unusually high tide (2.57 m), produced a high run-up and significant damage to coastal stabilized reaches (Smith et al. 2007, 2010). In contrast a comparable high swell struck in July (2009), however, in this case although the swell was as large as that of March 2007 and lasted longer, its direction was more southerly and it coincided with a neap tide. That swell lost significant energy as it was refracted across the continental shelf, and caused no infrastructure damage.

Observations made before, during and after the event confirm dramatic shoreline retreat of between 5 and 40 m and substantial property damage (Smith et al. 2007, 2010). Storm-return flow during the March 2007 event removed the nearshore bars, leaving the coast susceptible to erosion from lesser swell events (2–5 m H_s) during the ensuing austral winter. Catastrophic coastal erosion occurred at certain erosion hotspots (EHS: see: Thornton et al. 2007) where up to 100 m coastal retreat occurred locally. The nearshore bars had reformed by the beginning of 2008 (Smith et al. 2010).

The 2006–2007 erosion events highlighted the pitfalls of previous local efforts at coastal stabilization. Three examples have been chosen as illustrations. The first two illustrate the interaction between coastal stabilization works and the March 2007 swell and the third the effects of the follow-up erosion during lesser storms that occurred during the post-March 2007 large swell vulnerability window (when the coast had not yet recovered from the March 2007 erosion).

8.3.1 Salt Rock, Ballito: Constructed on a Bluff Stabilized by a Seawall

Historically the Salt Rock locality is a bluff fronted by a small coastal dune and beach (Fig. 8.1). Significant alteration of the inter tidal and supratidal shoreline was undertaken between 1957 and 1980 as the area was developed for recreation and housing. These modifications involved dune destruction and leveling and the construction of decorative seawalls on the bluff and beach (Fig. 8.2). A tidal pool was linked to the seawalls forming a bottle-neck to the northward flowing littoral drift. Consequently the coast in this area can no longer self-repair after storms and sediment supply across and around the bluff is constrained. This headland is bedrock-controlled and tipped by a dolerite outcrop projecting into the surf zone (Fig. 8.2). In addition to dune destruction and leveling, the Salt Rock bluff has been further stabilized by land reclamation so as to create a space for a caravan park, defended by a seawall (Fig. 8.2).

Fig. 8.2 Salt Rock, the above image was captured a month after the 2007 storm. The bottleneck section created between the defended beachfront properties and the landward wall of the tidal pool caused enhanced scouring of the beach

The March 2007 high swell was associated with a 0.3–0.4 m storm surge and a peak wave run-up in excess of 10 m was recorded (Smith et al. 2010). At Salt Rock, the shoreface is steep (water depth is −20 m at 80 m offshore) and thus deep water wave energy was not dissipated. Furthermore the coastal dune topography is steep and wave action could not be released landwards, resulting in a storm-return offshore flow active to depths of 20 m plus (Smith et al. 2010). The combination of the coastline's exposure and the SE swell propagation direction also caused a very strong northward-littoral drift. The bottle neck created by the Salt Rock bluff, accentuated by the decorative sea wall, impeded the northern littoral drift and reduced the coast's ability to dissipate high energy wave run up. This coastal irregularity also caused the development of a strong megarip in the southern lee of the Salt Rock Bluff and back-cutting at the head of the rip current resulted in strong scouring and property loss. The littoral current that did round the bluff caused scouring, especially at the northern end of the bottleneck (Fig. 8.2). Furthermore, the coastal defence structures acted to deflect and concentrate incoming wave energy at the weakest points of the structure, causing scour. Significant failure of the structure took place and up to 700 m^2 of property was lost from northern properties positioned at the end of the seawall.

Fig. 8.3 View of Santorini undercut by erosion. Note hard rocky headland to right which stopped the littoral drift and caused megarip-current action. Older defences, constructed to repair coastal erosion from an earlier erosion event were exposed in the eroded scarp

8.3.2 Santorini, Ballito: Constructed in a Headland-Bound Bay

Santorini is a multi complex residential building constructed in a headland-bound bay (Thomson's Bay), northern Ballito during the late 1980s (Fig. 8.1). Coastal stabilization at this site involved coastal dune destruction, ground filling and leveling. During the March '07 large swell Santorini was partially undercut by megarip-current head-cutting (Fig. 8.3).

The March 2007 high swell erosion (Fig. 8.4) revealed evidence of repairs from a previous phase of erosion when Santorini had to be "underpinned" in a bid to defend the structure. Accurate details were not made available to the authors but this previous bout of erosion probably took place during the late 1980s or early 1990s during or just after construction.

While the removal of the buildings would be deemed to be the most suitable response to the latest storm event, this would clearly be a massive undertaking with significant costs to its owners. Instead it is expected that further defence measures will be taken. Geofabric bags have subsequently been deployed on site and have further reduced the beach area. This structure remains at risk.

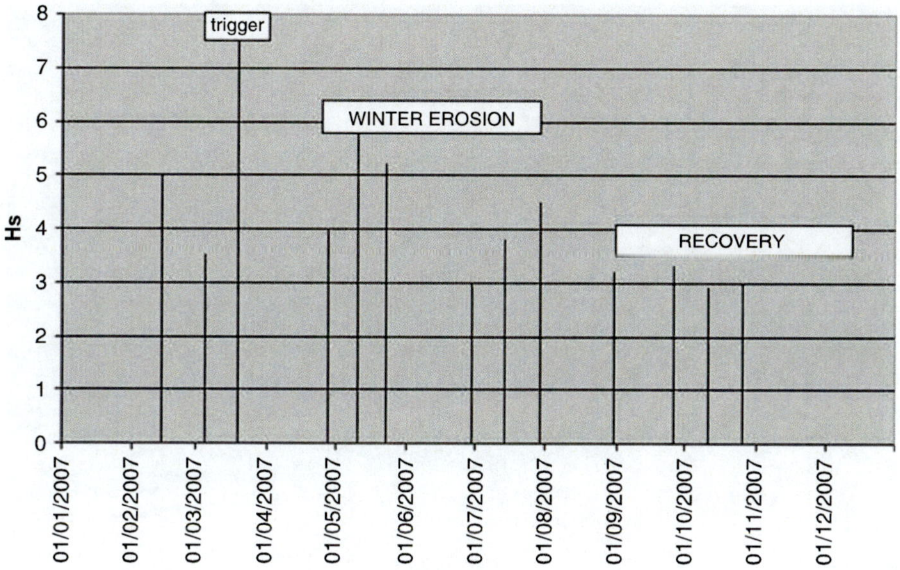

Fig 8.4 Peak swells >3 m Hs during 2007

8.3.3 Happy Wanderers: A Reversing Erosion Hotspot (EHS)

Happy Wanderers is a tourist resort located in the apex of an asymmetric headland-bound bay that tapers northward (Fig. 8.1). Here the coastal dune cordon has been leveled and a hotel complex and camping ground established. Happy Wanderers (Fig. 8.5) is constructed on former coastal dunes, at an elevation of about 3 m amsl, landward of a reversing EHS. Long-term coastal trends (1937–1983) indicate that the shoreline has fluctuated within a horizontal mobility envelope of 57 m (Cooper 1994). Significant erosion was experienced during September 2006 equinox and October 2006 LNC peak but no beachfront property was lost. Following this the coast recovered and no shoreline erosion was experienced during the March 2007 high swell event. However, significant and catastrophic erosion took place during the following austral winter when 80 m of retreat was recorded on the high tide shoreline (Fig. 8.5).

The March 2007 high swell struck during a time when the Happy Wanderer's beach was seasonally wide due to beach rotation. However, the March 2007 high swell-return flow removed the surfbars, which protect the beach, leaving the resort vulnerable to austral winter 2007 megarip activity (Smith et al. 2010). Similar scale erosion is known to have occurred in 1964 and 1991.

At Happy Wanderers erosion took place in two phases in 2007: (1) May-August and (2) September Equinox. The first phase culminated in a high swell at the end of August (Fig. 8.4). During this time large geofabric bags were laid down as a future marine hazard defence. Following the August storm the KZN coastline had

Fig. 8.5 Happy Wanderer's Resort which underwent catastrophic erosion during 2007. Photo taken during August event

seemingly recovered and surfbars had reformed. No more chronic erosion was experienced except at Happy Wanderers. On the September 2007 equinox, a 3 m swell (Fig. 8.4) struck the coast, causing catastrophic erosion and the resort infrastructure was almost lost.

8.4 What Was Learned

- Erosion was a significant infrastructural problem on stabilized coastal reaches. Rural beaches with intact dune cordons were able to self-repair following erosion. Natural beaches eroded during March 2007 and the 2007 austral winter storms had returned to near-normal beach widths by mid-late 2008 although dunes took longer to repair. At stabilized beaches, dune sand could not be naturally replaced due to dune cordon dysfunctionality and the dune buffer had to be replaced by geofabric solutions and in some cases hard seawalls. Structures were relocated when costs permitted.
- Developments that lay below the 5 m amsl contour where hardest hit by direct high swells and run-up as well as post-storm slumping of dunes.
- A number of coastal defences in KZN contributed to property destruction, rather than protection by reducing the coast's resilience (destroying dunes), focusing excess energy (e.g. via rip current formation) and interrupting longshore transport.

Coastal erosion is cyclic along the KZN coastline. In all instances winter erosion took place at known EHS localities and could therefore have been predicted and avoided. Erosion also appears to be linked to peaks in tidal cycles including the Lunar Nodal Cycle (Smith et al. 2010) and possibly the 4.4 year Perigean tidal cycle, indicating that future erosion vulnerability periods can be forecast.

Human nature is such that there appears to be a window of a few months during which authorities are prepared to listen and act on expert advice after an erosion event. It is the authors' experience that during this period authorities will listen to consulting scientists. After this period of grace, however, the traditional practice of coastal stabilization gradually resumes.

Although storm frequency does not appear to be increasing, storm intensity and wave heights are (van der Borch van Verwolde 2004; Elsner et al. 2008; Guastella and Rossouw 2009; Mavume et al. 2009). These phenomena, together with a declining coastal-sand budget and global warming-induced sea level rise, means that the probability of high water levels and large swells will increase and so, consequently will the frequency of coastal erosion events.

The 2006–2007 erosion event along the KZN coastline illustrates the outcomes of shoreline stabilization to facilitate urban coastal settlement and should act as a warning for future development.

References

Cooper JAG (1991a) Shoreline changes on the Natal coast: Tugela river mouth to Cape St Lucia. Natal town & regional planning commission reports, 76. The town & regional planning commission, P/B 9038, Pietermaritzburg, 3200, 57 pp

Cooper JAG (1991b) Shoreline changes on the Natal coast, Mkomazi river mouth to Tugela river mouth. Natal town & regional planning reports, 77. The town & regional planning commission, P/B 9038, Pietermaritzburg, 3200, 59 pp

Cooper JAG (1994) Shoreline changes on the Natal coast: Mtamvuna river mouth to the Mkomazi river mouth. Natal town & regional planning commission reports, 79. The town & regional planning commission, P/B 9038, Pietermaritzburg, 3200, 53 pp

Cooper JAG (2001) Gemorphological variability among microtidal estuaries from the wave-dominated South African coast. Geomorphology 40(1–2):99–122

Cooper JAG, Flores RM (1991) Shoreline deposits and diagenesis resulting from two Late Pleistocene stillstands near + 4 and + 5 m, Durban, South Africa. Mar Geol 97:325–343

Elsner JB, Kossin JP, Jagger TH (2008) The increasing intensity of the strongest tropical cyclones. Nature 455:92–95

Garland G, Moleko L (2000) Geomorphological impacts of Inanda on the Mgeni Estuary, north of Durban, South Africa. Bull Eng Geol Environ 59:119–126

Guastella LA, Rossouw J (2009) Coastal vulnerability: are coastal storms increasing in frequency and intensity along the South African coast? In: Abstracts, International Multipurpose Reefs (IMPR) conference, Jeffery's Bay, May 2009, p 10

Guastella LA, Smith AM, Mather AA, Bundy SC (2008) Post march 2007 marine storm erosion of KZN beaches. In: South African Marine Science Symposium: our changing seas, University of CapeTown, Cape Town 29th June–3rd July 2008, symposium guide & book of abstracts, p 44

Mather AA (2007) Linear and non-linear sea-level changes at Durban, South Africa. S Afr J Sci 103:509–512

Mather AA (2008) Sea level rise for the east coast of Southern Africa. COPDEC VII, 2008, Dubai, UAE

Mather AA, Stretch DD, Garland GG (2009) Southern African sea level: corrections influences and trends. Afr J Mar Sci 31:145–156

Mavume AF, Rydberg L, Rouault M, Lutjeharms JRE (2009) Climatology and landfall of tropical cyclones in the South-West Indian ocean. West Indian Ocean J Mar Sci 8(1):15–36

Moes H, Rossouw M (2008) Considerations for the utilization of wave power around South Africa. Workshop on ocean energy, Centre for Renewable and Sustainable Energy Studies, Stellenbosch, 21 Feb 2008, Abstracts

Schoonees JS (2000) Annual variation in the net longshore sediment transport rate. Coast Eng 40:141–160

Smith AM, Guastella LA, Bundy SC, Mather AA (2007) Combined marine storm and saros spring–high tide erosion event, march 19–20, 2007: a preliminary assessment. S Afr J Sci 103:274–276

Smith AM, Mather AA, Bundy SC, Cooper JAG, Guastella LA, Ramsay PJ, Theron A (2010) Contrasting styles of swell-driven coastal erosion: examples from KwaZulu-Natal, South Africa. Geol Mag 147:940–953

Thornton EB, MacMahan J, Sallenger AH Jr (2007) Rip currents, mega-cusps, and eroding dunes. Mar Geol 240:151–167

Tinley KL (1971) Determinants of coastal conservation: dynamics and diversity of the environment as exemplified by the Mocambique coast. In: Proceedings of symposium: nature conservation as a form of land use, South African Regional Committee for the Conservation and Utilisation of Soil, Gorongoza National Park, 13–17 Sept 1971, pp 125–53

Van der Borch van Verwolde E (2004) Characteristics of extreme wave events and the correlation between atmospheric conditions along the South African coast. Phd thesis, University of Cape Town, Cape Town, South Africa, pp 103 (unpubl)

Chapter 9
Presque Isle Breakwaters: Successful Failures?

Orrin H. Pilkey

Abstract Fifty-five breakwaters were constructed in 1992 along Presque Isle, a 7 mile-long spit extending into Lake Erie, Pennsylvania. A US Army Corps of Engineers Project, the breakwaters have halted the minor erosion along much of the peninsula and a good swimming beach has been retained. On the other hand, sand was predicted to continue to flow behind the breakwaters and it has not. This has required annual bulldozing of around 30,000 cubic yards per year to remove tombolos. Preservation of Gull Point, a critical natural area at the tip of the peninsula was also an important element of the original design and environmental impact statement but it has suffered severe erosion because the breakwaters blocked the point's sand supply. Engineering groups have declared the project to be a great success in spite of breakwater impacts on sand flow and the preservation of Gull Point. The Presque Isle project emphasizes the need for monitoring of project success or failure by independent groups.

9.1 Introduction

Presque Isle (Fig. 9.1) is a compound, re-curved sand spit of 3,200 acres that protrudes into Lake Erie (one of the US Great Lakes) from the shoreline of Pennsylvania. The spit has about seven miles of open-lake shoreline with a width that ranges from a few tens of yards at the narrow neck that connects it to the mainland to as much as a mile in its mid-section (Fig. 9.2). Presque Isle is a state park founded in 1921 (Delano 1991) and is also a National Natural Landmark established in 1967, a designation that is one step below a National Park.

O.H. Pilkey (✉)
Division of Earth and Ocean Sciences, Nicholas School of the Environment, Duke University, Durham, NC 27708, USA
e-mail: opilkey@duke.edu

J.A.G. Cooper and O.H. Pilkey (eds.), *Pitfalls of Shoreline Stabilization: Selected Case Studies*, Coastal Research Library 3, DOI 10.1007/978-94-007-4123-2_9,

Fig. 9.1 Oblique photo showing the entire Presque Isle spit taken shortly after the breakwater project was finished. A number of "near-tombolos" are visible behind the structures. (Public Domain Photo, USACE)

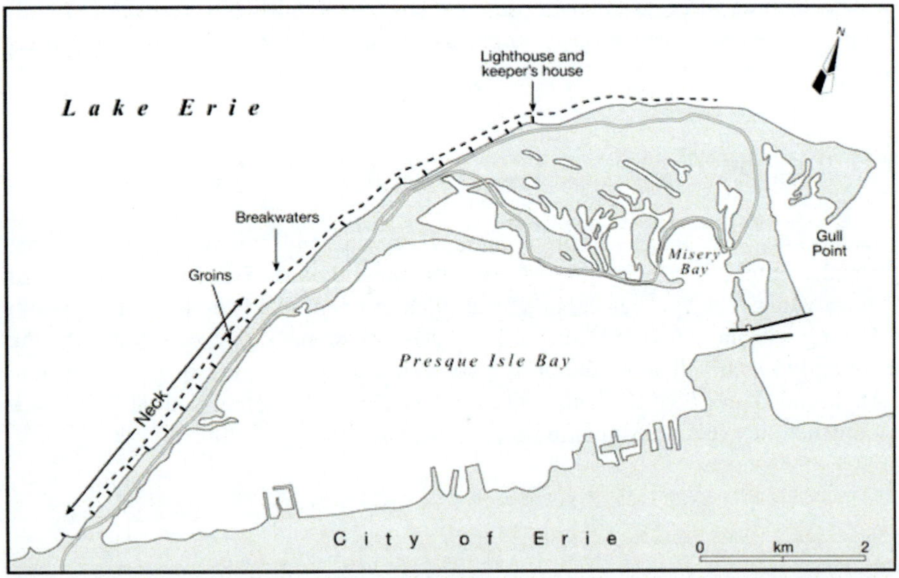

Fig. 9.2 Presque Isle is a 7-mile long "flying" spit extending into Lake Erie, Pennsylvania. The 55 offshore rock breakwaters shown here, each 150 ft long and spaced 350 ft apart, extend almost to Gull Point at the tip of the spit

Fig. 9.3 Gull Point is shown in this photo taken before breakwaters were emplaced in 1992. Preservation of the wetland environment of the point was one of the major design goals for the project. (Public Domain Photo, USACE)

At the tip of Presque Isle is Gull Point (Fig. 9.3), a broad 250 + acre forested area with large wetlands. According to National Park Service, Gull Point is one of the few areas in the world where one can study the succession of plant life from a sand and fresh water environment to a climax forest within a few hundred yards. Gull Point may also be one of the few significant remaining fueling and resting spots between the Great Lakes and the Atlantic Ocean for a number of species of migrating shore birds. This includes plovers, sand pipers, whimbrels, and sanderlings. Presque Isle also supports many of Pennsylvania's rarest plant species (Bissell 1993). For these and other subjective reasons, the preservation of Gull Point was considered to be important and thus was a central issue of the societal debate over shoreline engineering on Presque Isle.

The engineering/political/scientific controversy that preceded the construction of 55 offshore breakwaters along the spit is documented by Pilkey and Dixon 1996. The goal of this paper is to summarize the various events outlined by Pilkey and Dixon and to bring the story up to date. The Presque Isle story is one that illustrates how obviously unreasonable claims of environmental impacts can be used to clear the way for hard shoreline stabilization.

Presque Isle's sand supply through the nineteenth century came from eroding beaches and lakeshore bluffs to the west (Morang et al. 2011). This source has been much compromised by shorefront development and innumerable seawalls, groins and breakwaters (Carter et al. 1987). There are also a few groins at the base of the spit where it joins the mainland. Starting in 1975 the beach was nourished with a typical annual volume of 160,000 cubic yards and the sand moved down the peninsula basically reducing the erosion rate to a negligible level. Much of the sand used for pre-breakwater nourishment came from the west of the spit and contained a significant amount of gravel, which was not ideal for a swimming beach.

Ironically, the erosion problem had been "solved" by beach nourishment by the time the breakwater project was close to fruition. Actual shoreline retreat was quite minor and was not of primary concern to park managers. At that time, it seemed that the main recognized problem that faced Presque Isle was occasional winter storms that overwashed the spit, causing minor damage in parking lots. Like a snowball rolling down the hill, there was no way to stop the project in the late 1980s and early 1990s. No one in authority asked the question why.

9.2 The Solutions

The Buffalo District Corps of Engineers were the coastal experts that drove the political system to eventual hard stabilization of the shoreline of the spit. In 1968, the Corps (US Army Corps of Engineers 1973) came up with a number of possible solutions to the erosion problem on Presque Isle, a problem that really didn't exist.

The possible solutions were:

- Do nothing
- Construct additional groins and bulkheads
- Construct a continuous offshore breakwater
- Construct segmented offshore breakwaters
- Beach replenishment

The final solution was construction of 55 breakwaters (in addition to 3 breakwaters already in place) extending along most of the length of the spit. The coastal engineers of the Buffalo District of the Corps were very much impressed with the use of breakwaters in Japan where, at that time, more than 2,000 existed. But the Japanese shorelines with breakwaters usually were not sandy shorelines like Presque Isle.

Segmented breakwaters somewhat similar to the proposed Presque Isle breakwaters had been constructed at Winthrop Beach, Massachusetts (5 segments); Lorraine, Ohio (3 segments); Colonial Beach, Virginia (2 sites with 3 and 4 segments, respectively); and an open-ocean example on Holly Beach, Louisiana. None of these come close to the magnitude of the 55-segment breakwater chain at Presque Isle. The Corps of Engineers sold the segmented breakwater idea as a passive, permeable, non-intrusive and permanent solution to the problem.

9.3 The Breakwaters

The Corps design, finished in 1980 (U.S. Army Corps of Engineers 1986a, b, 1995) aligned the segmented breakwaters about 250–350 ft offshore (Fig. 9.4). Each breakwater was 150 ft long, 10 ft wide and 8 ft high above low lake water level and placed 350 ft apart. All told, more than 600,000 t of rocks were needed. Total construction cost was below $30 m and the fully inflated 50 year cost including beach nourishment was estimated at $150 m. Post-breakwater beach nourishment volumes typically are 32,000 cu yards per year.

Allowing natural sand transport to continue behind the breakwaters was an important part of the breakwater design. If sand flow did not continue, massive beach nourishment would be required and the all-important Gull Point would suffer severe erosion. The downdrift erosion problem related to offshore breakwaters is globally recognized. They are illegal in four ocean states and on all National Park Service shorelines.

The 55 rock breakwaters were completed in 1992. Erosion of Gull Point began immediately. Sand never did flow behind the breakwaters as predicted and it won't be long before Gull Point disappears.

The succession of errors and misstatements went something like this:

- In 1978, the Corps built three prototype breakwaters as an experiment. The breakwaters caused immediate severe downdrift erosion. The erosion was inexplicably attributed to a storm rather than to the breakwaters.
- The Corps stated that the breakwaters would slow but not stop the flow of sand down the length of the spit. This was not likely by any measure of the science of beaches. The flow of sand behind the breakwaters was unachievable, clearly and utterly impossible, and on average more than 30,000 cubic yards of sand is moved about annually to remove tombolos. This volume is about the same as the annual nourishment volume.
- The Corps claimed that a soft solution (beach nourishment) was not feasible, but after construction of the breakwaters as per the design, an annual nourishment program began using high quality offshore sand. The post-breakwaters annual nourishment volumes typically are 32,000 cu yards, 1/5 of the annual volume of pre-breakwater nourishments. The Corps claimed this as a benefit of the breakwaters but if sufficient sand had been pumped in to prevent the degradation of Gull Point, as per the project design, the required volume of sand would have been much larger.
- The Corps claimed that nourishment from an offshore source of high quality sand was too costly to use in place of the poor quality sand from an upland source that was being used pre-breakwater. Post-breakwaters however the offshore source of sand was used and the breakwaters were given credit for the improvement in sand quality.
- Scientists and engineers at the U.S. Army Corps of Engineers Coastal Engineering Research Center (CERC) in Biloxi, Mississippi, the Corps' coastal research arm (the ones who should know better), agreed that sand would continue to flow behind the breakwaters, an impossible conclusion (Daly and Pope 1986).

Fig. 9.4 Oblique aerial view of the eastern half of the spit showing the breakwaters and the incipient tombolos extending from the shore. Tombolo formation is quite extensive resulting in an ongoing bulldozing requirement moving around 30,000 cubic yards of sand annually. Photo Courtesy of the US Army corps of Engineers

- An outside engineering consulting firm (Moffat and Nichols 1988) hired by the state of Pennsylvania in 1988 said the project would not work as intended. The report was shelved.

A testimony to the often radically different philosophies held by the scientific and engineering communities (Mohr et al. 1999), the Presque Isle breakwaters have received resounding praise from professional engineers. The Michigan Society of Professional Engineers awarded the Corps its 1993 Outstanding Engineering Achievement Award for the breakwater project. In 1993, the National Society of Professional Engineers (Johnson 1993) recognized the breakwaters as one of the ten top U.S. engineering achievements of the year. In 2011, the American Shore and Beach Preservation Association (ASBPA) declared the beach to be a huge success, an example of successful beach engineering (Mohr 2011).

The National Society of Professional Engineers news release gives insight to the engineering view:

> One of the largest projects of its kind, … allows for the natural movement of littoral sediments to continue behind the breakwaters and for continued growth of the environmentally sensitive Gull Point area of the peninsula. The resulting shoreline is appealing and blends with the natural environment.

The engineering society simply accepted the Corps' assertion that sand transport would continue and seemed to have no appreciation of the rare beauty and incalculable value of a natural beach with an uninterrupted view of lake and horizon. The society also had little appreciation of coastal processes.

9.4 The End Result

In actuality, today the beach is a fine swimming beach, providing one doesn't stray too close to the boulder breakwaters. Continued moderate beach nourishment has been required to maintain the beach. Sand never did flow behind the breakwaters and frequent beach bulldozing, 30,000 cubic yards per year, is required to remove the tombolos connecting the breakwaters to the beach. Most importantly, Gull Point has been severely eroded. The treasured forest-wetland environment with its unique flora, preservation of which was one of the assumed main benefits of the breakwater project, will likely disappear altogether in the future. The shoreline at Gull Point, beyond the last breakwater had retreated by as much as 500 ft by 2009 (Fig. 9.5), and tangled trees and shrubs covered the beach and nearshore zone. The eroded sand has produced 43 acres of sand spit at the very tip of Gull Point but this is not the original wetland–forest environment.

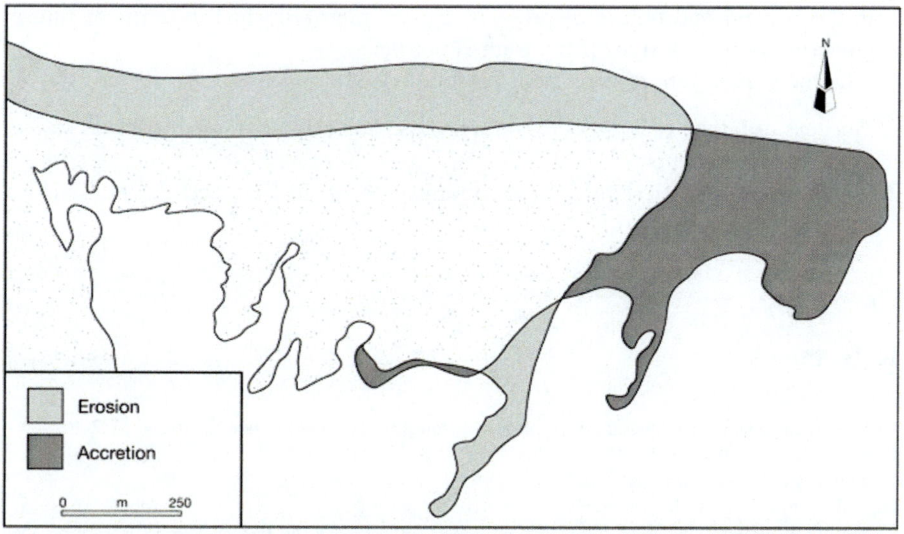

Fig. 9.5 Shown here are areas of post-breakwater erosion and deposition on Gull Point at the tip of Presque Isle. Lighter colored areas indicate erosion; darker areas are depositional. One of the primary concerns in the societal debate that preceded breakwater construction was preservation of the ecologically important Gull Point. The areas being lost are wetland/forest/beach environments critical to the fauna and flora of the wildlife refuge of the point

As with most societal debates about shoreline issues, our society has a short memory. The Corps stated after tombolos began to form that this was an indication that the breakwaters were 'too successful.' The superintendent of Presque Isle State Park was quoted as saying that the breakwaters were a positive addition to the park.

The Coastal Engineering Research Center (CERC) and the Buffalo District of the U.S. Army Corps of Engineers consistently expressed certainty and optimism about the breakwater design when such was not justified and made dismissive statements about alternative approaches. Also, the Corps of Engineers' publicly funded research, both on the CERC Research Center level and Buffalo District level, was all directed toward proving the viability of the breakwater alternative and not toward an objective comparison of various alternatives.

The bottom line is that 20 years after the completion of the project, success is measured by parameters other than the design parameters once claimed. The Corps of Engineers clearly stated in their project documents that the proposed breakwaters would not have disastrous impacts on the rare ecosystems at the Point.

9.5 Conclusions

There is much profit in engineering (and other aspects of life) to be gained from both the success and failure of projects. But nothing is gained in terms of future engineering project designs if the truth is not faced.

The most important lessons from Presque Isle are:

1. Success or failure of a project should be based on original project goals as stated in design documents.
2. Independent parties should do the monitoring of projects and declarations of success or failure.

References

Bissell JK (1993) Rare plants and rare plant communities of Presque Isle. Bartonia 57:2–8

Carter CH, Neal WJ, Haras WS, Pilkey OH (1987) Living with the Lake Erie Shore. Duke University Press, Durham, p 263

Daly WR, Pope J (1986) Detached breakwaters for shore protection: final report. Coastal Engineering Research Center, U.S. Army Corps of Engineers Waterways Experiment Station, Vicksburg, MS

Delano HL (1991) Presque Isle State Park, Erie County: a dynamic interface of water and land, Park Guide 21. Pennsylvania Geological Survey/Department of Environmental Resources/ Commonwealth of Pennsylvania, Harrisburg, 13p

Johnson D (1993) Corps protects eroding Lake Erie Shore. *Engineer Update*, vol 17 (Jan)

Moffat & Nichol Engineers (1988) Evaluation of segmented offshore breakwaters for beach stabilization at Presque Isle State Park, Erie, Pennsylvania. Report presented to Commonwealth of Pennsylvania Department of Environmental Resources. Moffat & Nichol Engineers, Raleigh, NC

Mohr M (2011) Preserving a historic resource. ASBPA best restored beaches 2011. Coastal Voice, American Shore and Beach Preservation Association

Mohr MC, Pope J, McClung JK (1999) Coastal response to a detached breakwater system; Presque Isle, Pennsylvania, USA. In: Proceedings, Coastal Sediments '99, American Society of Civil Engineers, 3, pp 2010–2025

Morang A, Mohr MC, Forgette CM (2011) Longshore sediment movement and supply along the U.S. shoreline of Lake Erie. J Coast Res 27:619–635

Pilkey OH, Dixon KL (1996) The Corps and the Shore. Island Press, Washington, DC, 273p

U.S. Army Corps of Engineers (1986a) Shoreline erosion control project, phase II, general design memorandum, detailed project design, Presque Isle Peninsula, Erie, Pennsylvania. Buffalo District/U.S. Army Corps of Engineers, Buffalo

U.S. Army Corps of Engineers (1986b) Shoreline erosion control project, Presque Isle Peninsula, Erie, Pennsylvania. Buffalo District/U.S. Army Corps of Engineers, Buffalo

U.S. Army Corps of Engineers, Buffalo District (1995) Cooperative beach erosion control project at Presque Isle Peninsula, Erie, PA, The Corps, http://books.google.com/books/about/Cooperative_beach_erosion_control_projec.html?id=Jb3qtgAACAAJ

U.S. Army Corps of Engineers, Buffalo District (1973) Review report on cooperative beach erosion control project at Presque Isle Peninsula, Erie, PA. Buffalo District/U.S. Army Corps of Engineers, Buffalo, 200p

Chapter 10
Armoring on Eroding Coasts Leads to Beach Narrowing and Loss on Oahu, Hawaii

Bradley M. Romine and Charles H. Fletcher

Abstract Coastal armoring (defined as any structure designed to prevent shoreline retreat that interacts with wave run-up at some point of the year) has, historically, been a typical response to managing the problem of beach erosion on the island of Oahu, Hawaii. By limiting the ability of an eroding shoreline to migrate landward, coastal armoring on Oahu has contributed to narrowing and complete loss of many kilometers of beach. In this paper, changes in beach width are analyzed along all armored and unarmored beaches on the island using historical shoreline positions mapped from orthorectified aerial photographs from as early as the late 1920s. Over the period of study, average beach width decreased by 11% ± 4% and nearly all (95%) documented beach loss was fronting armored coasts. Among armored beach sections, 72% of beaches are degraded, which includes 43% narrowed (28% significantly) and 29% (8.6 km) completely lost to erosion. Beaches fronting coastal armoring narrowed by −36% ± 5% or −0.10 ± 0.03 m/year, on average. In comparison, beach widths along unarmored coasts were relatively stable with slightly more than half (53%) of beaches experiencing any form of degradation. East and south Oahu have the highest proportion of armored coast (35% and 39%, respectively) and experienced the greatest percent of complete beach loss (14% and 12%, respectively). West and north coasts, with relatively little armoring (10% and 12% armored, respectively), experienced little complete beach loss (2% and 6%, respectively). However, beaches are still significantly narrowed compared to historical patterns on west and north coasts (61% and 70%, respectively). We find at these sites that cultivation of coastal vegetation may be a factor in beach narrowing on Oahu, along with beach erosion. Increased 'flanking' erosion (accelerated

B.M. Romine (✉) • C.H. Fletcher
Department of Geology and Geophysics, University of Hawaii at Manoa, 1680 East-west Road, Honolulu, HI 96822, USA
e-mail: romine@hawaii.edu; fletcher@soest.hawaii.edu

J.A.G. Cooper and O.H. Pilkey (eds.), *Pitfalls of Shoreline Stabilization: Selected Case Studies*, Coastal Research Library 3, DOI 10.1007/978-94-007-4123-2_10,

shoreline retreat adjacent to armored sections) is documented at several beaches, often requiring extension of armoring structures to protect abutting coastal properties, a process that leads to alongshore seawall proliferation.

10.1 Introduction

A recent study found that erosion dominates shoreline change on the beaches of Kauai, Oahu, and Maui. Since a strand plain of unconsolidated carbonate sand backs large segments of the Hawaii shoreline (Sherrod et al. 2007; Fletcher et al. 2012), one may assume there is adequate sediment on the backshore for an eroding beach migrating landward to develop a profile in equilibrium with nearshore conditions and underlying geology. However, on many Hawaii beaches, the response to beach erosion has been to armor the backshore to protect coastal properties, and thus impound this sand resource (Hwang 1981; Sea Engineering Inc 1988; Fletcher et al. 1997; Fletcher 1992; Makai Ocean Engineering and Sea Engineering 1991). In such cases, the water line continues to migrate landward while the backshore remains fixed – resulting in narrowing and eventually complete loss of the beach. Sediment that would otherwise be available to the littoral system is impounded behind seawalls, revetments, sand bags, and other designs; thereby depriving adjacent beaches and leading to a trend of increased erosion within the littoral cell. The narrowing effects of armoring on beach width are also documented in studies from other regions (e.g., Carter et al. 1986; Hall and Pilkey 1991; Komar and McDougal 1988; Kraus and McDougal 1996; McDonald and Patterson 1984; Tait and Griggs 1990).

'Healthy' Hawaii beaches are important to the local lifestyle and a vital attraction for the tourism-based economy. Fletcher et al. (1997) found that coastal armoring led to narrowing or complete loss along ~24% of beaches on the island of Oahu, Hawaii.

Seawalls and other armoring styles are often attributed with causing coastal erosion, yet in Hawaii we find that shoreline armoring is typically a response to pre-existing coastal erosion. Because of this, it is appropriate to ask two sets of questions. One, does armoring accelerate pre-existing erosion and does it initiate and or accelerate erosion on adjacent properties? Two, does armoring lead to other negative impacts such as beach loss or beach narrowing, which, although caused by erosion, we define as separate from erosion? Here, we primarily explore the latter through analysis of beach narrowing fronting coastal armoring. Evidence is also provided for 'flanking' erosion on beaches adjacent to coastal armoring.

10.2 Physical Setting

The Hawaiian Islands are comprised of eight high volcanic islands in the upper tropics of the north Pacific. Oahu, located between 21 and 22° north latitude, is the most populated of the main islands. The island is fringed by a Pleistocene

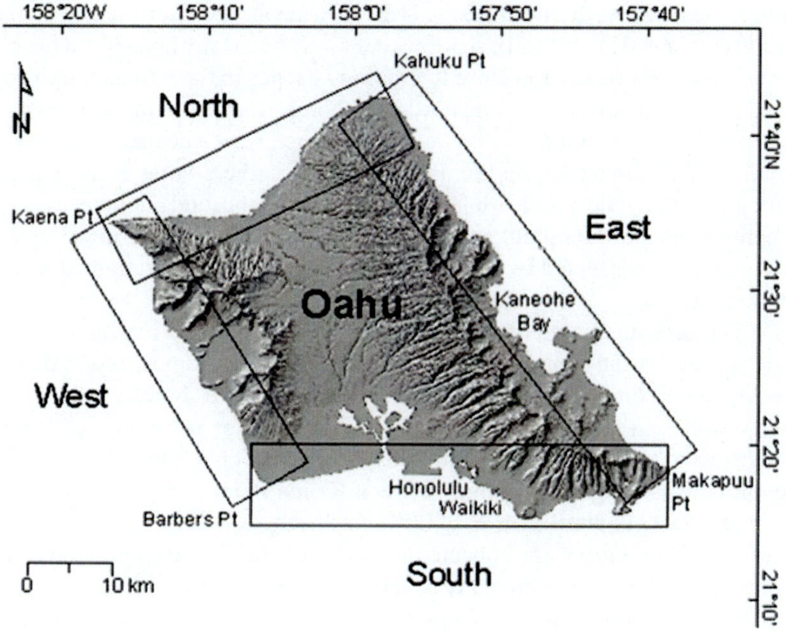

Fig. 10.1 Four regions of Oahu

reef platform cut by relict erosional features (e.g., channels, karst depressions) formed during periods of lower sea level (Fletcher et al. 2008). Hawaiian beaches are comprised primarily of calcareous sands. This sediment originated on the fringing reef platform through either direct organic precipitation in the reef ecosystem or through bioerosion of skeletal limestone. Sands may be stored in offshore channels and depressions, on low-lying coastal plains stranded by late-Holocene sea level fall (Fletcher and Jones 1996), or in the modern beach and dune system (Harney et al. 2000; Harney and Fletcher 2003). Hawaii beaches, like most carbonate beaches, are typically narrower than continental beaches due to limited sediment supply.

Located in the middle of the Pacific in a microtidal zone, wave energy is the predominant driver of shoreline processes in Hawaii. Large waves from North Pacific storms are common in winter months, typically affecting north and west - exposed shores. South-exposed shorelines are affected by smaller long-period swell from southern oceans in summer. Easterly trade winds and the waves they produce are common on leeward shores year-round but most frequent in summer months (Vitousek and Fletcher 2008).

The island is divided into four regions for analysis: east, south, west, and north (Fig. 10.1). East Oahu, from Kahuku Point in the north to Makapuu Point in the south, is moderately developed with single-family homes and a coastal highway lining most beaches. The east Oahu shoreline faces directly into the predominant easterly trade winds and is occasionally affected by large refracted northerly swells in winter. Beaches in the northeast (north of Kaneohe Bay) are typically narrow and fringed by a wide (~0.5 km), shallow reef platform. Many homes and the coastal highway were constructed too close to eroding beaches in the past century resulting in extensive coastal armoring along northeast shores. Beaches in the southeast (south of Kaneohe Bay) are wider, relative to the northeast, with a deeper fringing reef.

South Oahu, from Makapuu Point in the east to Barbers Point in the west, is the most densely populated and urbanized region of Oahu and includes the highly engineered shores of Honolulu and Waikiki. The south shore is fringed by a wide shallow reef and is affected by southerly swells in summer and refracted tradewind waves year-round.

West Oahu, from Barbers Point in the south to Kaena Point in the north, is the least developed of the four island regions. Single-family homes, beach parks, and undeveloped property line most beaches. Western, leeward shores receive refracted northerly waves in winter and refracted southerly waves in summer – leading to large seasonal changes in alongshore transport and beach width.

Development along north Oahu, between Kaena Point in the west and Kahuku Point in the east, is similar to east Oahu with single-family homes lining most beaches. Northern shores are impacted by large northerly waves in winter causing temporary seasonal erosion on many beaches. Relatively small, refracted tradewind waves are typical in summer.

10.3 Data and Methods

For our analysis, we use historical shoreline positions mapped from high-resolution (0.5 m pixel) orthorectified aerial photo mosaics following Fletcher et al. (2003, 2012), and Romine et al. (2009). Two shoreline proxies are utilized for beach width analysis: the Low Water Mark (LWM) and the vegetation line. The LWM or beach toe is the base of the foreshore and marks the seaward edge of the subaerial beach. The vegetation line marks the landward edge of the beach and is located at the seaward extent of interannual vegetation growth (vegetation that survives annual high run-up of waves) or at the base of armoring structures (e.g., sea wall). Beach width is defined as the distance between the LWM and vegetation line (or armoring) (Fig. 10.2).

We use survey-quality vertical aerial photographs with sufficient spatial resolution (<0.5 m) and tonal contrast to identify shoreline features. New imagery was acquired for the Oahu shoreline in 2005–2008 including synchronous position and orientation (POS) navigation data from an on-board aircraft global positioning system and inertial and mobilization unit (IMU). The POS data is used with a high-resolution digital elevation model (DEM; 5 m horizontal, sub-meter vertical) to rectify and mosaic the imagery. Typically, one historical air photo set meeting minimum quality standards is available for each decade going back to the late 1920s or late 1940s. Historical air photos are orthorectified and mosaicked using ground control points collected from more recent ortho imagery. The orthorectification process typically produces mosaics with root mean square (RMS) positional errors <2 m.

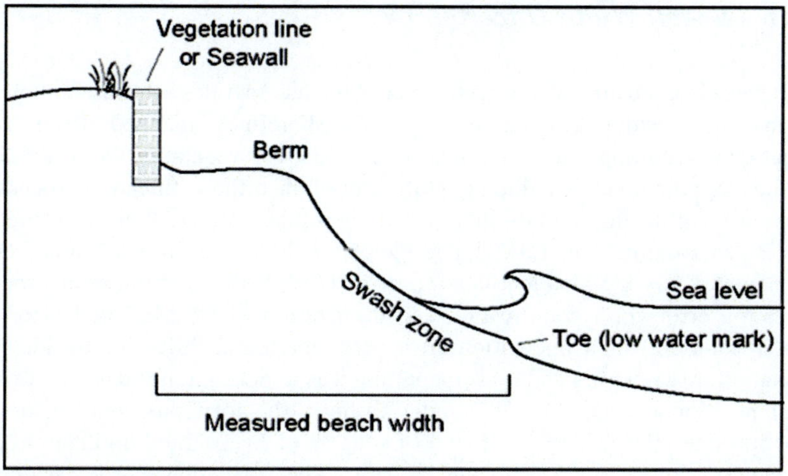

Fig. 10.2 Beach width is the distance between the beach toe (*low water mark*) and vegetation line (or armoring) (Modified from Fletcher et al. (1997))

Due to limited availability of historical air photos, we attempt to locate and utilize all available imagery. We do not remove historical shorelines from a time series based on records of large storms or waves. Rather, we account for fluctuations in shoreline position due to waves and storms in our uncertainty analysis (see: Uncertainties). However, historical shorelines may be removed from the time series in special cases. Some Oahu beaches have been artificially altered to the extent that the physics of the beach system have been permanently changed. Examples include removal of beach sand by mining operations, artificial beach fills, and construction of coastal engineering structures such as groins or sea walls. In these cases, shorelines prior to such alterations are removed from the time series and beach changes are analyzed only for the recent configuration of the beach. LWM and vegetation line positions are measured at regularly-spaced (roughly 20 m) shore-normal transects cast from an arbitrary offshore baseline.

For this study we define coastal armoring as any structure coming in contact with wave run-up and thereby interfering with natural coastal processes at any point of the year. Typically, these are designed to prevent coastal recession and retain sand. This includes rubble or stone revetments (with or without mortar); cement, brick, or stone walls; and wood or metal bulkheads. We also include landscaping or retaining walls that have transitioned into shoreline armoring on receding coasts. Armoring structures typically have little or no intra-annual vegetation growth (e.g., tall shrubs or trees) on the seaward side indicating the wall is impacted by wave run-up.

Coastal armoring is mapped using the most-recent (2005–2008) orthophoto mosaics. Locations are verified with high-resolution (~10 cm resolution) original air photo images and site visits. For this study we map only shore-parallel armoring structures on beaches or former locations of beach (i.e., where the beach was lost in the time span of analysis). Armoring on rocky shoreline or along engineered shorelines that never had beach in the time span of this study are not included in this study.

10.3.1 Beach Width Uncertainties

LWM shoreline positions are highly variable due to tides, storms, and waves resulting in positional uncertainties with shorelines mapped from aerial photographs. Additional uncertainties for LWMs and vegetation lines also arise from the mapping process including RMS error of the orthorectification process and on-screen identification and digitization of shoreline features. Following Fletcher et al. (2003), Romine et al. (2009), and Fletcher et al. (2012), five positional errors are calculated for LWMs: rectification error (*Er,* RMS error of ortho process), digitization error (*Ed,* identification and digitization of LWM), pixel error (*Ep,* spatial resolution,) tidal fluctuation error (*Etd,* horizontal shifts due to tides) and seasonal error (*Es,* waves and tides,); combined as a root sum of squares to arrive at a total positional error, *Etp*. In similar fashion, the total positional error of a vegetation line (*Eveg*) is the root sum of squares of *Er*, *Ep*, and digitization error for vegetation lines (*Evid*, estimated at 2 m). The vegetation line is assumed to mark the annual high wash of waves and is, therefore, not prone to shorter-term (intra-annual) fluctuations. Thus, *Es* and *Etd* are not included when calculating positional uncertainties for vegetation lines.

Beach width is the difference between vegetation line distance and LWM distance along a transect. However, calculating the uncertainty of the beach width as the root sum of squares of *Etp* and *Eveg* overestimates the error. We may omit the rectification errors (*Er*) for both the LWM and vegetation line because we are no longer concerned with geographic position; only the net distance between the vegetation line and LWM. Any errors due to rectification between the shoreline features are assumed to be negligible at those distances (<100 m). Therefore, a more accurate estimate of the beach width error, $Ev - t$, is:

$$(Ed^2 + 2^* Ep^2 + Etd^2 + Es^2 + Evid^2)^{0.5}$$

10.3.2 Calculating Beach Width Changes

Beach width change rates and net beach width change are calculated at each transect using weighted least squares (WLS) linear regression to fit a trend line to the time series of measured beach widths. Beach width uncertainties are applied as weights ($1/Ev - t^2$). Thus, beach widths with higher uncertainty values have less influence on the trend line. This method is similar to recent studies (Romine et al. 2009; Hapke et al. 2010; Fletcher et al. 2012) – only, beach width data is used instead of shoreline positions. The annual rate of beach width change (m/year) is the slope of the trend line (Fig. 10.3). The net change in beach width is the difference between the estimated beach width values at the end points of the WLS trend line (at the earliest and most recent shoreline times). Uncertainties of estimated

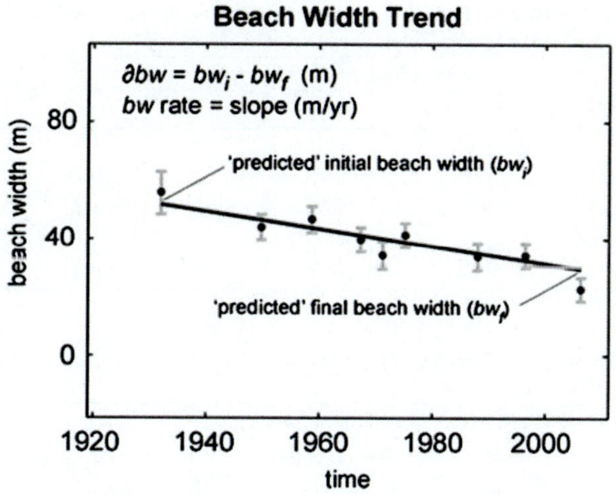

Fig. 10.3 Calculating beach width change with weighted least squares (WLS)

beach widths from the regression line are calculated at 1-sigma (standard deviation) to be consistent with 1-sigma positional uncertainties calculated for measured beach widths. Uncertainty of the net change in beach width is the root sum of squares of the uncertainties of the initial and final beach widths.

Regional average beach widths, average beach width changes, and average beach width rates are the average of values from all transects in a beach section. Following equation 9 of Hapke et al. (2010), the uncertainty of regional averages is estimated using an effective number of independent uncertainty observations (n*), calculated using a spatially-lagged (along-shore) autocorrelation of the uncertainty values.

A beach width trend (narrowing or widening) is considered significant if the net change is greater than the uncertainty (@ 1-sigma). A section of beach is considered completely lost to erosion if no beach remains (beach width = 0 m) at the most recent shoreline time(s) and beach was present at the earliest shoreline time(s). The total percent of 'degraded' beach is the sum of percents of beach lost and beach narrowed. To avoid reporting some beach width rate uncertainties as ±0.0 m/year, we report rates and uncertainties to the nearest cm/year (±0.00 m/year) even though our measurements at individual transects may not provide this high level of precision.

Shoreline change rates are calculated at select locations to compare rates before and after installation of coastal armoring. For this, we use historical shoreline positions (LWMs) and the method of single-transect WLS rate calculation from Fletcher et al. (2012).

10.4 Results

Beach width changes are measured at 5,332 shore-normal transects spaced roughly 20 m along 107 km of Oahu beaches from 1928 or 1949 to near-present (2005–2008) (Table 10.1). Approximately 29 km or 27% of the total extent of

Table 10.1 Length and percent of armored and unarmored beach on Oahu (measured from recent air photos and ground surveys)

	Beach studied, total		Armored beach			Unarmored beach		
Region	Transects	(km)	Transects	(km)	(%)	Transects	(km)	(%)
East	2,101	42.0	734	14.7	35	1,367	27.3	65
South	1,316	26.3	512	10.2	39	804	16.1	61
West	628	12.6	61	1.2	10	567	11.3	90
North	1,287	25.7	157	3.1	12	1,130	22.6	88
Total	5,332	106.6	1,464	29.3	27	3,868	77.4	73

Table 10.2 Beach width trends for Oahu (all beaches, armored beaches, and unarmored beaches; 1928 or 1949 to near present)

All beaches (armored and unarmored)							
	Lost		Narrowed (%)		Degraded (%)[b]	Widened (%)	
Region	(km)	(%)	Total (%)	Significant (%)[a]	Total (%)	Total (%)	Significant (%)[a]
East	5.7	14	42	17	55	45	18
South	3.1	12	38	22	49	50	25
West	0	0	60	41	61	39	23
North	0.2	1	69	46	70	30	12
Total	9.1	8	49	28	58	42	19
Armored beaches							
	Lost		Narrowed (%)		Degraded (%)[b]	Widened(%)	
Region	(km)	(%)	Total (%)	Significant (%)[a]	Total (%)	Total (%)	Significant (%)[a]
East	5.6	38	36	20	74	26	10
South	2.8	27	40	27	67	33	17
West	0	2	80	59	82	18	2
North	0.2	6	70	54	76	24	11
Total	8.6	29	43	28	72	28	12
Unarmored beaches							
	Lost		Narrowed (%)		Degraded (%)[b]	Widened (%)	
Region	(km)	(%)	Total (%)	Significant (%)[a]	Total (%)	Total (%)	Significant (%)[a]
East	0.1	0	45	15	45	55	23
South	0.4	2	36	18	38	61	29
West	0	0	58	40	58	42	25
North	0	0	69	45	69	31	12
Total	0.5	1	52	29	53	47	21

[a]Percent of transects where narrowing or widening is greater than 1-sigma uncertainty
[b]Degraded total equals percent lost plus total percent narrowed

Oahu beaches (or locations of former beaches) are armored. Over 9 km or 8% of Oahu beach was completely lost to erosion in the time span of analysis – nearly all of it (95%) fronting artificial coastal armoring (Table 10.2). A majority or 58% of Oahu beaches are degraded (narrowed or lost) including 49% narrowed (28% significantly) and 8% completely lost. Of the 49% of narrowed beaches, roughly

Table 10.3 Average beach width changes for Oahu (all beaches, armored beaches, and unarmored beaches)

All beaches (armored and unarmored)

Region	Initial average beach width (m)[a]	Final average beach width (m)[a]	Average beach width change (m)[a]	(%)[a]	Average beach width change rate (m/year)[b]
East	19.4 ± 1.0	18.4 ± 0.8	−1.0 ± 1.4	−5% ± 7	−0.02 ± 0.05
South	18.2 ± 0.7	16.4 ± 0.4	−1.8 ± 0.7	−10% ± 4	−0.02 ± 0.02
West	35.5 ± 2.3	32.3 ± 1.6	−3.1 ± 2.8	−9% ± 8	−0.03 ± 0.12
North	33.2 ± 1.4	27.5 ± 1.2	−5.7 ± 1.9	−17% ± 6	−0.07 ± 0.07
Total	24.3 ± 0.7	21.8 ± 0.6	−2.6 ± 0.9	−11% ± 4	−0.03 ± 0.03

Armored Beaches

Region	Initial average beach width (m)[a]	Final average beach width (m)[a]	Average beach width change (m)[a]	(%)[a]	Average beach width change rate (m/year)[b]
East	15.3 ± 1.1	8.7 ± 1.0	−6.6 ± 1.5	−43% ± 10	−0.09 ± 0.07
South	21.3 ± 1.0	14.5 ± 0.3	−6.9 ± 1.1	−32% ± 5	−0.09 ± 0.03
West	39.3 ± 1.8	24.9 ± 1.2	−14.4 ± 2.1	−37% ± 5	−0.18 ± 0.08
North	29.3 ± 1.2	20.6 ± 1.0	−8.7 ± 1.5	−30% ± 5	−0.11 ± 0.05
Total	19.9 ± 0.7	12.7 ± 0.6	−7.2 ± 0.9	−36% ± 5	−0.10 ± 0.03

Unarmored Beaches

Region	Initial average beach width (m)[a]	Final average beach width (m)[a]	Average beach width change (m)[a]	(%)[a]	Average beach width change rate (m/year)[b]
East	21.7 ± 1.0	23.6 ± 0.6	1.9 ± 1.2	9% ± 6	0.02 ± 0.03
South	16.2 ± 0.6	17.7 ± 0.7	1.5 ± 0.9	9% ± 5	0.02 ± 0.03
West	35.1 ± 2.4	33.1 ± 1.7	−1.9 ± 2.8	−6% ± 8	−0.01 ± 0.12
North	33.7 ± 1.5	28.5 ± 1.2	−5.3 ± 1.9	−16% ± 6	−0.07 ± 0.07
Total	26.0 ± 0.8	25.2 ± 0.7	−0.8 ± 1.1	−3% ± 4	−0.01 ± 0.03

[a] ±1-sigma uncertainty, calculated using effective number of independent observations (n*), see text

[b] ±95%Cl, calculated using effective number of independent observations (n*), see text

one-quarter (24%) is attributed to armoring. Island-wide, average beach width decreased by 11% ± 4% (2.6 ± 0.9 m) at a rate of −0.03 ± 0.03 m/year (Table 10.3). Forty-two percent of beaches widened (19% significantly), overall, with most of the widening (82%) occurring along unarmored beaches.

Looking at beach width changes on armored and unarmored beaches separately, we find the majority, or 72%, of armored beaches are degraded, including 43% narrowed (28% significantly) and 29% completely lost to erosion. The average width of beaches fronting coastal armoring decreased by 36% ± 5% (7.2 ± 0.9 m) at a rate of −0.10 ± 0.03 m/year.

Beach widths along unarmored coasts were roughly stable, overall, with 52% of unarmored beaches narrowed (28% significantly) and 47% widened (21% significantly). Complete beach loss was documented at only 1% of unarmored beaches where sandy shoreline was replaced by natural rock shoreline. Average beach width on unarmored beaches remained approximately the same at 26.0 ± 0.8 m at the beginning of historical data and 25.2 ± 0.7 m near the present (−3% ± 4%).

10.5 Discussion

Coastal armoring on eroding beaches of Oahu has resulted in beach narrowing and loss as beaches that are prevented from migrating upland are unable to access coastal plain sands that are trapped behind structures. In addition, increased erosion due to 'flanking' is observed adjacent to several armored sections on Oahu, often resulting in further construction of armoring to protect abutting property, a process that leads to alongshore proliferation of seawalls. Here we provide analysis on a regional scale and present several case studies documenting the effects of coastal armoring on Oahu beaches.

10.5.1 East Oahu

Of the four island regions, the relatively narrow (average ~ 18 m) beaches of east Oahu suffered the most damage from beach erosion and coastal armoring (Fig. 10.4). Roughly 35% or 14.7 km of east Oahu beaches are armored. The average beach width fronting coastal armoring decreased from 15.3 ± 1.1 m to 8.7 ± 1.0 m (−43% ± 10%), suggesting that many of the remaining narrowed beaches fronting armoring likely become unusable at high tide. Nearly 6 km or 14% of east Oahu beaches were completely lost to erosion; nearly all of it (98%) fronting coastal armoring. Seventy-four percent of armored beaches on the east side are degraded including 38% lost and 36% narrowed (20% significantly). Forty-five percent of east Oahu beaches widened (18% significantly), of which 80% occurred on unarmored coasts.

While erosion and narrowing is a problem on many east Oahu beaches, the region also has some of the longest extents of accreting beaches in Hawaii (Fletcher et al. 2012). As a result, widths of east Oahu beaches remained approximately stable, as a whole, with an average change of −5% ± 7%. Beach widths on unarmored beaches on east Oahu increased by 9% ± 6% or roughly 2 m. However, it is interesting to note that Kailua Beach, which is accreting along most of its length, actually narrowed as seaward growth of vegetation outpaced the prograding beach.

The highest proportion of armoring, narrowing, and beach loss on any segment of the Oahu shoreline is found between Laie and Kaaawa on the northeast coast. Flanking erosion north of armoring at Makalii Point has resulted in shoreline recession of over 40 m since 1967, loss of beachfront property, and is threatening to undermine beachfront homes (Figs. 10.5 and 10.6).

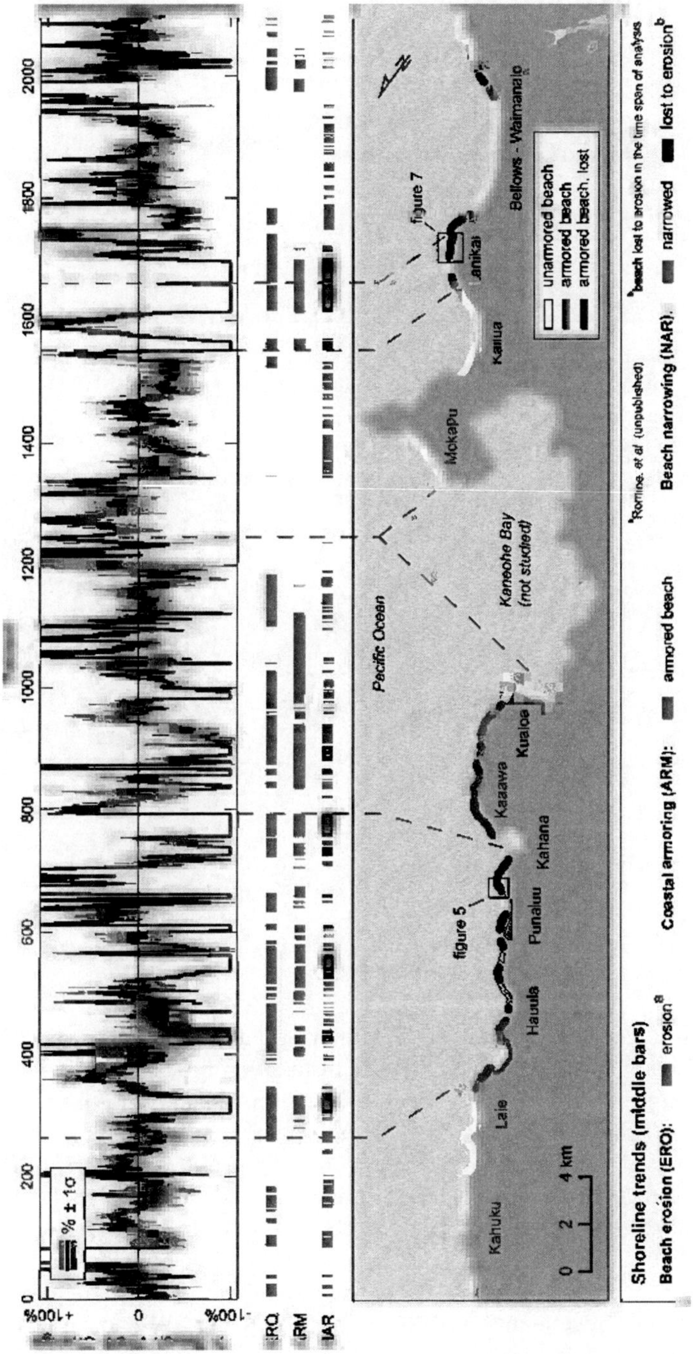

Fig. 10.4 East Oahu beach width percent changes (plot, 1928 or 1949 to near present), shoreline trends (*middle bars*), and coastal armoring (map)

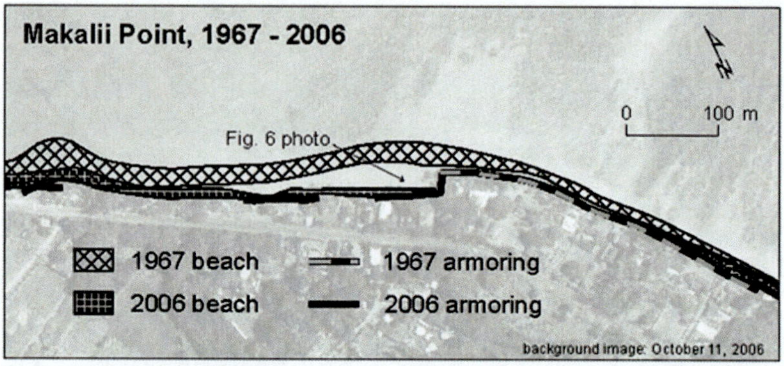

Fig. 10.5 Beach loss and flanking erosion at Makalii Point, east Oahu (1967–2006, location shown in Fig. 10.4). The unmarked area between the 1967 and 2005 beach was vegetated sand, which has since been lost to erosion

Fig. 10.6 Flanking erosion at Makalii Point (Photo location shown in Fig. 10.5; photo date, March 15, 2011)

There is strong evidence that coastal armoring has contributed to accelerated flanking erosion at Makalii Point following installation of armoring in the 1960s. Shoreline change rates calculated for the beach immediately north of armoring installed by 1967 show statistically significant increases in erosion rates (at the 95% confidence interval) when comparing rates from 1928 to 1967 and 1967 to 2005. As an example, directly adjacent to the armoring (within Fig. 10.6 photo) the rate changed from 0.5 + 0.4 m/year (accretion) to -1.0 ± 0.5 m/year (erosion) following installation of armoring. Erosion also increased fronting the northern half of the 1967 armoring, though not to the degree measured on the flanking unarmored beach. Low rubble revetments were recently (2000s) installed to protect homes on the north side of the point.

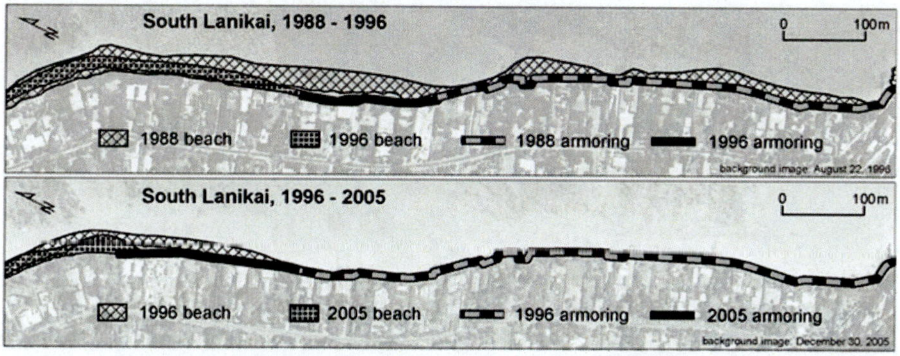

Fig. 10.7 Beach loss and flanking erosion at south Lanikai (1988–2005, location shown in Fig. 10.4)

At south Lanikai beach a trend of accretion reversed in the late 1970s. In the late 1980s, in response to the erosion, seawalls were constructed along much of the southern end of the beach to protect coastal properties (Fig. 10.7). By the mid-1990s the beach at the southern end of Lanikai had been completely lost to erosion and armoring proliferated to the north ~ 150 m in response to the northward-moving beach loss. By 2005 the beach had completely disappeared along the southern half of Lanikai. Recent beach surveys at south Lanikai indicate that flanking erosion continues to move north.

Comparisons of shoreline change rates at south Lanikai indicate that accelerated erosion due to the flanking process followed installation of the first armoring in the 1980s. Shoreline change rates are compared for the periods 1975–1988 (from the beginning of the erosion trend at south Lanikai to the first installation of coastal armoring) and 1988–2005 (after the first installation of coastal armoring). Rates along roughly 700 m of the beach flanking the north end of the armoring became more erosional and in most cases switched from accretion to erosion following installation of the armoring. However, none of the rate changes are statistically significant due largely to the limited number of historical shorelines available for the two measurement periods (three shorelines, each).

10.5.2 South Oahu

Along south Oahu (Fig. 10.8), analysis of beach width changes and its relation to shore-parallel coastal armoring is complicated by extensive use of other types of coastal engineering including groins, breakwalls, dredging, and fill – especially along beaches of Hawaii Kai to Kahala and Waikiki. As mentioned previously, beach changes are calculated for the modern configuration of the shoreline following major engineering efforts.

Thirty-nine percent (10.2 km) of beaches along south Oahu are armored; the highest percent of the four Oahu regions. Looking at south Oahu beaches as a

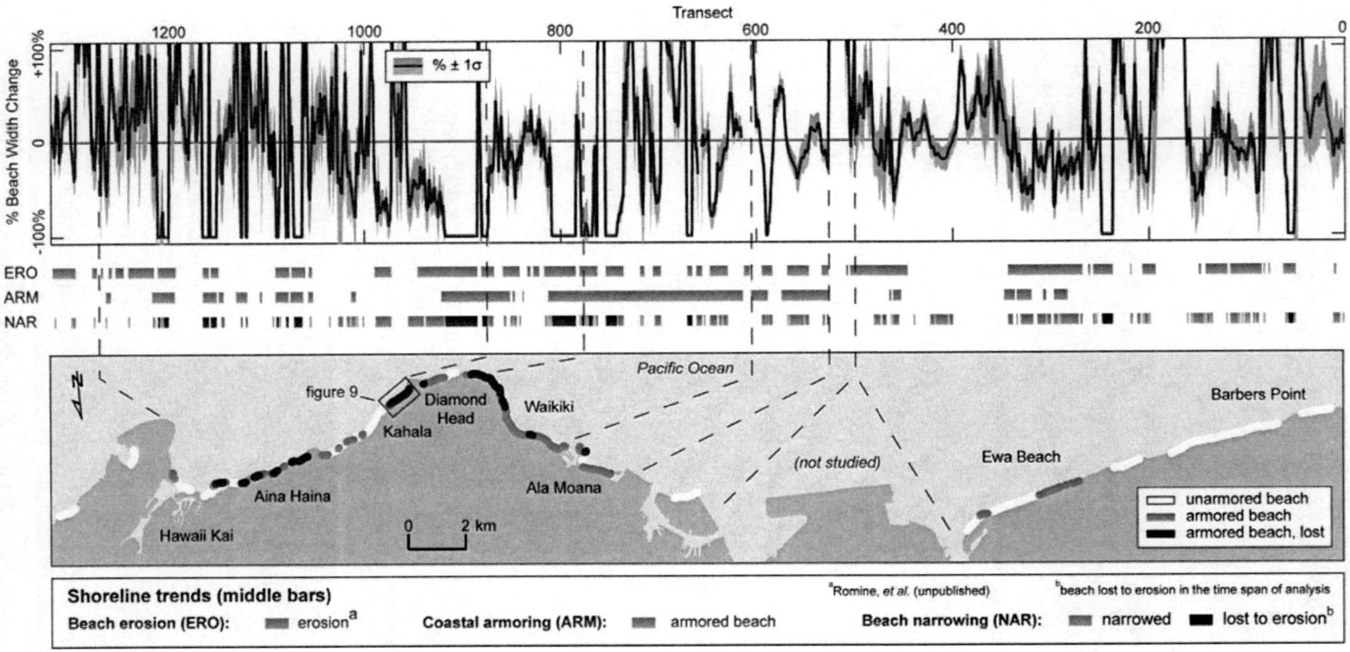

Fig. 10.8 South Oahu beach width percent changes (plot, 1928 or 1949 to near present), shoreline trends (*middle bars*), and coastal armoring (map)

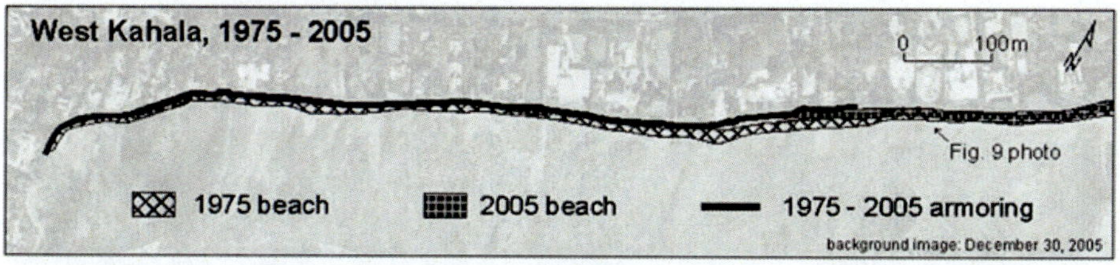

Fig. 10.9 Beach loss at Kahala, south Oahu (1975–2005, location shown in Fig. 10.8)

whole, roughly half of the beaches are degraded (22% significantly) and half widened. Twelve percent (3.1 km) of south Oahu beaches were completely lost to erosion. Average beach width along south Oahu decreased by 10% ± 4% or 1.8 ± 0.7 m.

Comparing armored and unarmored beaches we find that the majority (67%) of armored beaches along south Oahu are degraded with 40% narrowed (27% significantly) and 27% lost, while the majority, or 61%, of unarmored beaches have widened over the period (29% significantly). Beach width decreased by 32% ± 5% (6.9 ± 1.1 m) on armored beaches and beach widths increased by 9% ± 5% (1.5 ± 0.9 m) on unarmored beaches.

Areas of significant narrowing fronting coastal armoring include the Kahala shoreline where the beach has been completely lost to erosion. Beach width changes for the rest of Maunalua Bay (Hawaii Kai – Kahala) and Waikiki are highly variable alongshore. This is likely related to numerous groins and other shore-perpendicular structures that interrupt alongshore sediment transport leading to updrift impoundment and downdrift erosion. Nearly the entire length of the Waikiki and Ala Moana shoreline is armored. The greatest extent of beach loss in this section is at the eastern end of Waikiki adjacent to Diamond Head.

At the west end of Kahala Beach, roughly 900 m of beach was completely lost to erosion fronting coastal armoring (Fig. 10.9). Historical changes in the extent of armoring along west Kahala are difficult to discern from air photos due to dense cultivated vegetation along seaward property lines. It appears that most or all of the armoring was constructed prior to 1975 with extensions along a few adjacent properties in recent years in response to flanking erosion (Fig. 10.10). Analysis of changes in erosion rates on flanking beaches is not provided for this region due to the difficulty in mapping armored locations from historical air photos and limited shoreline data following the installation of armoring.

10.5.3 West Oahu

The west Oahu coast (Fig. 10.11) is the least armored of the four Oahu regions with armoring along only 1.2 km or 10% of beaches. However, the beaches are highly

Fig. 10.10 Flanking erosion and temporary armoring (*sand bags*), west Kahala Beach (location shown in Fig. 10.9; photo date, March 21, 2011)

erosional (Fletcher et al. 2012) and coastal armoring has contributed to beach narrowing. As a whole, 61% of west Oahu beaches are degraded, including 41% significantly narrowed; while 39% of beaches widened (23% significantly). Complete beach loss was noted at only a handful of transects. West Oahu has the widest initial and final average beach widths, though beaches narrowed by 9% ± 8% (3.1 ± 2.8 m).

Of the 10% of beaches armored along west Oahu, 82% are degraded with 80% narrowed (59% significantly) and only 2% completely lost. The average beach width fronting coastal armoring decreased by 37% ± 5% (14.4 ± 2.1 m). The majority or 58% of unarmored beach also narrowed (40% significantly), while 42% widened (25% significantly). The average change in beach width was not significant along unarmored beaches at −6% ± 8% (−1.9 ± 2.8 m).

The shoreline at the north end of Maili has retreated over 100 m due to chronic erosion and removal of sand by mining operations in the mid-1900s (Hwang 1981) (Fig. 10.12). In spite of the shoreline recession, substantial beach still remains at north Maili. Coastal armoring has only been constructed along a short section (~ 50 m) to protect a public restroom. The beach is preserved as the vegetation line is allowed to erode into a lightly-developed beach park, which has acted as a buffer between the receding beach and the coastal highway.

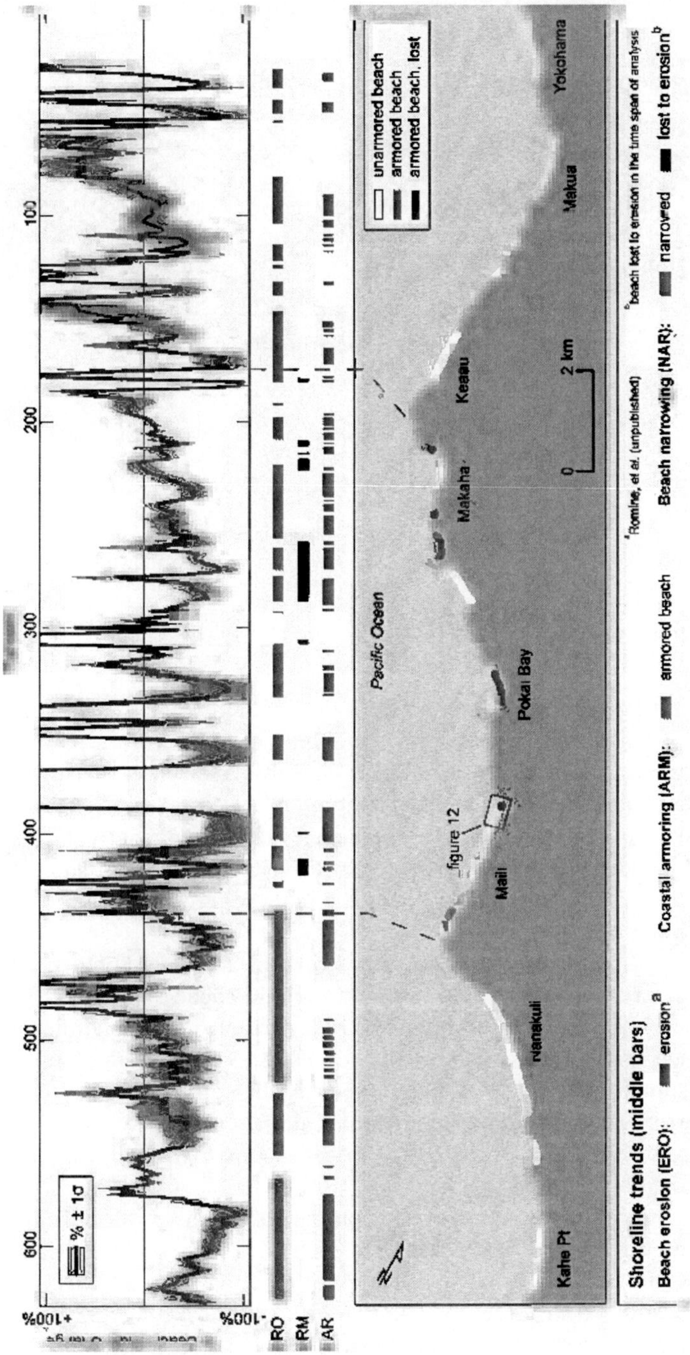

Fig. 10.11 West Oahu beach width percent changes (plot, 1928 or 1949 to near present), shoreline trends (*middle bars*), and coastal armoring (map)

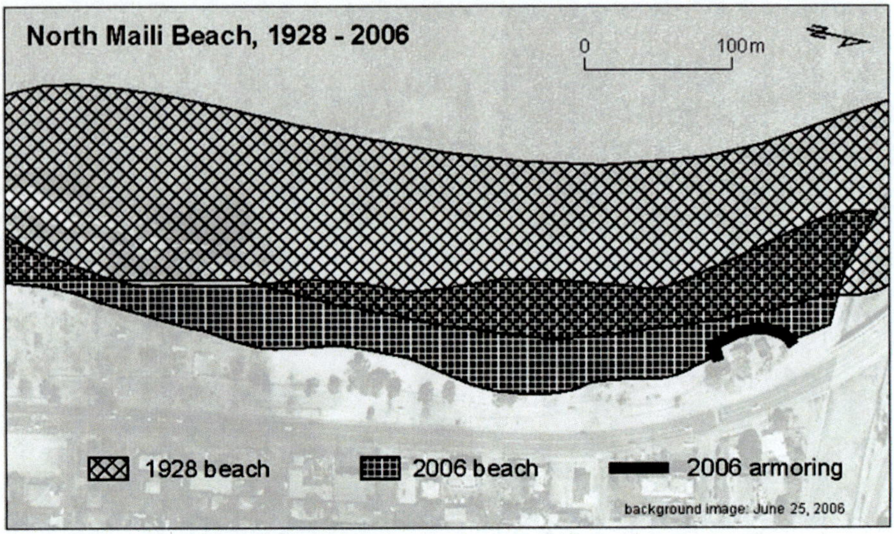

Fig. 10.12 In spite of shoreline recession of over 100 m, substantial beach remains along the (mostly) unarmored northern end of Maili Beach (1928–2006, location shown in Fig. 10.11)

10.5.4 North Oahu

Over 3 km or 12% of north Oahu beaches are armored (Fig. 10.13). Only about 200 m (1%) of north Oahu beaches was completely lost to erosion – all of which was at the northern end of Haleiwa fronting sea walls. As a whole, narrowing is the dominant trend of beach width change along north Oahu beaches, with 69% narrowed (46% significantly) and 30% widened (12% significantly) – the lowest percentage widened of the four Oahu regions. On average, north shore beaches narrowed by 17% ± 6% or 5.7 ± 1.9 m – the highest percent and net decrease of the four Oahu regions.

Significant narrowing is found on both armored and unarmored north Oahu beaches, though narrowing was greater on armored beaches. Seventy-six percent of armored beaches are degraded including 70% narrowed (54% significantly) and 6% lost. Beach widths decreased by 30% ± 5% or 8.7 ± 1.5 m along armored beaches. The majority or 69% of unarmored beaches also narrowed, though the amount of narrowing was less than along armored sections with average decrease in beach width of 16% ± 6% or 5.3 ± 1.9 m – the most narrowing on unarmored beaches of the four regions.

Beaches are narrowed along most of a continuous beach between Mokuleia and Waialua, including armored and unarmored sections. Near-complete beach loss is observed in 2006 air photos of a small embayment at Mokuleia (Fig. 10.14). Armoring, constructed in the early 1970s, was extended in the 1980s and more recently to protect coastal properties threatened by flanking erosion. Continued narrowing has resulted in complete beach loss fronting armoring in the middle of the bay as observed in a site visit in March of 2011 (Fig. 10.15).

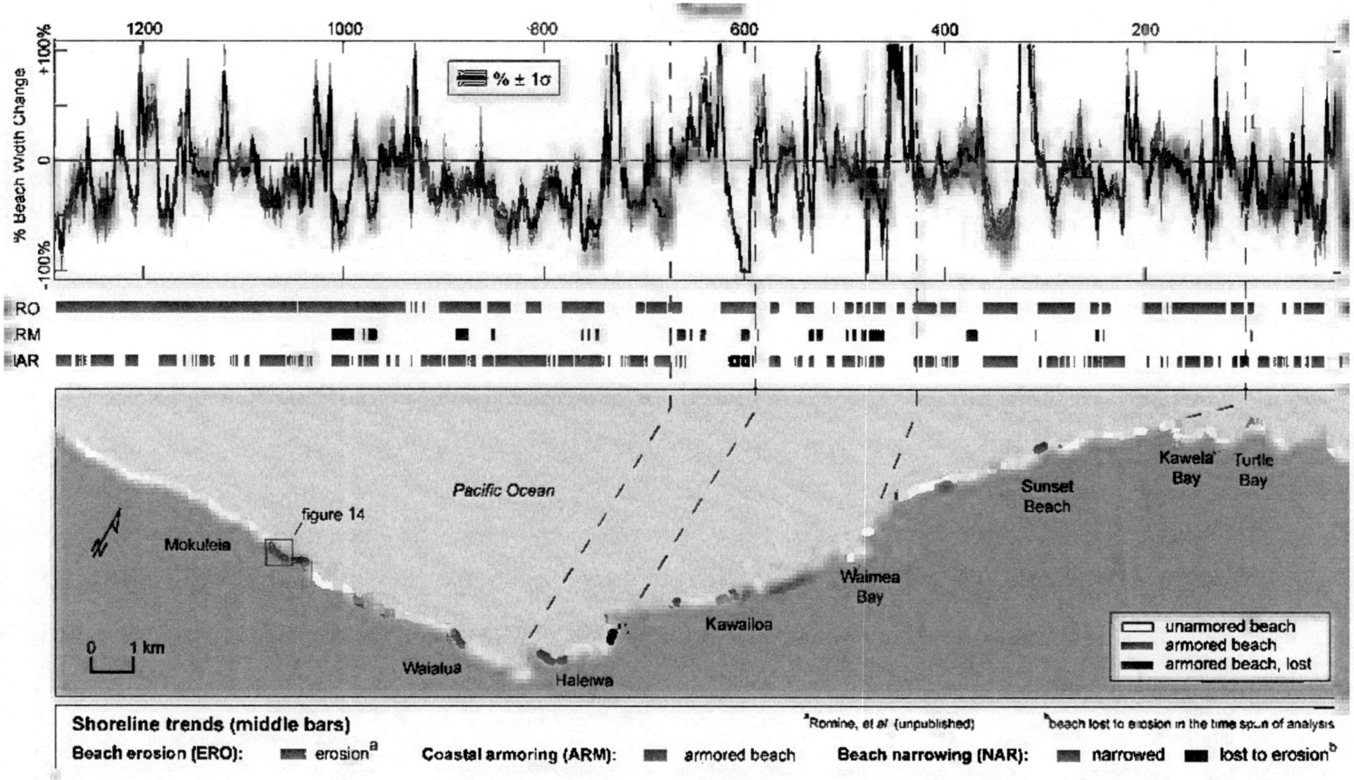

Fig. 10.13 North Oahu beach width percent changes (plot, 1928 or 1949 to near present), shoreline trends (*middle bars*), and coastal armoring (map)

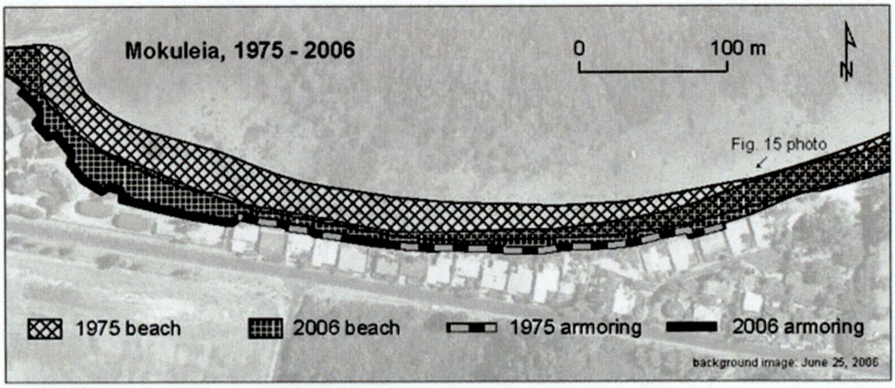

Fig. 10.14 Beach narrowing and flanking erosion at Mokuleia, north Oahu, as of June, 2006 (1975–2006, location shown in Fig. 10.13)

Fig. 10.15 Beach loss at Mokuleia, north Oahu (location shown in Fig. 10.14; photo date March 22, 2011)

Unlike Makalii Point and Lanikai, beach erosion rates flanking the west side of the 1975 armoring at Mokuleia appear to have slowed following installation of the armoring. Rates fronting the armoring and along roughly 100 m of the eastern flanking beach suggest accelerating erosion following installation of the armoring. As with Lanikai, none of the rate changes are statistically significant.

10.5.5 Island-Wide

Over the period of study, average beach width decreased by 11% ± 4% and nearly all (95%) documented beach loss was fronting armored coasts. Among armored beach sections, 72% of beaches are degraded, which includes 43% narrowed (28% significantly) and 29% (8.6 km) completely lost to erosion. Beaches fronting coastal armoring narrowed by −36% ± 5% or −0.10 ± 0.03 m/year, on average. In comparison, beach widths along unarmored coasts were relatively stable with slightly more than half (53%) of beaches experiencing any form of degradation.

As mentioned in the introduction, we examine two questions regarding the effects of coastal armoring on eroding coasts on Oahu. One, does armoring accelerate pre-existing erosion and does it initiate and or accelerate erosion on adjacent properties? Two, does armoring lead to other negative impacts such as beach loss or beach narrowing, which we define as separate from erosion? Analysis of shoreline change rates preceding and following installation of armoring suggests accelerated erosion on flanking beaches at several locations on Oahu after installation of armoring. However, the statistical significance of some of these rate changes is questionable due largely to limited shoreline data. Also, the argument could be made that the evidence is somewhat circumstantial. It is not possible through our analysis to conclude what proportion of the documented rate accelerations are due to the influence of coastal armoring or unrelated coastal dynamics. In response to question two, our analysis has clearly shown that armoring beaches in response to preexisting erosion leads to increased beach narrowing and loss by fixing the landward edge of the beach (vegetation line) and preventing it from receding with the seaward edge (beach toe).

These results support the findings of Fletcher et al. (1997) that construction of coastal armoring on eroding beaches of Oahu has contributed to beach narrowing and loss. However, the cause of narrowing along the majority of unarmored coasts of west and north Oahu (58% and 69%, respectively) is not clear. The north and west shores are dominated by beach erosion (Fletcher et al. 2012) so some narrowing is expected. However, the relatively high percentage of narrowing on unarmored beaches suggests that movement or stabilization of vegetation lines by means other than coastal armoring may be a factor. Cultivation of vegetation along the seaward edge of coastal properties is common practice and in some cases may be an attempt at 'soft armoring' to protect property from seasonal or chronic erosion – perhaps contributing to narrowing along these coasts. Therefore, the vegetation line does not necessarily denote the stable landward edge of the beach on all coasts and may be governed by more than erosion and accretion.

Another possible cause of narrowing is that interannual run-up interaction with a seawall, which would not be identified by our methodology, is responsible for a trend of narrowing. An example of this might include non-recovered sand loss related to wave reflection off seawalls during particularly high swell events such as in 1969 and 1998. Such intermittent losses, if significant, could contribute to decreased sand availability and, thus, beach narrowing.

Historical shoreline studies are typically hindered due to limited data (often <10 shorelines). By utilizing all available beach data with WLS regression, rather than an end-point analysis (only two data points), our analysis provides a more statistically defensible analysis of beach width change for highly variable coastal regions like Hawaii.

Sea level rise is likely to accelerate in coming decades (Vermeer and Rahmstorf 2009) and is almost certain to increase erosion and beach loss along Hawaii shores. With this study we have documented the negative effects of armoring eroding beaches and identified 'hotspots' of beach erosion and narrowing – data that may assist coastal resource managers in protecting beaches for future generations through improved management practices.

10.6 Conclusions

Coastal armoring has been a typical response to beach erosion on Oahu, Hawaii. To better understand the effects of armoring on eroding beaches, changes in beach width are compared among armored and unarmored beaches using historical shorelines mapped from aerial photographs. The results from this study show that armoring has contributed to beach narrowing and loss as receding beaches are prevented from migrating upland and sediment is trapped behind structures. Evidence is also provided for increased 'flanking erosion' on select beaches adjacent to coastal armoring by increased shoreline erosion rates following installation of armoring.

Over 27% of Oahu beaches (or former locations of beach) are armored and the majority, or 72%, of armored beaches are degraded (including 43% narrowed and 29% completely lost to erosion). Virtually all beach loss documented in this study (95%) occurred fronting coastal armoring. The remaining beaches fronting coastal armoring narrowed by 36% ± 5%. In contrast, beach widths along unarmored sections were much more stable with percents of degraded and widened beaches roughly even (53% vs. 47%), little or no change in average beach width change (−3% ± 4%), and little beach loss (1%).

The most armored regions of Oahu, the east and south sides (35% and 39% armored, respectively), suffered the greatest percents of beach loss (14% and 12% lost, respectively). Many of the remaining beaches along armored sections of east Oahu are narrowed to the extent that they likely become unusable at high tide (average beach width 8.7 ± 1.0 m). In comparison, the relatively unarmored west and north regions (10% and 12% armored, respectively) experienced little beach loss (0% and 1% lost, respectively). Like south and east Oahu, beaches along armored sections of the west and north shores are highly degraded (82% and 76%, respectively). In all four coastal regions of Oahu the majority of the beach fronting armoring was degraded (between 67% and 82%). Along south and east

Oahu the majority of unarmored beaches widened (55% and 61%, respectively). Sixty-nine percent of unarmored beaches on north Oahu narrowed (45% significantly) indicating that the common practice of stabilizing seaward property lines by cultivating vegetation may be contributing to narrowing.

Acknowledgments This work was supported by grants from the State of Hawaii Department of Land and Natural Resources, City and County of Honolulu, Hawaii Sea Grant College Program, U.S. Army Corps of Engineers, Harold K.L. Castle Foundation, and Hawaii Coastal Zone Management Program. We thank Matthew Barbee, Chyn Lim, Tiffany Anderson, Haunani Kane, Matthew Dyer, Amanda Vinson, and Craig Sentor of the University of Hawaii Coastal Geology Group for their support on this project. Thank you to Neil Frazer of the University of Hawaii Geology and Geophysics Department for his advice on statistical methods. Thank you to Chris Conger and Dolan Eversole of Hawaii Sea Grant and Jessica Podoski of the U.S. Army Corps of Engineers for their advice on defining and mapping coastal armoring.

This paper is funded in part by a grant/cooperative agreement from the National Oceanic and Atmospheric Administration, Project R/IR-4, which is sponsored by the University of Hawaii Sea Grant College Program, SOEST, under Institutional Grant No. NA09OAR4170060 from NOAA Office of Sea Grant, Department of Commerce. The views expressed herein are those of the author(s) and do not necessarily reflect the views of NOAA or any of its subagencies. UNIHI-SEAGRANT-BC-09-02.

Literature Cited

Carter CH, Monroe CB, Guy DE (1986) Lake Erie shore erosion; the effect of beach width and shore protection structures. J Coast Res 2:17–23

Fletcher CH (1992) Sea-level trends and physical consequences: applications to the U.S. shore. Earth Sci Rev 33:1–36

Fletcher CH, Jones AT (1996) Sea-level highstand recorded in Holocene shoreline deposits on Oahu, Hawaii. J Sediment Res 66(3):632–641

Fletcher CH, Mullane RA, Richmond BM (1997) Beach loss along armored shorelines on Oahu, Hawaiian Islands. J Coast Res 13(1):209–215

Fletcher CH, Rooney JJB, Barbee M, Lim S-C, Richmond BM (2003) Mapping shoreline change using digital orthophotogrammetry on Maui, HI. J Coast Res (Special Issue 38):106–124

Fletcher CH, Bochicchio C, Conger CL, Engels M, Feirstein EJ, Grossman EE, Grigg R, Harney JN, Rooney JB, Sherman CE, Vitousek S, Rubin K, Murray-Wallace CV (2008) Geology of Hawaii reefs. In: Riegl B, Dodge R (eds) Coral reefs of the U.S.A. Springer, Berlin, pp 435–488

Fletcher CH, Romine BM, Genz AS, Barbee MM, Dyer M, Anderson TR, Lim SC, Vitousek S, Bochicchio C, Richmond BM (2012) National assessment of shoreline change: historical shoreline changes in the Hawaiian Islands. U.S. Geological Survey Open-File Report 2011–1051

Hall MJ, Pilkey OH (1991) Effects of hard stabilization on dry beach width for New Jersey. J Coast Res 7(3):771–785

Hapke CJ, Himmelstoss EA, Kratzmann MG, List JH, Thieler ER (2010) National assessment of shoreline change: historical shoreline changes along the New England and Mid-Atlantic coasts. U.S. Geological Survey Open-File Report 2010–1118

Harney JN, Fletcher CH (2003) A budget of carbonate framework and sediment production, Kailua Bay, Oahu, Hawaii. J Sediment Res 73(6):856–868

Harney JN, Grossman EE, Richmond BM, Fletcher CH (2000) Age and composition of carbonate shoreface sediments, Kailua Bay, Oahu, Hawaii. Coral Reefs 19:141–154

Hwang D (1981) Beach changes on Oahu as revealed by aerial photographs. Prepared by the Urban Regional Planning Program and the Hawaii Institute of Geophysics and Sea Grant Program, University of Hawaii for the Dept. of Planning and Economic Development, State of Hawaii, Technical Supplement 22. Hawaii Coastal Zone Management Program, Honolulu

Komar PD, McDougal WG (1988) Coastal erosion and engineering structures: the Oregon experience. J Coast Res (Special Issue 4):77–92

Kraus NC, McDougal WG (1996) The effects of seawalls on the beach: part I, an updated literature review. J Coast Res 12(3):691–701

Makai Ocean Engineering, Sea Engineering (1991) Aerial photograph analysis of coastal erosion on the Islands of Kauai, Molokai, Lanai, Maui and Hawaii. Office of State Planning/Coastal Zone Management Program, Honolulu

McDonald HV, Patterson DC (1984) Beach response to coastal works, Gold Coast, Australia. In: Proceedings of the international conference on coastal engineering, No 19 '84, Houston, TX, pp 1522–1538

Romine BM, Fletcher CH, Frazer LN, Genz AS, Barbee MM, Lim S-C (2009) Historical shoreline change, Southeast Oahu, Hawaii: applying polynomial models to calculate shoreline change rates. J Coast Res 25(6):1236–1253

Sea Engineering Inc (1988) Oahu shoreline study. City and county of Honolulu Department of Land Utilization

Sherrod DR, Sinton JM, Watkins SE, Brunt KM (2007) Geologic map of the State of Hawaii. U.S. Geological Survey, Reston

Tait JF, Griggs GB (1990) Beach response to the presence of a seawall. Shore Beach 58:11–28

Vermeer M, Rahmstorf S (2009) Global sea level linked to global temperature. Proc Natl Acad Sci 106(51):21527–21532

Vitousek S, Fletcher CH (2008) Maximum annually recurring wave heights in Hawaii. Pac Sci 62 (4):541–553

Chapter 11
Compromising Reef Island Shoreline Dynamics: Legacies of the Engineering Paradigm in the Maldives

Paul S. Kench

Abstract Located in the central Indian Ocean, the Maldives archipelago consists of 25 atolls and oceanic reef platforms that contain 1,200 low-lying reef islands. These islands are among the most dynamic landforms on earth. Island instability and the pressures of high population densities have resulted in the proliferation of engineered structures to combat erosion and maintain island shorelines. In many instances the introduction of hard-engineered structures has exacerbated island erosion and degraded ecological processes. Reasons for these negative environmental consequences relate to the appropriateness of the design and placement of these structures. The materials used and the mode of construction employed by many small island nations contravene most standard measures of sound engineering design. Sound design is also constrained by the absence of environmental information on local coastal processes (e.g. waves, currents). This paper summarizes field observations from the Maldives to highlight the natural dynamics of small island shorelines. Island dynamics are examined in light of standard engineering structures and shown to destabilize island landforms. Such physical responses necessitate reconsideration of classic concepts of island instability and erosion. As a result management solutions are often inappropriate with respect to natural coastal processes and dynamics of small island shorelines. It is proposed that island maintenance will be best achieved by ensuring that management solutions safeguard the integrity of natural geomorphic processes. This approach requires the replacement of the prevailing paradigm of islands as 'static landforms' with the recognition and incorporation in planning of each island's natural dynamism. This approach places an emphasis on understanding the natural processes of small islands and provides new challenges for managers to seek planning alternatives to conventional *ad hoc* engineering solutions.

P.S. Kench (✉)
School of Environment, The University of Auckland, Private Bag 92019, Auckland, New Zealand
e-mail: p.kench@auckland.ac.nz

J.A.G. Cooper and O.H. Pilkey (eds.), *Pitfalls of Shoreline Stabilization: Selected Case Studies*, Coastal Research Library 3, DOI 10.1007/978-94-007-4123-2_11,

11.1 Introduction

Located in the central Indian Ocean, the Maldives archipelago consists of 25 atolls and oceanic reef platforms that stretch 886 km from Ihavandhippolhu in the north (6°57′ N) to Addu atoll (0°34′ S) just south of the equator (Fig. 11.1). The reef platforms and atolls of the Maldives support more than 1,200 reef islands, which are wave deposited accumulations of carbonate sands and gravels derived from the surrounding reef platforms (Kench 2011). The reef islands of the Maldives are small and range in area from less than 1 ha to a maximum of 595 ha, although the modal size is approximately 2.5 ha in area. The islands are also low lying. While the maximum natural land levels approach 6 m above mean sea level (MSL) in places, the mean elevation of land is approximately 1 m above MSL.

Low-lying reef islands, such as those found in the Maldives, are physically dynamic landforms, changing their size, shape, elevation and position on reef platforms in response to extreme events (Kench et al. 2006a, 2008), short-term and seasonal adjustments in wind, wave and current patterns (Flood 1986; Kench and Brander 2006), and medium to long-term shifts in sea level (Leatherman 1997; Kench and Cowell 2001). A range of anthropogenic activities also promote shoreline change and instability.

The Maldivian reef islands are host to a population of 325,000, who are located on 198 islands in the archipelago. The population density on islands varies considerably from small rural villages of less than 300 to the capital island Malé, which is host to more than 102,000 people, with a population density in excess of 1,020 people per hectare. A further 102 islands are designated exclusively as resorts, which provide one of the principal sources of external income to the nation. The limited land area of reef islands and their morphological instability pose significant management problems for island communities. Island change is commonly misconstrued as shoreline erosion, which is perceived as one of the most pervasive environmental problems in small island nations. For example, in the Maldives, 41 of the inhabited islands reported severe erosion in 2009 (Shaig 2011). Furthermore, erosion issues are expected to become more prominent with future sea level rise and climatic variability (Leatherman 1997). Indeed, a number of projections suggest islands will be entirely eroded from reef surfaces in the near future, rendering the inhabitants of small island states the first environmental refugees of climate change (Khan et al. 2002; Barnett and Adger 2003; Dickinson 2009).

Confronted with unstable island shorelines and perceived erosion, management responses have relied upon structural engineering solutions. Increasing population pressure, and the need to provide services, have also resulted in a wide range of engineered projects such as causeways, dredged boat channels and reclamation to the coastal zone that interact with reef-top processes. Introduction of such structures is known to result in a number of adverse environmental effects including accelerated island erosion and reef degradation (e.g. Maragos 1993). However, few studies have examined how structures interact with reef platform processes that affect island shorelines.

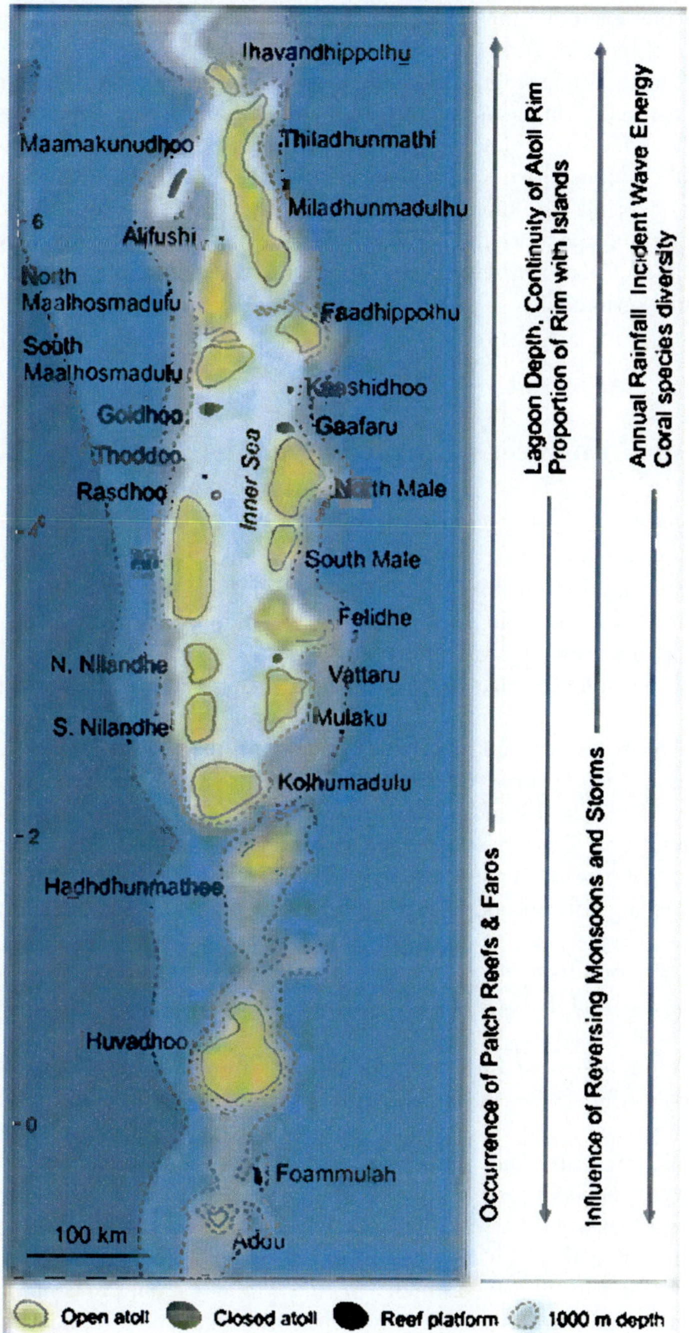

Fig. 11.1 Location and setting of the Maldives archipelago, central Indian Ocean

This chapter examines the use of shoreline management strategies adopted in the Maldives archipelago. First, it reviews the type and prevalence of engineering approaches to shoreline stabilization and discusses some of the common problems that have emerged from the use of such structures. Second, the chapter considers the broad scale processes regime and natural dynamics of reef island shorelines. It highlights the unique process regime of island shorelines and evaluates the interaction of engineering structures with the nearshore process regime. Findings are discussed with respect to the validity of using existing standard engineering structures in reef platform settings and future directions for island shoreline management are considered.

11.2 Use of Engineering Structures in the Maldives

A range of management strategies have been adopted in the Maldives to combat coastal erosion that have typically included the standard suite of engineering solutions to stabilize shorelines to maintain island size and sediment volume (Fig. 11.2). Development pressure on small islands has also led to the introduction of a range of structures (causeways, boat channels, reclamation) into the coastal environment that interact with environmental processes (Fig. 11.2). For example, Table 11.1 presents the number and type of structures found on a survey of 45 of the inhabited and resort islands in the Maldives. The data show that more than 75% of surveyed inhabited islands have constructed shore-attached sea walls (Fig. 11.2a, b) and approximately 90% have infrastructure associated with artificial boat harbors (quaywalls, breakwater etc. Fig. 11.2c, d). The data also show that resorts have adopted different shoreline management options. In particular, a greater proportion of resorts have adopted nearshore breakwaters and groynes (Fig. 11.2e, f). In general, resorts have fewer sea walls, although they are still found on 50% of resort islands. Resorts also have much lower proportions of boat harbors and associated structures, instead adopting piled jetties for boat access.

Introduction of these structures is known to have a high failure rate in the Maldives (Fig. 11.3). In particular, resort owners increasingly express frustration that groynes and other measures appear to have little effect on erosion or shoreline stability and they continually seek and trial new approaches (Kench 2010). For example, a number of resorts are actively exploring the use of offshore breakwaters as an alternative to shore-attached structures. The use of structures, and their failure, has also been implicated in the generation of additional island instability problems in the Maldives. The reasons for failure of structures and their broader environmental impacts are explored in the following sections.

Fig. 11.2 Range of structures used to manage shoreline erosion. (**a**) Coral block seawall used as breakwater for harbor. (**b**) Coral block quay wall in harbor. (**c**) Concrete wall over stacked sandbag seawall. (**d**) Tetrapod seawall, Malé. (**e**) Arrangement of groynes on island shoreline. (**f**) and (**g**) Breakwaters on reef edge. (**h**) Sand pumping to reclaim land. A similar technique is adopted to nourish beaches

Table 11.1 Summary of engineering structures used on 45 inhabited and resort islands in the Maldives

Type of coastal structure or modification	Number of structures on Inhabited Islands (n = 30)	Number of structures on resort Islands (n = 15)
Seawall	22 [73%]	8 [53%]
Nearshore breakwater	3 [10%]	10 [67%]
Revetment	3 [10%]	–
Groynes	10 [33%]	11 [73%]
Reclamation	10 [33%]	–
Quay wall	30 [91%]	4 [27%]
Harbor breakwater	26 [87%]	2 [13%]
Entrance channel protection	23 [77%]	2 [13%]
Land reclamation	30 [91%]	4 [27%]
Bridge/causeway	2 [7%]	4 [27%]

Source: Shaig (2011), and field observations

11.3 Failure of Engineered Structures

Field inspection of numerous structures indicates that they: physically fail; have a short life span; do not stop the erosion problem, and; they often exacerbate island erosion and can impact on reef productivity (Kench et al. 2003). Coastal structures in the Maldives fail for a range of reasons. First, failure is associated with the design and the mode of construction. In the Maldives major infrastructure projects are commonly associated with international aid programs or government initiatives. In such cases structures undergo a more formal design and construction process (Kench 2010). However, outside the major urban centre structures are designed and built *ad hoc* by local communities or resorts. In such cases the technical capacity to design structures is lacking and consequently, designs and structures commonly contravene most standard measures of engineering 'best practice' (Kench 2001, 2010). For example, considerations such as design height, length, how structures are tied to islands; the use of filter cloth; grading of aggregates, structural foundations; and the size and type of materials used and construction methods are common elements that can compromise the structural integrity of structures in the Maldives.

Second, structural solutions have been adopted that simply copy designs transposed from developed countries. Such structural designs have been applied without special regard for the unique character of the nearshore and coastal process regime of small reef islands. In particular, designs are developed in a vacuum of environmental information on wave conditions, currents, beach dynamics and sediment flux (as also highlighted by Kench 2001). This system has evolved due to the lack of coastal process information. Therefore, the choice of structure, its physical design and construction materials are selected without detailed knowledge of the coastal processes and in particular extreme

Fig. 11.3 Examples of common failings of coastal structures in reef island settings. In most cases structures do not comply with the design considerations for structures. (**a**) and (**b**) Undermining and collapse of cement sand bag seawall due to overtopping and lack of geotextile. (**c**) and (**d**) Slumping of land behind seawall and collapse of structure. (**e**)–(**h**) Overtopping, breaching and failure of walls

forces that occur at the shoreline. As a consequence, many structures have failed due to overtopping or the size of engineering units is too small to withstand the hydrodynamic forces impacting the structure.

Third, In the Maldives coastal structures (seawalls, revetments, groynes and breakwaters) have been built from a wide variety of materials. In general, more robust materials have been employed in major Government infrastructure projects and resort developments. Recent structures have been built using concrete slab, cement sand bags or large cast engineering units (CEUs) such as tetrapods (e.g. Fig. 11.2d). However, older structures have been constructed from coral block (e.g. Figs. 11.2a and 11.3f, g). Structures sponsored or constructed by local communities consist of a more diverse array of materials which include concrete, cement sand bags, coral rock, metal drums and in isolated instances, wooden structures. Coral fragment shape and density mean that individual clasts are frequently too small to remain stable under moderate to high energy conditions. Furthermore, cements are either of poor quality or not used.

Fourth, apart from major infrastructure projects, poor construction practices compound inappropriate designs and undermine structures. In part, this is also a function of a lack of technical capacity. For example, except where erected on beach rock, toe protection is rarely considered and appropriate foundation materials are rarely used. The absence of suitable natural structures to which seawalls and groynes can be fixed, leads to structures liable to flanking at the lateral extent of seawalls or on the inshore end of groynes. Filter cloth is seldom used behind seawalls and revetments. Consequently, in many cases water and sediments can move freely around, and occasionally over and through seawalls.

The net effect of failed structures in many cases is exacerbation of erosion problems. This leads to promulgation of engineering structures with some examples of entire island shorelines being armored.

11.4 Environmental Effects of Engineered Structures

Current scientific understanding of the impacts of coastal structures on coastal processes and geomorphology emanates largely from studies on continental coastlines (e.g. Sawaragi 1967; Kraus 1988; Weggel 1988; Kraus and McDougal 1996; Morton 1998). It is necessary to consider the relevance of these impacts in coral reef settings, which have unique differences in the configuration of the coast, the coastal process regime and shoreline dynamics. These differences are summarized in the context of the Maldives before consideration is given to the potential interaction of structures with reef island process systems and potential environmental impacts.

11.4.1 Nearshore Process Regime

Erosion or change in island shorelines is controlled by a combination of wave and current processes that transport sediment (sand) around island shorelines. Understanding of the controls on island erosion, therefore, must consider the wave processes and the influence of climate in modifying nearshore wave and current processes.

The reef island process regime and shoreline dynamics in the Maldives are modulated by marked seasonal reversals in monsoon conditions, which are characterized by strong reversals in wind direction that are confined to a narrow range of wind angles. Analysis of 30-years of wind data from Hulule Island since 1964 indicates that the Maldives experience southwest to northwest winds (~225–315°) from April to November during the *Hulhangu* monsoon, with a mean wind speed of 5.0 m s^{-1}. This is also known as the westerly monsoon. In contrast, the *Iruvai* monsoon, from December to March is characterized by winds from the northeast-east (~45–90°) with a mean wind speed of 4.8 m s^{-1} (Kench et al. 2009a). This period is also known as the northeast monsoon.

Detailed experiments of wave and current processes in the west and northeast monsoon conditions on a number of reef platforms in South Maalhosmadulu atoll have provided the first insights into the coastal process regime of island shorelines (Kench et al. 2006b, 2009a, b). First, swell wave energy is able to refract/diffract around reef platforms and influence the entire perimeter of reef island shorelines. Second, the magnitude of energy impacting an island varies depending on location in an atoll (e.g. exposed atoll rim, lagoonal setting or leeward atoll). Third, wave energy is unevenly distributed around the entire 360° perimeter of reef islands. Windward shorelines receive greater input of energy through a combination of swell and wind-wave energy. In contrast, leeward shorelines receive lower total energy input in each season as wind-wave energy is effectively dissipated on windward reef surfaces. Fourth, the Maldives is characterized by a microtidal regime. These tides act to modulate the amount of wave energy that leaks onto reef surfaces, with greatest wave energy accessing reef island shorelines at higher tidal stages. Fifth, changes in wind direction, driven by monsoons, promote changes in the areas of a reef island shoreline that receive greatest wave energy.

Monsoonally forced changes in incident wave conditions have been shown to modulate reef platform circulation patterns between seasons. As reported by Kench et al. (2003, 2009b) nearshore currents around Maldivian reef islands were found to be unidirectional in most instances. Furthermore, circulation patterns showed distinct changes between monsoon seasons. In general, nearshore flows are toward the east under westerly monsoon conditions and toward the south-southwest under northeast monsoon conditions (Fig. 11.4). The precise nature of current patterns on reefs differed between reef islands. Such differences reflect a number of factors: the shape of each platform, which controls localized wave refraction patterns; and, direction and character of waves impacting reefs (swell versus wind waves), which is a function of the boundary wave climate, position of each reef platform within the atoll and their proximity to neighboring reefs and gaps in the atoll periphery (Kench et al. 2009a).

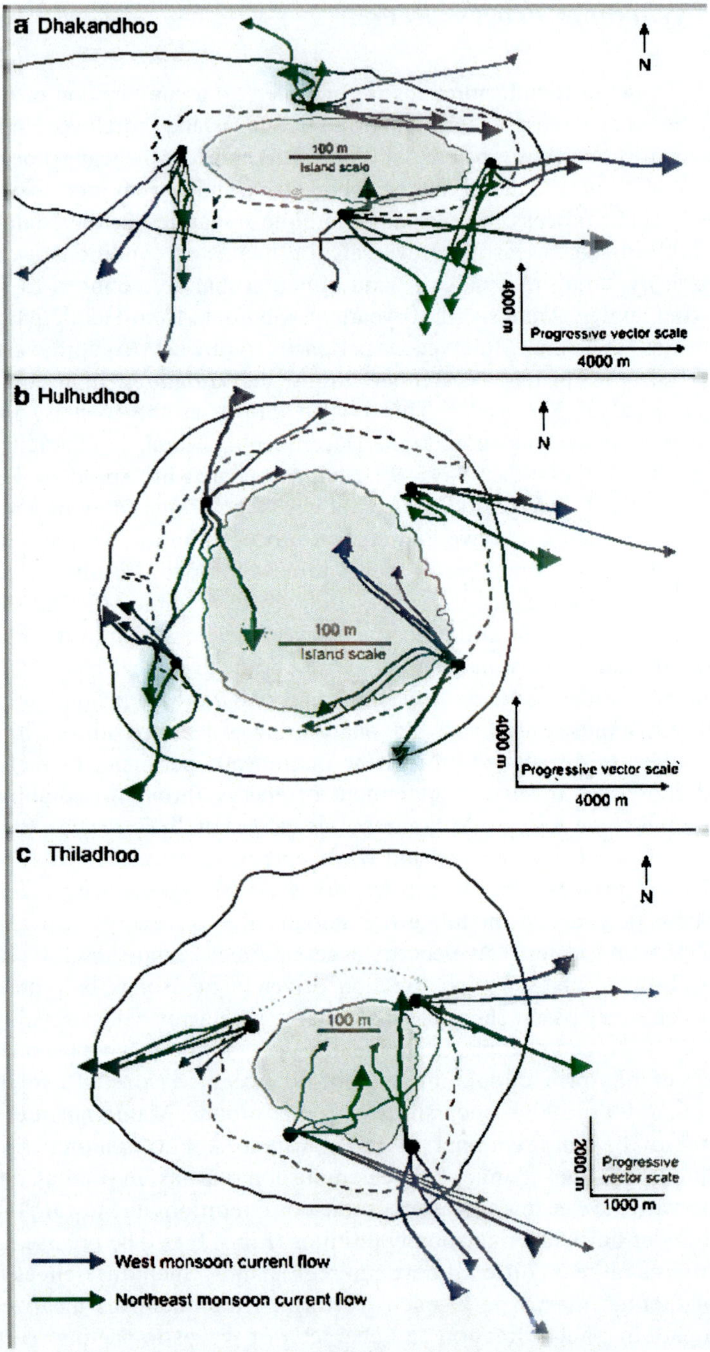

Fig. 11.4 Summary nearshore current patterns around three reef islands in South Maalhosmadulu atoll. Each *colored arrow* represents 12.5 h of continuous current flow (Source: Kench et al. 2009b)

11.4.2 Shoreline Dynamics

Monsoonally-forced changes in nearshore wave and current patterns control seasonal morphological responses of island beaches. Waves and currents provide the process mechanism that drives change in shoreline position between seasons. Kench and Brander (2006) examined the morphological sensitivity of 13 islands in the Maldives to predictable changes in wind, wave and circulation conditions controlled by the reversing monsoons. Results showed large (up to 53 m of beach change) and rapid excursions of beach material around island shorelines in response to monsoonally forced changes in nearshore circulation (Fig. 11.4). Of note, beach changes involved the alongshore reorganization of beach material, rather than the off/onshore exchange common on continental shorelines. In particular, unidirectional shoreline currents transfer sediment alongshore to leeward depocentres and govern the reorganization of mobile beach materials around the study islands (Fig. 11.5). Beaches show the greatest degree of morphological change at locations where current patterns exhibit the largest changes between monsoon seasons.

This study also showed that seasonal oscillations in planform position of the mobile beach were balanced at the annual scale. The magnitude of seasonal morphological change, and sensitivity of islands to change, were found to vary between islands as a function of reef platform shape, which controls wave refraction patterns. Kench and Brander (2006) proposed the 'island oscillation index' to predict the sensitivity of islands to changes in wave climate. Analyses showed circular islands were most sensitive to changes in incident wave processes (Fig. 11.5).

11.4.3 Implications of Process Regime for Shoreline Management

Collectively the process and morphodynamic observations of islands in South Maalhosmadulu atoll provide the first detailed process linkages between waves, currents and consequent shoreline dynamics on reef platform islands in the Maldives. These findings provide insights into the morphodynamic behavior of reef islands that are likely to have implications in other reef platform settings and which can be synthesized to improve the information base with which to make better informed coastal management decisions. A conceptual model of the process signature and island behavior can be constructed. Key features of the shoreline process regime for reef islands include:

- Unlike continental shorelines, reef islands can have a convex planform coastline that encompasses 360°, that is exposed to incident wave and current processes around the entire island perimeter.
- Process studies show that reef island shorelines are dominated by alongshore current processes. Alongshore currents promote alongshore reorganization of sediments around island shorelines that control beach change.

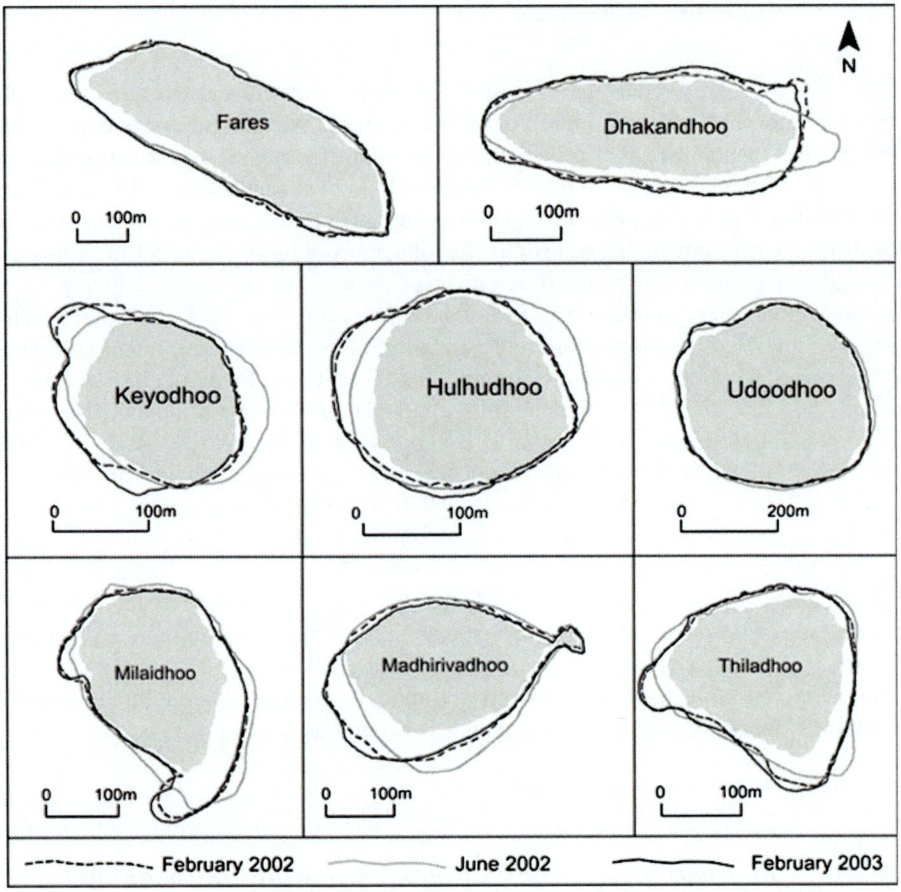

Fig. 11.5 Seasonal change in eight reef island shorelines, South Maalhosmadulu atoll. Surveys undertaken using GPS (Source: Kench and Brander 2006)

- The nearshore current patterns that control sediment transport are influenced by seasonal variations in wind and wave patterns which can lead to total reversal in circulation around islands.
- Island shorelines are morphologically very dynamic and exhibit large changes in position in response to changes in wave energy and current energy. Unlike shoreline dynamics on continental coastlines the sweepzone on reef islands involves the alongshore reorganization of beach sediments rather than the cross-shore exchange of sediments.
- Shoreline dynamics indicate there are sectors of island shorelines that act as deposition zones and other sectors that act as sediment transfer zones.
- The dynamics of island shoreline change is dependent upon the shape of the reef platform, position in the atoll and magnitude of change in wave energy between seasons. Alongshore process and morphological change signatures are likely to

dominate on reef platforms where wave refraction around the reef structure can take place, but is not likely to occur on islands situated on extensive linear atoll reef rims where wave refraction is precluded by the reef extent and/or presence of a backing lagoon.

- Islands are dynamic landforms that can migrate on platforms over decadal timescales as a consequence of incremental shifts in the balance of seasonal oscillations in beach behavior.

11.4.4 Interaction of Coastal Processes and Structures on Maldivian Reef Islands

Observational insights into the coastal processes and shoreline dynamics of reef islands have important implications for how coastal structures can interact with processes and potentially promote adverse environmental impacts.

11.4.4.1 Process Interactions with Shore Perpendicular Structures

Shore perpendicular structures are common in the Maldives and include groynes and structures associated with boat harbors (Fig. 11.2). The insertion of shore perpendicular structures can have significant impacts that compromise nearshore current processes, sediment transport and promote further island erosion and instability (summarized in Figs. 11.6 and 11.7). First such structures directly interfere with alongshore current patterns. In particular, currents are deflected away from the shoreline on the updrift side of structures and current shadows (eddies) can form on the downdrift side of groynes. Localized zones of accelerated and decelerated flows induce erosion and deposition. Second, shore perpendicular structures alter sediment transport and deposition that include: the trapping of sediment on the updrift side of structures promoting shoreline accretion, and; reduction in transfer of sediment to the downdrift side of structures depleting the sediment volume on this shoreline and potentially promoting erosion. These effects are similar to those commonly reported from continental coastlines. However, such modifications to the sediment transport system are compounded on small reef islands where the sediment budget is limited in volume and undergoes periodic and reversing fluxes, and can undermine natural shoreline dynamics. Perpendicular structures can unevenly partition the sediment reservoir into discrete cells. This can promote shoreline erosion by reducing the volume of sediment in one part of the shoreline sediment budget. Where shore perpendicular structures are located in the transfer pathway of the alongshore reorganization of sediment, the downdrift sector of the island shoreline is depleted of sediment. In the subsequent monsoon period the small volume of sediment, trapped on this downdrift coast, is rapidly remobilized and transported away from the shoreline exposing the vegetated shoreline

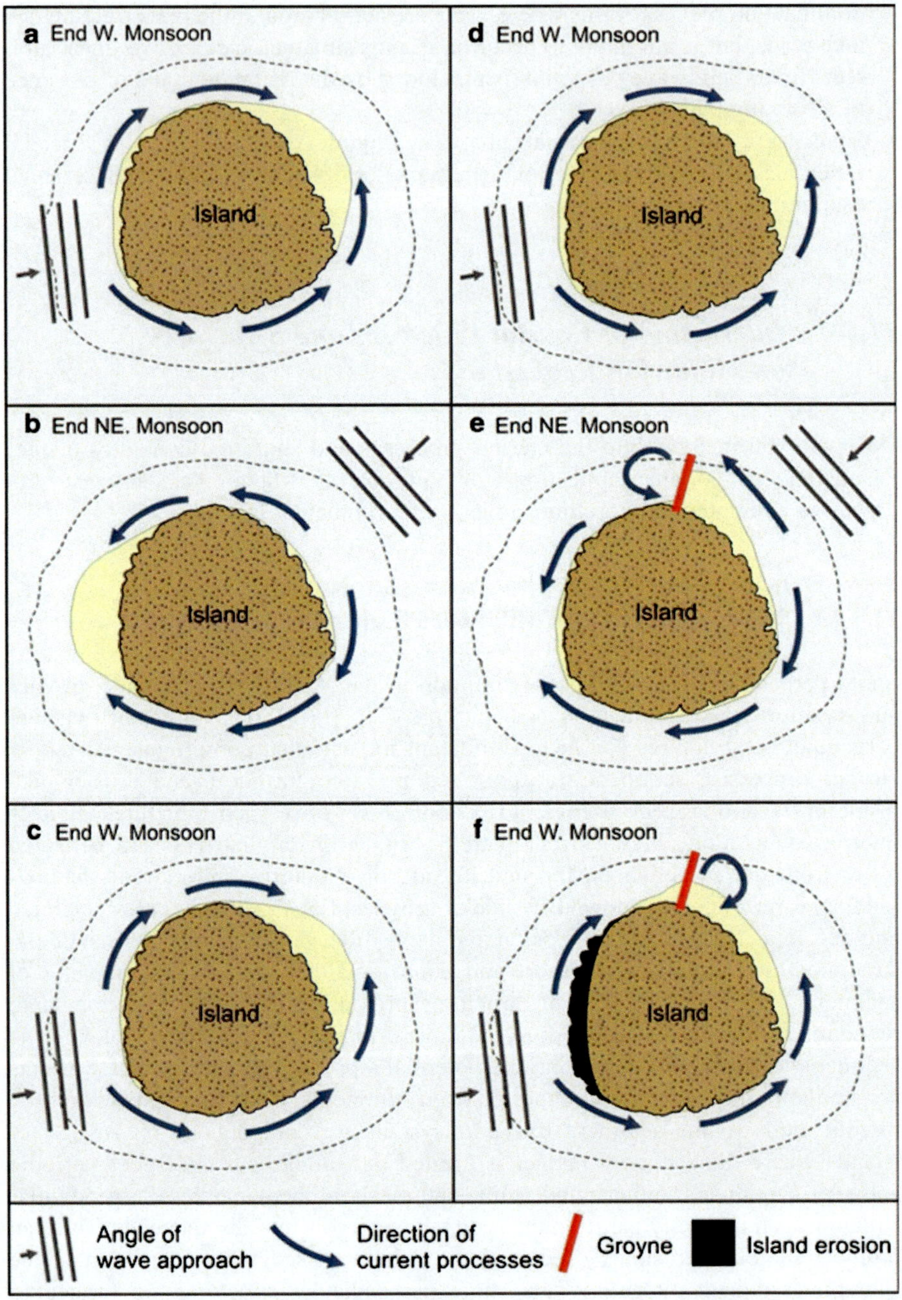

Fig. 11.6 Natural seasonal shoreline dynamics on circular reef platforms (**a–c**). Potential impacts on shoreline dynamics and shoreline erosion following insertion of a shore perpendicular structure (groyne, **d–f**). Note: potential aggregate effect may be to force migration of the island on the reef platform

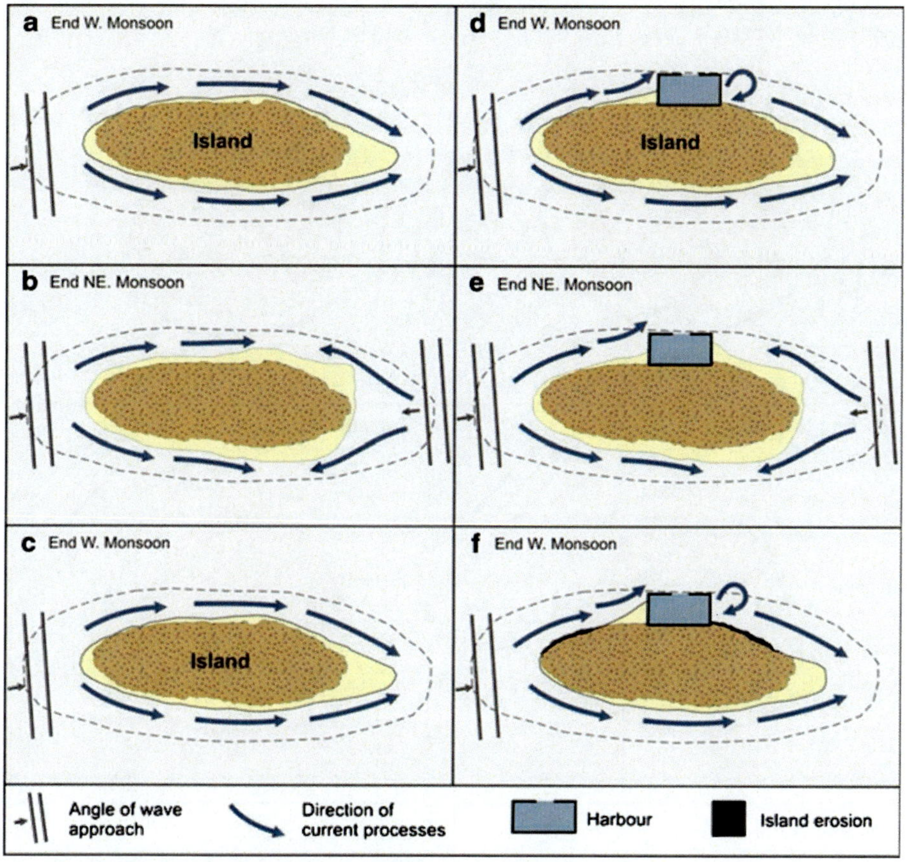

Fig. 11.7 Natural seasonal shoreline dynamics on an elongate reef island (**a–c**). Potential impacts on shoreline dynamics following insertion of boat harbor (**d–f**)

to prolonged wave attack during the remainder of the seasonal cycle. This can exacerbate shoreline erosion on the sector of the island which has a net sediment deficit. This outcome is expected on islands where the shoreline undergoes significant movement between seasons. A possible outcome of these effects over timescales of a decade is to force island migration in a different trajectory to the natural direction of island movement.

11.4.4.2 Interaction of Processes with Shore Parallel Structures (Seawalls)

Seawalls are another ubiquitous erosion management strategy adopted in the Maldives (Fig. 11.8). In the presence of ample sediment supply beaches will form in front of rock structures. Indeed, on many reef island shorelines, beachrock (a naturally occurring rock that forms through cementation of beach sand) parallels

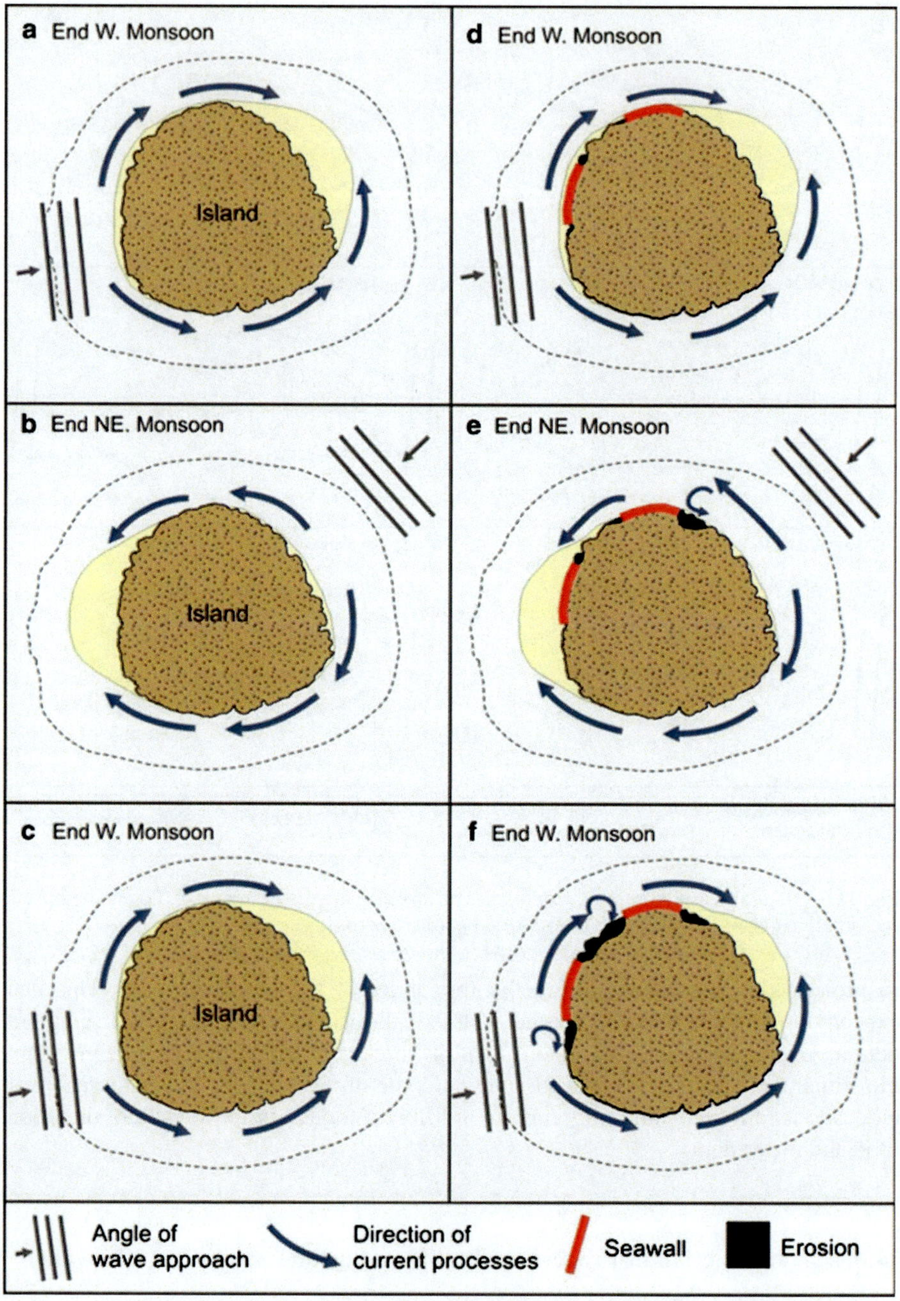

Fig. 11.8 Natural seasonal shoreline dynamics on a circular reef island (**a–c**). Potential impacts on shoreline dynamics following insertion of shore parallel structure (**d–f**)

the shoreline. In such cases this rock can be seasonally covered and uncovered with beach material. Therefore, given a sufficient supply of sediment and assuming a shore parallel structure does not protrude from the shoreline it is possible that such structures have minimal impact on nearshore processes and on sediment transport and deposition. However, a range of impacts of seawalls can be observed in the Maldives. Most of these effects are associated with poor design and construction and include the same range of effects that have been well documented in other coastal settings (e.g. flanking, collapse, overtopping, sediment leakage; Fig. 11.8). However, there are additional considerations related to the unique process characteristics of island shorelines. As documented in examples from South Maalhosmadulu atoll island beaches can migrate substantial distances alongshore (and around island shorelines) between seasons and the extent of beach migration varies between islands of different shape and exposure (Fig. 11.5). However, such dynamism implies that island shorelines and structures can be seasonally exposed to wave action. During periods when structures are exposed turbulence at the terminal ends of structures and strong alongshore current gradients can accelerate shoreline erosion of unprotected coastlines (Fig. 11.8) and can promote a crenulate shoreline configuration.

Placement loss as a consequence of seawall construction can entirely undermine the natural behavior of islands. Islands are dynamic landforms that can migrate on platforms over decadal timescales as a consequence of incremental shifts in the balance of seasonal oscillations in beach behavior. Seawalls lock sediment behind them, which is unable to contribute to the normal range of seasonal beach dynamics. However, each season natural processes require a finite volume of sediment to satisfy the balance of energy and capacity of currents to transport sediment. If sediment is locked behind a seawall other sectors of the island shoreline can erode to compensate for losses elsewhere in the shoreline. As reef islands are small this leads to cannibalization of shorelines and accelerated shoreline erosion. Consequently, this can change the trajectory of island migration on reef platforms and exacerbate local erosion issues.

11.4.4.3 Interaction of Processes with Offshore Breakwaters

Recently the use of offshore breakwaters has been proposed as an alternative to shoreline structures to protect shorelines from wave attack and erosion. The principle of offshore breakwaters is to present a structure that breaks waves offshore and refracts waves across the reef surface (Fig. 11.9), in such a manner that opposing nearshore currents are generated that transport sand to the lee of the breakwater forming a beach. In extreme cases breakwaters built above high tide level can filter all wave energy from a shoreline (e.g. tetrapod walls surrounding Malé).

There have been no studied examples of breakwaters on the reef platform of reef islands that document whether manipulation of processes in this manner does occur. However, observational evidence indicates that where breakwaters have been

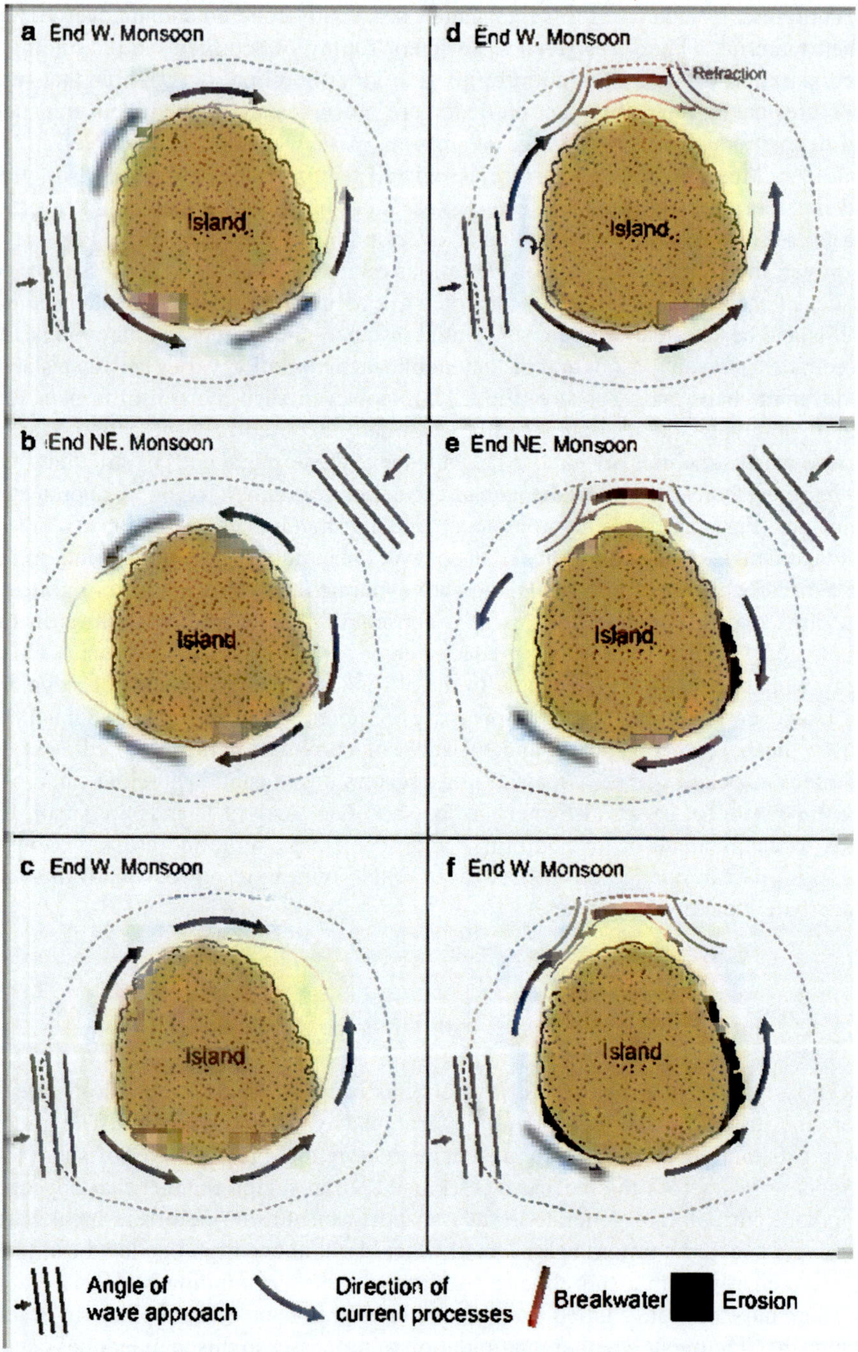

Fig. 11.9 Natural seasonal shoreline dynamics on a circular reef island (**a–c**). Potential impacts on shoreline dynamics following insertion of a shore detached breakwater (**d–f**)

inserted shoreline instability and erosion continue to be a problem. Based on the conceptual models of island shoreline processes there are a number of factors that are likely to contribute to the ineffectiveness of breakwaters and exacerbate coastline instability.

The insertion of a breakwater on one sector of an island shoreline implies that the structure is able to partition the shoreline into discrete process zones that do not interact. Furthermore, this suggests nearshore currents and beach dynamics can be manipulated on one part of the island shoreline in isolation from the remaining shoreline. However, reef island shorelines are integrated current and sediment transport systems. Consequently, modification of processes on one sector will have alongshore impacts on the island shoreline. In particular, if beach accretion is able to be manipulated in the lee of a breakwater this sand is most likely to be drawn from alongshore zones which will deplete the sediment reserves, disrupt natural shoreline dynamics and promote erosion (Fig. 11.9).

Most applications of breakwaters have occurred on linear continental shorelines where cross-shore processes dominate. However, reef islands are dominated by alongshore current processes. To be locally effective in contributing to shoreline deposition the breakwater must overcome the normal nearshore process regime and replace it with a localized counter current (Fig. 11.9). Given the wave environment of reef surfaces (see following paragraph) such an outcome is doubtful and undesirable. However, if such a change is achieved the nearshore process regime of the entire shoreline will be affected promoting island instability.

Wave processes on coral reef flats are characterized by a mixed surfzone environment consisting of a combination of broken swell, reformed waves and short period waves. It is unclear whether the refraction that is necessary to alter shoreline current patterns can be achieved.

The wave and current characteristics of reef platforms can alter between seasons. In extreme cases the nearshore current processes reverse and sectors of the reef platform undergo marked changes in the magnitude of incident wave energy. Consequently, a breakwater can be seasonally ineffective and will only modify processes during those periods when waves propagate onto the reef platform and interact with the breakwater.

11.5 Discussion and Conclusions

There is little doubt that properly designed shoreline protection structures are able to maintain the position of a shoreline in reef settings. Furthermore, where such structures completely encircle islands (e.g. Malé) they can be effective in maintaining the size and relative position of islands on reef platforms surfaces. However, field observations and consideration of the natural coastal processes of islands (outlined above) suggest that such approaches have the potential to completely undermine natural coastal dynamics. In such cases island communities and managers are committing to a future of maintenance of shoreline structures. Furthermore, as illustrated in this study, many structures are not properly designed, which can lead to localized impacts associated with process interactions with structures and breakdown of alongshore processes that further destabilize islands on their reef surfaces. Collectively, these outcomes (artificially fossilizing island shorelines with structures, or enhancing erosion problems) further increase the vulnerability of island communities to natural variability and future global climatic change.

Field observations of island change (Kench and Brander 2006; Kench et al. 2006a) and initial attempts to model reef island morphodynamics (Kench and Cowell 2001) highlight significant differences in the stability of reef island shorelines to continental coastlines. Over short timescales (days to years) islands can exhibit large morphological adjustments in response to extreme events or seasonal to interannual variations in climatic processes (Bayliss-Smith 1988; Kench and Brander 2006; Kench et al. 2006b). These short-term changes are expressed as alongshore variations in storage of sediment around the perimeter of islands. At medium timescales (decades) islands have been found to be highly mobile changing their position on reef platforms in response to changing climate, wave regimes or sea level (Flood 1986; Webb and Kench 2010; Dawson and Smithers 2010). Such changes reflect the aggregate adjustment of islands to variations in short-term processes and result in net movement of the vegetated core of islands. Stoddart (1965) in Belize and Webb and Kench (2010) in the central Pacific present evidence for the migration of islands away from the exposed reef margin under conditions of increasing sea level. Field studies have also confirmed modeling predictions that islands can exhibit a range of styles of morphological change that include whole island migration, washover and vertical island building, as well as island narrowing (erosion) (Kench et al. 2006b, 2008).

Collectively, geomorphic studies indicate that reef islands have an inherent physical resilience. This resilience is dependent on the ability of islands to change their morphology and position on reef surfaces. Consequently, the secret to long-term resilience of islands relies on the maintenance of geomorphic processes so that sediment can be remobilized and island morphology can adjust to a new process regime.

This geomorphic understanding of islands and island dynamics necessitates reconsideration of conventional notions of 'erosion' and island stability with respect to reef island management. Island migration and washover are mechanisms that promote island morphological adjustment, but may maintain island size and sediment volume. Indeed, washover processes can enhance future resilience through vertical land building. Net loss of land (erosion) is only one possible outcome of island morphological change.

The examples of the use and impacts of engineered structures presented in this study highlight how structures directly compromise the natural processes of island dynamics (inhibit washover, impede alongshore sediment exchange) and can force island migration along alternate spatial trajectories than natural processes dictate.

The limited land area of reef islands and their morphological instability pose significant management problems for island communities. How can communities coexist with dynamic islands? The challenge for managers is to recognize that islands are not static landforms. A paradigm shift in management approach is necessary that is based on the intrinsic capacity of the island system to adjust its morphology. Such an approach should comprise a broader suite of options than simply engineered structures and should also involve land use and planning options.

It must be recognized that there is a spectrum of different island types. From a geomorphic perspective islands range from small platform islands to large linear islands on the periphery of atolls. From a human dimension the spectrum ranges from urban centers with high population densities (e.g. Malé) to those that are uninhabited. In heavily urbanized islands the natural processes that afford intrinsic geomorphic resilience have been compromised. In such settings ongoing engineering and structural solutions are likely to be the preferred option to sustain viable socioeconomic functions. However, in many non-urban islands and uninhabited islands the integrity of geomorphic processes has been maintained. In these islands alternate solutions to island maintenance should be sought. These solutions should be founded on a more complete understanding of the range of behaviors and process regimes of islands across the spectrum of island types.

New solutions for the physical management of island shorelines should be underpinned by a number of key considerations. First, management actions should endeavor to protect or enhance the intrinsic morphodynamic resilience of islands. Second, and as a consequence, increased effort is necessary to define/refine the understanding of nearshore processes and natural island dynamics. Third, the use of traditional engineering solutions should be adopted with extreme caution and only be used where they do not compromise natural geomorphic processes.

References

Barnett J, Adger WN (2003) Climate dangers and atoll countries. Clim Chang 61:321–337

Bayliss-Smith TP (1988) The role of hurricanes in the development of reef islands, Ontong Java Atoll, Solomon Islands. Geogr J 154:377–391

Dawson JL, Smithers SG (2010) Shoreline and beach volume change between 1967 and 2007 at Raine Island, Great Barrier Reef, Australia. Glob Planet Change. doi:10.1016/j.gloplacha.2010.01.026

Dickinson WR (2009) Pacific atoll living: how long already and until when? GSA Today 19:4–10

Flood PG (1986) Sensitivity of coral cays to climatic variations, southern Great Barrier Reef, Australia. Coral Reefs 5:13–18

Kench PS (2001) Maldives erosion assessment and management. Report prepared for UNDP as part of the Republic of Maldives climate change enabling activity project MDV/95/G32/A/1 G/99

Kench PS (2010) Coastal monitoring programme: final report. Report to the Environment Protection Agency, Ministry of Housing, Transport and Environment, Republic of Maldives. Maldives Environmental Management Project, 103 p

Kench PS (2011) Maldives. In: Hopley D (ed) Encyclopedia of modern coral reefs: structure form and process. Springer, Dordrecht

Kench PS, Brander RW (2006) Morphological sensitivity of reef islands to seasonal climate oscillations: South Maalhosmadulu atoll, Maldives. J Geophys Res F01001. doi:10.1029/2005JF000323

Kench PS, Cowell PJ (2001) The morphological response of Atoll Islands to sea-level rise. Part 2: application of the modified Shoreline Translation Model (STM). J Coast Res SI 34:645–656

Kench PS, Parnell KE, Brander R (2003) A process-based assessment of engineered structures on reef islands of the Maldives. In: Paper 74 in Kench PS, Hume T (eds) Proceedings of the coasts and ports Australasian conference, Auckland, 10 Sept 2003

Kench PS, Brander RW, Parnell KE, McLean RF (2006a) Wave energy gradients across a Maldivian atoll: implications for island geomorphology. Geomorphology 81:1–17

Kench PS, McLean RF, Brander RF, Nichol SL, Smithers SG, Ford MR, Parnell KE, Aslam M (2006b) Geological effects of tsunami on mid-ocean atoll islands: the Maldives before and after the Sumatran tsunami. Geology 34:177–180

Kench PS, Nichol SL, Smithers SG, McLean RF, Brander RW (2008) Tsunami as agents of geomorphic change in mid ocean reef islands. Geomorphology 95:361–383

Kench PS, Brander RW, Parnell KE, O'Callaghan JM (2009a) Seasonal variations in wave characteristics around a coral reef island, South Maalhosmadulu atoll, Maldives. Mar Geol 262:116–129

Kench PS, Parnell KE, Brander RW (2009b) Monsoonally influenced circulation around coral reef islands: a control on seasonal dynamics of reef island beaches. Mar Geol 266:91–108

Khan TMA, Quadir DA, Murty TS, Kabir A, Aktar F, Sarker MA (2002) Relative sea level changes in Maldives and vulnerability of land due to abnormal coastal inundation. Mar Geodesy 25:133–143

Kraus NC (1988) The effects of seawalls on the beach: an extended literature review. J Coast Res SI 4:1–29

Kraus NC, McDougal WG (1996) The effects of seawalls on the beach: part 1, an updated literature review. J Coast Res 12:691–701

Leatherman SP (1997) Sea-level rise and small island states: an overview. J Coast Res SI 24:1–16

Maragos JE (1993) Impact of coastal construction on coral reefs in the U.S.-affiliated Pacific Islands. Coast Manag 21:235–269

Morton RA (1998) Interactions of storms, seawalls, and beaches of the Texas coast. J Coast Res SI 4:115–134

Sawaragi T (1967) Scouring due to wave action at the toe of permeable coastal structure. In: Proceedings of 10th coastal engineering conference, American Society of Civil Engineers, Tokyo, pp 1036–1047

Shaig A (2011) Survey of climate change adaptation measures in Maldives: integration of climate change risks into resilient Island planning in the Maldives project. Ministry of Housing and Environment, Government of Maldives

Stoddart DR (1965) British Honduras cays and the low wooded island problem. Inst Br Geogr Trans 36:131–147

Webb AP, Kench PS (2010) The dynamic response of reef islands to sea-level rise: evidence from multi-decadal analysis of island change in the Central Pacific. Glob Planet Change 72:234–246

Weggel JR (1988) Seawalls: the need for research, dimensional considerations and a suggested classification. J Coast Res SI 4:29–39

Chapter 12
"Alternative" Shoreline Erosion Control Devices: A Review

Orrin H. Pilkey and J. Andrew G. Cooper

Abstract A variety of patented approaches have been devised in efforts to halt shoreline erosion. Commonly termed 'alternative' or 'innovative' technologies, these are typically variations on the traditional approaches. A categorization of these approaches is presented that identifies devices placed in the water and devices placed on the beach. These categories are further subdivided. Despite their innovative nature and the claims of their inventors and promoters, these devices suffer from a variety of weaknesses when deployed in the real world. We present a non-exhaustive list of 110 devices for which US patents were awarded since 1970.

The view of success of 'alternative' devices often differs between reports made by the developer and those of the end-user and only in a few cases have objective assessments been made. Using a variety of sources we review experiences with artificial surfing reefs and beach drainage systems. We conclude that 'alternative' devices offer the same range of shortcomings as traditional shoreline stabilization approaches because of the inherent inability to control such a dynamic sedimentary environment and the failure to address the underlying causes of shoreline recession (sea level rise, sediment supply, other engineering structures, and the presence of infrastructure in the active coastal zone).

O.H. Pilkey (✉)
Division of Earth and Ocean Sciences, Nicholas School of the Environment, Duke University, Durham, NC 27708, USA
e-mail: opilkey@duke.edu

J.A.G. Cooper
School of Environmental Sciences, University of Ulster, Coleraine BT52 1SA, Northern Ireland, UK
e-mail: jag.cooper@ulster.ac.uk

J.A.G. Cooper and O.H. Pilkey (eds.), *Pitfalls of Shoreline Stabilization: Selected Case Studies*, Coastal Research Library 3, DOI 10.1007/978-94-007-4123-2_12,

12.1 Introduction

Coastal managers faced with shoreline retreat (coastal erosion) problems, typically seek a mitigation remedy. At the broadest level, the decision usually involves a choice between soft stabilization (beach nourishment), hard stabilization (seawalls or other structures), and retreat from the shoreline. This initial decision is made difficult because holding the shoreline in place, especially by hard stabilization, interferes with the coastal sedimentary system and creates additional problems including loss of beaches. Soft stabilization or beach nourishment avoids this problem in the first instance or is used to remedy the problems caused by hard stabilization, but it is costly and temporary. In addition, in a time of rising sea level it will become more expensive as more sand and more frequent nourishments are required to hold the beach in place. Purely from the standpoint of maintenance of the coastal ecosystem and preservation of beaches for future generations, the best approach is moving development back from the retreating shoreline. But there are other considerations besides the beach. What will be the fate of the buildings lined up along a beach, especially high-rise buildings which are, in all practicality, immovable? The retreat option is usually beyond the scope of the coastal manager's authority and this restricts the response options to some kind of engineered approach.

Preservation of buildings and infrastructure has been deemed the highest priority in many, if not most, of the world's coastal communities and, as a consequence, hard stabilization is a common erosion response. The hard engineering of shorelines, usually in relation to construction of harbors, has been carried out for more than 3,000 years. Over centuries of engineering, the principles of seawall, groin, and offshore breakwater construction in surf zone conditions have been developed. Today, as an unprecedented rush to the shore is occurring and at a time when the sea level is rising and expected to do so at ever-increasing rates, engineering of shorelines is becoming an ever more widespread societal endeavor.

In this paper, we focus on a category of engineering structures known as *alternative (non-traditional)* or *innovative erosion control technologies*. Those who harbor skepticism about such structures sometimes refer to them as *snake-oil devices*,[1] an uncomplimentary term derived, in large part, from the often exaggerated claims commonly made by manufacturers. Non-traditional ideas abound concerning how to halt shoreline retreat and hold shorelines in place. Most of these ideas have led to patented inventions, particularly in the last three decades. Many such devices have not been tried or have been installed along only a few shorelines, and thus exist mainly on paper. Coastal managers and coastal politicians are faced with a choice from an array of coastal engineering devices being promoted by their inventors or patent holders. The claims of the promoters regarding the positive

[1] Snake oil was once advocated as a solution to a variety of medical ailments by unscrupulous salesmen in the American west.

benefits of their devices and the lack of objective assessment of their performance data make decision-making difficult for coastal managers.

Perusal of the technical and commercial literature about these structures usually reveals little monitoring of them post-emplacement, except for that done by the engineering firm that installed them. The State of Florida, in recognition of this problem, has a limited device-monitoring-and-evaluation program (Woodruff 2006), and an evaluation of 'alternative' stabilization methods was carried out in Puget Sound (Gerstel and Brown 2006) but, for the most part, coastal communities are on their own when attempting to judge the claims of manufacturers or installers regarding the efficacy of their devices.

In addition to protecting resources at risk, community-level coastal managers must usually assess the potential impact of any erosion response on environmental and economic issues, such as turtle and beach bird nesting, the sand supply of adjacent beaches, beach quality, and the tourist industry, but also must consider the role of continued sea level rise in the future.

The devices listed and briefly described here represent a cross section of the available alternative technology for shoreline stabilization. Of the hundreds of alternative devices, most follow the standard basic principles of sea walls, groins, and breakwaters that attempt to trap sand or in some way reduce wave energy. However, they differ from "standard" coastal engineering structures either in configuration and/or the type of materials used.

12.2 US Patents – An Overview of 'Alternative' Devices

A non-exhaustive analysis of US patent records reveals at least 110 patents, since 1970, that fall into the category of beach erosion mitigation. The oldest such patent we found (for a 'sand and water break') dates from 1881. The list below includes 6 from the 1970s, 30 from the 1980s, 46 from the 1990s, and 27 from the 2000s. The list is certainly not complete, because such devices may be classified in different ways, but it provides an impression of the types of devices being contemplated.

The 110 US patent applications are a sometimes wild collection of ideas ranging from the ingenious to the ridiculous. The former include instant and temporary seawalls to be put up within a few hours as a storm approaches and the latter include a sand-trapping device so efficient that it is said to form a small protective barrier island just offshore. Classification of these patents is difficult because many do not fall conveniently into recognizable categories. However, roughly 40% can be considered breakwaters in that they are emplaced parallel to the shoreline but are at least partially submerged most of the time. Broadly-defined seawalls of various types make up another 25% of the total, and groins an additional 10%. The remainder includes various kinds of mats, drainage systems, artificial seaweed, and a variety of nearshore current manipulators with pumps.

The various patented devices include ones described as permeable, adjustable, and even biodegradable. A few are designed to be emplaced immediately before storms and removed after storm passage. Many include an element of sand trapping that is envisaged as being accomplished by reducing the size or the velocity of the backwash in the surf zone. One patented device is supposed to reduce backwash by removing water from the swash zone, and another proposes to reduce longshore transport by pumping water against the current. Other devices include:

- Prefabricated seawalls deemed to be easy to construct and remove
- Seawalls that protect other seawalls
- Erosion mitigation devices that incorporate wave or tidal power generators
- "Fluid dynamic repellers" that create turbulence that dissipates wave energy
- Fish net groins
- Permeable, removable, adjustable, and biodegradable groins
- Structures that imitate kelp and coral reefs
- Reefs, breakwaters, seawalls, mats, and groins made of used tires
- Various breakwaters that break, block or realign waves
- Artificial surfing reefs (breakwaters)
- Sand-wetting devices that reduce wind sand loss
- Various beach dewatering systems, and
- Repeated removal of sand from shallow shore-parallel troughs expected to be refilled naturally in future tidal cycles

Few patent descriptions indicate any concern for or recognition of the impact on the quality of beaches or other coastal issues. Attention seems rather to focus on shoreline stabilization and preservation of oceanfront buildings and infrastructure. The ecosystem function of a beach was not mentioned in any patent with the occasional exception of turtle nesting but in all cases (when it was mentioned) the impact was declared to be negligible.

The likelihood of survival of most of these structures in major storms under open-ocean conditions is very low. Most design descriptions did not recognize the wide variety of wave conditions to which a particular structure might be subjected, and limitations related to where the device was to be installed (for example, fetch-limited bay shorelines versus open-ocean shorelines) were rarely stated. Putting it another way, "one solution fits all" is a common underlying assumption. Understandably, given the motivations behind filing a patent, optimism prevails in the descriptions of these devices.

12.3 A Classification of 'Alternative' Devices

While the 'alternative' devices portfolio contains an eclectic mix of approaches, most are derivations of or modifications of the traditional shoreline engineering approaches. The structures are categorized by placement location (in the water or on the beach) and functional similarity to well-recognized engineering structures such as seawalls, breakwaters and groins. There is some overlap, as some of the alternative devices do not necessarily conform to the criteria for any particular

Fig. 12.1 Beach prisms on the Maryland shoreline of Chesapeake Bay

category. They are subdivided into devices placed in the water (breakwaters and artificial seaweed) and devices placed on the beach (groins, seawalls, dewatering systems, dune stabilizing systems and other devices).

An internet survey of currently or recently available devices, mostly US in origin, was undertaken. The survey was not intended to be exhaustive but includes examples from nearly every available type of alternative structure. The essential characteristics and problems associated with each category of devices are described in the following tables. The left-hand column shows the commercial name (if one exists), a very brief description of the device, and the name of the company that installs it, if available. The middle column lists some locations where the device has been installed, and the right-hand column notes some of the claims made by the manufacturer.

12.3.1 Devices Placed in the Water

12.3.1.1 Breakwaters

The devices described in this category are shore-parallel structures placed offshore, either submerged or floating. Their intended purpose is to modify the incoming wave so as to create a "wave shadow" on the beach causing sand deposition (Fig. 12.1).

Problems commonly associated with breakwaters are that they may cause scour in the vicinity of the device and may increase downdrift erosion by removing material from the littoral current. They can impact water quality because of reduced water circulation and they can be a hazard to swimmers or boaters.

Table 12.1 lists 22 variations on the offshore breakwater theme. They vary in design, material and in the claimed secondary effects (e.g. surfing, benthic habitat).

12.3.1.2 Artificial Seaweed

These features involve some type of "imitation kelp" anchored to the seafloor. They are intended to slow waves and reduce energy, causing sand carried by the waves to be deposited. They are also intended to slow return wave energy, so that sand carried off the shore by return flow is deposited nearshore (Table 12.2).

The devices are placed in shallow water and may be hazardous to swimmers and boaters. Many are not suitably anchored and do not withstand storms. When they are dislodged they create debris on the beach.

Table 12.2 lists four variations on the artificial seaweed theme showing a wide divergence in materials.

12.3.2 Devices Placed on the Beach/Dunes

12.3.2.1 Groins

Groins are constructed perpendicular or at a high angle to the shoreline. They commonly are located on the inter- or supra-tidal beach but occasionally extend into the subtidal zone. Their primary purpose is to trap sediment that is moving alongshore in the littoral current. Groins cause erosion of downdrift beaches, they often create rip currents that are hazardous to swimmers and cause loss of sediment offshore, and they act as a barrier to activities on beaches, for example walking.

Table 12.3 provides details of seven variations on the groin theme. These vary in material and design.

12.3.2.2 Seawalls

Seawalls are walls placed at the base of a bluff, at the edge of shoreline property or at the landward edge of a beach. They are designed to protect land from the impact of waves.

Seawalls cause both active and passive erosion of the front beach. By preventing erosion they cut off the local sediment supply while waves that hit the wall are reflected downward, scouring the toe of the wall. In the passive mode, seawalls provide a barrier to the landward movement of beaches preventing them from adopting a storm profile and inhibiting a natural response to sea level rise.

The alternative devices surveyed include six variations on the seawall theme (Table 12.4).

Table 12.1 Name, description, installation and manufacturer's claims for 22 devices in the offshore breakwaters category

Device	Installation examples	Manufacturer's claims
Artificial surfing reefs – sand bag offshore breakwaters	Gold Coast, Australia Mount Maunganui, New Zealand	Reduces waves onshore acting as a breakwater Realigns wave crests and/or spreads wave energy to reduce wave driven currents Enhances surfing, providing several types of waves for surfers with different skill levels
Atlas shoreline protection system – Stacked timber, laid horizontally, held together by steel supports. Arranged in a sawtooth pattern on the nearshore, parallel to beach		Prohibits erosion & allows for accretion inward & outward of system Long life, low maintenance
Beach cones – Concrete donut 6″ high, 2′ across, 40″ across the bottom weighing 92 lb each	1992 – Shell Island in Lower Plaquemines, LA	Provide hard bottom stabilization for sand accretion No loss during Hurricane Andrew of an installation that included 300 cones and 13–72 cu. yds. of sand Average accretion 6′, max gain 3′
Beach prisms – Concrete blocks, with a triangular cross-section. Each unit is 6′ high, 12′ long and 84″ wide with a concave, openwork front face	1988 – Chesapeake Bay	Openwork face allows more water to flow through, which reduces scour
Beach protector tire mat – Tires anchored to each other & to the seafloor in a section 30–60′ wide & at least 1 mile long. Can be shorter if between 2 promontories & close to end of one of the promontories		Slows the return of "sand laden" waves
Beachsaver reef – Interlocking, concrete units, triangular in cross-section. Each is 10′ long, 6′ high, 16′ wide. The front face is ridged to reduce wave reflection, with a slotted opening at the top (Breakwaters International, Inc.)	1993 – Avalon, NJ 1994 – Cape May Pt. & Belmar/ Spring Lake, NJ	Water flows through slotted openings at top, sand is suspended & carried forward by incoming waves Stabilizes beach nourishment, requiring less sand for renourishment Attracts wildlife
Burns beach erosion device – Concrete block (5′ × 2′ × 8″) with rubber tire strips (1″–2″ wide) attached to the top of the block. Acts as artificial seaweed		Dissipates wave energy reducing offshore transport of sand Allows for greater accretion of sand during storm conditions May provide protection for turtles & substrate for crustaceans

(continued)

Table 12.1 (continued)

Device	Installation examples	Manufacturer's claims
Flow and erosion control system – breakwater – zigzag louvered 6–10-foot long, 4-foot high panels (Sandi Technologies)		Traps sand by slowing backwash Widens beach – four times cheaper than beach nourishment Reduces rip tides and storm impacts
Menger submerged reef – Triangular in cross-section; welded iron frame covered with steel screen mesh & concrete. Submerged offshore by filling with sand		Prevents sand from washing seaward by slowing wave energy Units can withstand severe weather changes, because the materials expand and contract Re-usable; not permanent Environmentally friendly because it can be moved with ease
MOTO – Primary function is to harness wave energy but also acts as a breakwater to reduce coastal erosion. Installation – 3 toroids, 10′ in diameter, weighing 4 tons each, placed at least 20′ deep		Waves lose power by creating energy, thus reducing erosion Provides clean energy and fresh water
Pep (Prefabricated Erosion Prevention) reef – Concrete units, triangular in cross-section, 6′ high, weighing 20 tons. Placed 2–4′ below surface at low tide (Designed by American Coastal Engineering, West Palm Beach, FL)	1988 – Palm Beach, FL (Privately funded) 1992–1993 – Palm Beach, FL (removed in 1995 because of increased erosion) 1996 – Vero Beach, FL	Builds trough and bar areas beyond the foreshore which shifts the foreshore outward Stabilizes the shoreline Reduces wave energy 40–70% Can be relocated easily Shelter and habitat for animals
Reef balls – Concrete balls with holes; mimic natural coral heads; sometimes integrated onto articulating concrete mats to create a breakwater in nearshore waters; range from 1.5′ × 1′ (30–45 lb) to 6′6″ × 3″ (4,000–6,000 lb) (The Reef Beach Co., Ltd.)	1996 – Turks & Caicos 1998 – Dominican Republic 2003 – Dade County, FL	Reduces wave energy that reaches the shore in area of chronic erosion Protects beach from erosion or builds up eroded beach Serves as an artificial reef structure providing hard bottom substrate for attachment of corals Provides shelter and habitat for fish
Reef mitigation gardens – Method of encouraging biological colonization of nearshore hardgrounds (Surfbreak Engineering Sciences, Inc.)		Especially applicable off East Florida where abundant hardgrounds occur

(continued)

Table 12.1 (continued)

Device	Installation examples	Manufacturer's claims
Seabox – 7.5 ton concrete box can be deployed as floating breakwater, reef breakwater, semi-submerged (on lower beach), as seawalls on upper beach or buried within frontal dunes (Seament Shoreline Systems)	Colonial Beach, VA Carolina Yacht Club, Wrightsville Beach, NC	Modular long-lasting concrete box Easily transported and installed Can be moved readily by truck, train, barge Fills with sand in most situations
Sealift – Shoreline breakwater, triangular in cross-section, placed beyond foreshore where it is shallow at low tide. Angled, so as to slow wave energy	1990 – proposed for Palm Beach	Pollution free installation Waves lose much of their destructive power Reduces long-term erosion Compresses the configuration of wave cells
Shore guard – high energy structure – High grade carbon steel coated with environmentally friendly coating; install 7′ or deeper (Seabull Marine, Inc. Shoreline Erosion Reversal Systems)		Patented zigzag design supports Mother Nature in supplying nourishment to build up shoreline Reinforces the natural balance of Mother Nature
Shoreprotector – Submerged sand fence, placed 400′ offshore. Made of openwork steel frame with 4 baffles on each side; 7′ tall, 16′ wide at base, weighing 650 lb	1975 – Virginia Beach, Va. Removed due to failure. Installation cost: $108,000 Removal cost: approx. $67,700	Flexible design Removable
Surge breaker – Permanent, steel reinforced "prisms", 4′ high, 4′ wide and 6′ deep; placed in 3–8′ deep water. Can be joined by steel cables for higher energy environments. Recommend installing 2 systems parallel to each other	*1976 – Highlands Park, IL 1979 – Bayou State Park, FL 1984 – Kuala Regional Park, Oahu, HI	Simulates offshore sandbars and reefs Stimulates accretion, reduces erosion Will work on any beach Will withstand extreme weather *In 8 months beach accreted 50′ in width in IL
Temple beach system – Reinforced concrete, triangular in cross-section, placed at mean low-tide/ 12-18″ below high tide, parallel to the shore. Metal rods are used to anchor		Does not interfere with boaters, bathers or turtles Mitigates against storm damage Moves high & low tide line an average of 200′ outward Protects beach nourishment
Waveblock – Modularized, permeable, steel reinforced concrete. Structure is an angled tower		Absorbs wave energy before it reaches the shoreline

(continued)

Table 12.1 (continued)

Device	Installation examples	Manufacturer's claims
Waveshield – Floating system made of steel; each unit is 80′ long, 20′ wide & 18′ high, weighing 40 tons. Unit of 3 compartments. Best in 25–30′ deep water		Provides protection against wave damage and erosion Breaks 8–10′ roller waves Economical, simple & easy to make Can be floated to any location, thereby avoiding high accretion on the landward side
Wave wedge – Concrete, interlocking units; triangular in cross-section, weighing 5,000 lb. Three slots/ holes on the front face	1985 – Michiana, MI	Slots on front absorb energy Builds up foreshore & sandy beach Restores sand lost during storms

Table 12.2 Name, description, installation and manufacturer's claims for four devices in the artificial seaweed category

Device	Installation examples	Manufacturer's claims
Cegrass – Synthetic seaweed made of foamed polypropylene, attached to open grid mat, held to seafloor by ballasts. The length of the mat is tailored to the environment	1985 – Germany, to fix scour caused by pipeline – Italy – Wetlands in Europe	Reduces nearshore current velocities, thereby sand is dropped in sandbars which build up to 1.6 m high Reduces offshore sand movement and scour
Coil system – 9-gauge wire, 24–30″ in diameter, intertwined with smaller wire, attached to the ocean floor. Installed between inlets, 500′ to 2000′ from shore, in grid system 100′ between units, which are placed at an angle to the shore		Sand is captured within the coil grid and returned to the shore by tides & wind Coils interrupt ocean currents, allowing for sand to be trapped while currents pass through If properly emplaced, there will be no sand loss to adjacent beaches
Seabee – A series of six-sided concrete blocks, weighing 35 lb to 1 ton, with holes (honeycomb design), placed on slope in the nearshore. 20% of construction material is recycled ash	1989, Tidewater Community College – Portsmouth, VA. Monitored by VIMS	20″ of sand and silt collected between 1989 and 1996 on Tidewater Test Site Reduces energy of wave run-up, causing sand to be deposited
Seascape – Synthetic seaweed. Plastic filaments attached to a bag which is filled with sand to anchor the device	1981 – Cape Hatteras, NC 1983/1984 – Barbados	Controls shoreline erosion Fronds reduce current flow, sand is dropped

Table 12.3 Name, description, installation and manufacturer's claims for seven devices in the groins category

Device	Installation examples	Manufacturer's claims
Brush fence – Christmas trees or discarded lumber laid out in a "crib" fashion; 4′ wide, 72′ long	Jefferson Park, LA	Wave stilling device; decreases wave energies, capturing suspended sediment; protects the shoreline
Holmberg undercurrent stabilizer – A series of concrete-filled tubes buried at angles to the shore. "Interlocked network of geotextile forms injected with concrete." Site-specific design with longshore & offshore components laid perpendicular & parallel to the waterline. Accretion template which builds the submerged nearshore profile	1982-Manasota Key, FL 1983 – Michigan near Buffalo; Captiva, FL and Ogden Dunes, IN 2000 – Najmah Beach, Ras Tanura, Saudi Arabia USACE has permitted approx. 100 installations to date	Energy dissipator; slows currents so that inlets & jetties don't divert sand Nearshore sand stays nearshore Sand coming from offshore no longer transported downshore by littoral currents, therefore beaches accrete. This induces nearshore shoaling
NuShore beach reclamation system – Porous net groin system (Benedict Engineering Co., Inc.)	Okaloosa County, FL	Accumulate sand on the dry portion of the beach Intercept cross-shore transport without significantly restricting long-shore transport Removable netting material
Parker sand web system – Series of fish nets (50–100′ apart) perpendicular to the shore, strung from the high tide line, into the water. Nets are made of heavy nylon material. Work similar to a groin, trap suspended sand (Parker Beach Restoration, Inc.)	1987 – Pelican Bay Beach, FL. Had to be removed after 20 days because the installation did not have a permit 2001 – Naples, FL (not successful)	Nets cause a build-up of sand Promoting onshore movement of sand
Sealogs – (Sediment Shoreline Systems) – Attached approximately 3 foot concrete "logs" forming mats on beach surface	1994 – Colonial Beach, VA	Scour protection at seawall base Boat launch ramp on beach
Shoreline construction corp. groin – Low profile sill and groin system. Sill placed at an angle to the shore; acts as an artificial bar. Groin, perpendicular to the shore on either end of the sill & in the middle. The groin directs the flow of the sediment & water & reduces currents		System is at or below the water level, so waves can still overtop which eliminates scouring, flanking and reflection Eventually the whole system is covered by sand
Stabilito – Plastic groin/artificial ripple, 5 m long, 1.8 m wide, 60 cm high; placed perpendicular to shore on a submerged beach or dune		Slows & "elevates" currents, thereby creating sand ripples Stabilizes coastlines, riverside erosion and dunes

Table 12.4 Name, description, installation and manufacturer's claims for six devices in the seawall category

Device	Installation examples	Manufacturer's claims
High energy return wall – Concave seawall that causes wind and water to work against each other, thereby flattening the sea surface. Individual sections are 33′ by 44′ at base. Wall is 30′. Perforations in "splash pad" allow for water to pass through & sand to be deposited on back side of wall		Reduces toe scour common with traditional seawalls Causes beach accretion
L wall bulkhead and T groins – Primarily used as a seawall but can also be used as a groin; 4-ton concrete L-shaped units (Shoreline Systems, Inc.)	Several locations in Chesapeake Bay, VA	Modular Easy installation and removal Long-lasting Can be used in conjunction with sealogs and seaboxes
Marine bin walls – Steel bin filled with "granular material" to withstand freezing & thawing. Placed at shoreline or base of bluff	Protects homes	Best suited for marine construction
Ravens retaining wall – Aluminum, corrugated retaining wall placed at the water's edge at the base of a bluff	Unspecified	Protects property from slippage & erosion by tides
Wave buster – Seawall with angled top to reduce wave reflection. Associated drainfield above & behind bluff to reduce hydrostatic pressure. Base secured with geotextile bed	Great Lakes-unspecified	Deflects water up & back without reflecting the waves-reduces toe scour
Z-wall – Low-lying concrete wall placed in a saw-tooth pattern at the base of a bluff or, ideally, offshore, submerged halfway	1973, Buttersville Park, Luddington, MI	Reduces erosion & encourages the build-up of sand in front of the wall Redirects wave energy so that sand is dropped

12.3.2.3 Dewatering

These installations extract water from the beach allowing for more percolation of water from incoming waves and reducing backwash. As the groundwater is pumped out, it is funneled to the ocean or collected as a resource.

Like all structures on beaches, such devices can be damaged or destroyed during storms. In Nantucket, the system broke down during every major storm. The pipes pose a hazard for swimmers and other beach users. On turtle nesting beaches, they must be turned off during the nesting season because groundwater extraction affects the temperature of the sand.

Three beach dewatering devices are tabulated in Table 12.5.

Table 12.5 Name, description, installation and manufacturer’s claims for three devices in the beach dewatering category

Device	Installation examples	Manufacturer’s claims
HDSI – Buried wells extract groundwater, thereby leaving an unsaturated zone. Waves run up & the water percolates below ground, depositing sand		Easier to install & more cost effective than traditional dewatering devices Not susceptible to storm damage Environmentally friendly, even to turtles Can be operated at variable rates
Pressure Equalizing Modules (PEM) – Beach dewatering with vertical pipes that have slits cut into the walls of them (ECO Shore International)	Several installations in Jutland, Denmark Old Skagen Lonstrup Skodbjerge	Turtle friendly Can be used in conjunction with groins and offshore breakwaters Invisible installation Easily installed and removed using light equipment on the beach
Stabeach – System includes a pump placed on the high tide beach with drain pipes attached. The pipes run underground & discharge into the ocean	1988–Sailfish Pt., FL 1994–Englewood, FL 1996–Nantucket, MA	Builds beaches while reducing erosion – less water washes back to the ocean in return flow, so less sand is carried with it Installation causes relatively little disturbance to the beach

12.3.2.4 Bluff/Dune Stabilization

These are low-lying barriers placed on the beach to prevent erosion of the back-beach topography, whether a dune or bluff. They also include structures that aim to trap wind-blown sand to build artificial dunes. Revetments protect only the land behind the structure and have little influence on the beach which may continue to erode. They may, however, act like seawalls in cutting off sediment supply from inland. A range of such interventions is listed in Table 12.6.

12.3.3 Miscellaneous Devices

A number of devices that are not easily categorized are described in Table 12.7.

Table 12.6 Name, description, installation and manufacturer's claims for six devices in the dune and bluff stabilization category

Device	Installation examples	Manufacturer's claims
Biodune sand gel – Spray gel-mixture of 97% beach sand & water with non-toxic biodegradable aqueous polymer gel	St. Augustine, FL Melbourne Beach, FL Ft. Fisher, NC	Stabilizes dunes Doesn't deter marine turtles Withstood 3 years of storms (dunes lost elevations, but were not undercut) Damage can still be caused by walkover Does not impede growth of vegetation
Dune guard – Similar to sand fencing but made of polymer grid attached to poles	Avalon, NJ	Captures wind-blown sand Especially suited for storms Lasts longer than ordinary sand fencing, partially because it resists weathering Can resist 9-ton force
Fabric fence – Sand fence made from yarn impregnated & coated with foam vinyl plastic, attached to poles & placed at the high tide mark or base of the dune line. Rolls are 150′ long, 46″ high		Highly visible Easy to install Stable & weather resistant
Nicolon geotubes – textile tube made from woven polyester; 30′ in circumference & variable lengths. Bags are filled with sand and placed in a trench at the toe of a dune	1995-Atlantic City, NJ	Stabilizes dunes & prevents landward erosion Can also be used as a groin
Soukup rubber tire revetment – Tires placed in a 16–18″ deep, 15′ wide trench, lined with filter cloth on the low-tide dry beach. Tires are covered with the sand that is dug out		Tires act as a more stable sandbag Stabilize the shoreline behind the revetment
Subsurface dune restoration – A dune is created by burying sandbags on a re-contoured slope. Vegetation is then established to protect the dune	Caledon Shores 1997-Long Island, NY	Dissipates storm wave energy which reduces erosion Designed for a 25 year storm Also allows for percolation of waves which builds up sand on the surface
Subsurface dune stabilization (Advanced Coastal Technologies, LLC) Three sloped, wedge-shaped, sand-filled geotextile tubes underlying frontal dune system, covered by 3–5 ft of sand	1986 – Satellite Beach, FL 1988 – Long Boat Key, FL	Soft solution Turtle friendly Wedge shape provides gradual wave force dissipation Rapid deployment capability Subject to puncture and tearing by waveborne debris
Triton marine mattress – Stone filled mattresses used for bluff or dune stabilization	Trinidad Boston Harbor, MA	Stabilization of bluffs & dunes Protection from scour

Table 12.7 Name, description, installation and manufacturer's claims for five diverse types of devices in the 'miscellaneous' category

Device	Installation examples	Manufacturer's claims
Beachbuilder technique – Elastomer coated industrial fabric, 25′ wide, anchored from the high beach to the tide line. Uses the energy of waves to build the beach (maximum winter buildup) by preventing the removal of sand during wave retreat		Restricts the return flow of sand carried by a retreating wave "Accretion concentration of 60 cu yd/ft in less than 4 days"
Beachbuilder (Project Renaissance, public domain) Perforated pipes through which compressed air is pumped		Produce curtains of air bubbles to slow water and cause sediment deposition Principle same as snow fence
Biorock – Use of electrical current in the water to precipitate calcium carbonate (Biorock, Inc.)	1996 – Maldives 2002 – Bali	Low voltage electrolysis of seawater to grow limestone structures and accelerate growth of coral reefs, oysters, seagrasses and salt marshes Could act as breakwater like a natural coral reef
Stabler disks – Concrete disks, 4′ in diameter, attached to pilings & placed at the storm high tide line (Sold by Erosion Control Corp., Livingston, NJ)	1993-Spring Lake, NJ 1996-Myrtle Beach, SC	Protects beaches & dunes by reducing storm wave energy Waves are slowed, sand is dropped & disks are covered
WhisprWave – Floating plastic breakwater (Wave Dispersion Technologies, Inc.)	Iraq Dubai California	Reduces wave energy on the shoreline Used mostly for security

12.4 Impacts of Alternative Devices

Despite the lack of acknowledgement of the negative impacts of these alternative devices, it is obvious that any artificial device placed on or near a beach will interfere with the natural dynamic equilibrium that controls beach behavior. Different devices interfere or impact to different degrees and in different ways. The alternative erosion control devices and some of their potential impacts on the beach environment are categorized in Table 12.8.

Of particular concern is the device's performance during storms. During storm conditions, even the most robust infrastructure is threatened with damage and the debris from these coastal devices is often left strewn along the coast. This debris creates dangerous conditions for everyone, from beachcombers to swimmers.

The pollution problem related to these structures is dependent upon original water quality. If a device creates standing or very slow moving water and the local

Table 12.8 A tabulation of the likely impacts of the 'alternative' devices surveyed

Device	Harms beach access	Erosion of downdrift beaches	Erosion of fronting beaches	Potential hazard to water-based recreation	Impact on water quality	Impact on turtle nesting	Impact on beach fauna and flora	Impairs aesthetics
In the water								
Artificial Surfing Reef		X		X				X
Atlas Shoreline Protection System	X	X		X	X	X	X	X
Beach Cones		X		X	X	X		
Beach Prisms		X		X	X	X		X
Beach Protector Tire Mat	X	X		X	X	X	X	X
Beachsaver Reef		X		X	X	X	X	X
Biorock		X			X		X	
Burns Beach Erosion Device		X		X	X	X		
Cegrass		X		X	X	X		
Coil System		X		X	X			
Flow & Erosion Control System (FECS)		X		X		X	X	X
Menger Submerged Reef		X		X	X	X	X	
MOTO		X						
PEP Reef		X		X	X	X	X	X

Device	Harms beach access	Erosion of downdrift beaches	Erosion of fronting beaches	Potential hazard to water-based recreation	Impact on water quality	Impact on turtle nesting	Impact on fauna and flora	Impairs aesthetics
On the beach								
Beachbuilder Technique	X			X		X	X	X
Biodune Sand Gel			X				X	
Brush Fence		X		X	X			X

(continued)

Table 12.8 (continued)

Device	Harms beach access	Erosion of downdrift beaches	Erosion of fronting beaches	Potential hazard to water-based recreation	Impact on water quality	Impact on turtle nesting	Impact on fauna and flora	Impairs aesthetics
Dune Guard			X					X
Fabric Fence			X					X
Geo-Textile Low Profile Stabilization System	X	X		X		X	X	X
HDSI	X					X	X	X
High Energy Return Wall		X	X	X				X
Holmberg Undercurrent Stabilizer		X	X	X	X	X	X	X
L Wall Bulkhead	X	X	X	X		X	X	X
Marine Bin Walls		X	X	X				X
Nicolon Geotube			X					X
Parker Sand Web	X	X						X

Device	Harms beach access	Erosion of downdrift beaches	Erosion of fronting beaches	Potential hazard to water-based recreation	Impact on water quality	Impact on turtle nesting	Impact on shellfish resource	Impairs aesthetics
On the beach								
Pressure Equalizing Modules (PEM)	X	X		X		X	X	X
Ravens Retaining Wall		X	X	X				X
Sealogs (ramps)	X	X	X			X		X
Shoreline Construction Corp. groin		X	X	X	X			
Soukup Rubber Tire Revetment			X	X	X			X
Stabeach	X	X		X		X	X	X

(continued)

Table 12.8 (continued)

Device	Harms beach access	Erosion of downdrift beaches	Erosion of fronting beaches	Potential hazard to water-based recreation	Impact on water quality	Impact on turtle nesting	Impact on shellfish resource	Impairs aesthetics
Stabilito		X	X					X
Stabler Disks	X			X		X		
Subsurface Dune Restoration System			X					
Subsurface Dune Stabilization		X	X			X	X	X
Triton Marine Mattress			X					X
T-wall Groins	X	X	X	X		X	X	X
Wave Buster		X	X	X				X
Z-wall		X	X	X				X

Device	Harms beach access	Erosion of downdrift beaches	Erosion of fronting beaches	Potential hazard to water-based recreation	Impact on water quality	Impact on turtle nesting	Impact on fauna and flora	Impairs aesthetics
In the water								
Reef Ball		X		X	X		X	
Reef Mitigation Gardens		X						
Seabox	X	X	X	X	X	X	X	X
Sealift		X		X	X	X	X	X
Seascape		X		X	X	X		
Shoreprotector		X		X	X	X		
Surge Breaker		X		X	X	X		
Temple Beach System		X	X	X	X	X	X	X
Waveblock		X		X	X			
Waveshield		X		X				X
Wave Wedge		X		X	X	X		

water is already polluted, additional pollution may result. This is a possible impact of all breakwaters. Since most erosion control involves holding sand in place or causing it to deposit, essentially all devices will create downdrift sand loss. Damage to fauna and flora includes harming the biota of the sand, the plants and animals that live within beach sand and are an essential part of the whole ecosystem.

Although objective analysis of the performance of alternative devices is typically lacking, some impression of the performance of a few is presented below. These are based upon firstly, results from the State of Florida's monitoring program and secondly from an analysis of secondary sources which we present for artificial surfing reefs and beach drainage systems.

12.5 Florida Monitoring Program

In light of the profusion of new technologies being offered, the State of Florida passed a law in 1989 to encourage development of new methods of shoreline stabilization and to test them along its shores. The law (Sec. 29, 89–175 – Rule 62B-41.0075) was intended to encourage development of new and innovative approaches to deal with the widespread erosion problems of the state whose economy depends heavily on beaches and associated tourism. Before this program on innovative erosion control technology was created, state officials had no basis on which to assess the performance of proposed devices, and no established basis for ascertaining the performance of a device once in the water. As is the case elsewhere, they were at the mercy of the companies that proposed such devices.

The Florida Department of Environmental Protection summarized the state's experience with innovative technology in a historical overview that included pre-1989 experience (Woodruff 2006). The results of that analysis are summarised below.

12.5.1 Artificial Seaweed

Installations: Three, from 1983 to 1984. All were declared to be ineffective. The required monitoring in two cases was too short or not carried out at all.

Problems: Problems included an inadequate anchoring system, buoyancy loss by individual fronds, and unknown environmental impact.

12.5.2 Net Groins

Installations: Four, in 1987, 2000, 2001 and 2005. Third party reviews showed that success criteria were not met and there was significant downdrift beach loss.

Problems: Significant hazard potential for swimmers, surfers and jet skis because of possible entanglement with nets.

12.5.3 Beach Scraping

Installation: Three, in 1985 (2) and 2004, were considered ineffective or inconclusive. In this category are a variety of mostly post-storm scraping approaches. One ineffective approach involved **Beach Builder Screws**, large augers intended to bring sand ashore from beyond low tide (Florida Department of Environmental Protection 2008).

Problems: Beach scraping is not permitted except with an emergency permit. It can be a form of beach erosion, with possible impact on dunes.

12.5.4 Beach Dewatering

Installations: One installation, 1985 – StaBeach System.

Problems: Beach was stable but inadequate information to determine effect of installation.

12.5.5 Physical Structures (Geotextile Groins)

Installations: 1. **Protect Tube II**, 1989: beach remained stable 2 years; 2. **Longard Tubes**, 1992: performed well but tubes damaged and settled; 3. **Undercurrent Stabilizers**, 1984: no beneficial effects.

Problems: Long-term groin effects expected, plus concern about inhibition of turtle nesting.

12.5.6 Proprietary Reef Structures or Thin Line Submerged Breakwaters

Installations: **P.E.P. Reef** (Palm Beach, FL), **Beach Saver** (Avalon, NJ), **Beach Beam** (Maryland), **Beach Prism** (Bennetts Point, MD), **Campbell Module** (Sea Island, GA). All are concrete, segmented, prefabricated offshore breakwaters with more-or-less triangular cross sections.

Installations: **P.E.P. Reef** – 2 installations: 1. Palm Beach, 1993, installation removed after 2 years; reef segments used to make groins for nourished beach retention. 2. Vero Beach, 1996, minimal effect.

Problems: Relatively little impact on beach volume; swimmer hazard; Palm Beach structure had enhanced seaward sand loss.

The Florida analysis is unique in having made an objective assessment of the field performance of several 'alternative' devices. As can be seen from the analysis, none was regarded as an unqualified success.

12.6 Case Studies

Probably the highest profile 'alternative' devices at the present time are artificial surfing reefs and beach dewatering systems. Both approaches have seen numerous installations worldwide and sufficient material exists to compare the manufacturers' claims with practical experience and public opinion. In the following section we review the application of both technologies and the outcomes as perceived by the public. These are then compared to the manufacturers' claims.

12.6.1 Artificial Surfing Reefs

A US patent for artificial surfing reefs (ASRs) was first filed in 1991 and granted in 1993. Early installations for surfing reefs include Cable Station, Western Australia (Bancroft 1999) and Pratte's Reef, California (2000). Subsequent structures have been built in Queensland (Australia), New Zealand, England and India using a variety of designs and materials. Most of the available literature on artificial surfing reefs derives from the manufacturers and designers themselves (e.g., Black (1998); Hutt and Mead (1998); Hutt et al. (1998); Mead and Black 1999; Black 2001; Mead et al. 2010) and much is aspirational and theoretical in character, reporting the outcomes of modelling studies and describing the *expected* outcomes. A few studies, however, do describe monitoring of artificial surfing reefs for a short period after installation. The authors of many of those studies also have interests in the construction of artificial surfing reefs. For example, Bancroft (1999) assessed the Cable Station reef as "working to design specifications and is performing as well, or better than, was predicted," but few reports have described the results of monitoring of ASRs over several years. In several cases this is because they have not been completed to specification, thus hampering analysis of their performance. The 4-year monitoring study reported by Jackson et al. (2005) for Narrowneck Reef (Gold Coast, Queensland) was inconclusive regarding the influence of the reef on shoreline position and surfing but did prompt a number of design changes during and after installation, as unexpected physical changes in the seabed occurred. A subsequent report on 7 years' monitoring (Jackson et al. 2007) was much more emphatic in concluding that the reef had been a success in retaining the nourished beach, in providing a substrate for development of a diverse marine community and improved surfing conditions. The report did conclude that there were only a few occasions when the ideal swell simulated by modeling had been reproduced in the field.

An artificial surfing reef at El Segundo, California, was constructed to compensate for a surf break that was destroyed by construction of a 900-foot groin by the Chevron oil company in 1984. The Surfrider Foundation was given a 10-year permit in which to make an artificial surfing reef work or take it out. That point was reached in 2008 when it was acknowledged that the efforts had failed. A spokesperson for Surfrider, the proponents of the scheme, stated eventually that "For eight plus years, Pratte's Reef was as useless as useless gets". The sandbags that formed the artificial reef were removed from the seabed in 2008 (http://www.surfline.com/surf-news/after-years-of-unspectacular-closeouts-prattes-reef-is-removed-from-el-segundo-sandbagged_19261/photos).

The rationale behind installation of artificial surfing reefs often refers to them as multipurpose, offshore and adjustable (e.g. Black 2001), and several potential benefits are commonly cited to justify their installation. The priority and importance of each of these benefits, however, appear to evolve according to the way the installation pans out. Turner et al. (2000), for example, described the Gold Coast reef's aims as primarily to stabilise and enhance the beach by promoting beach widening and, secondly, to enhance local surfing conditions. The diverse marine environment created by the reef was identified subsequently as a significant value (Jackson et al. 2005).

Studies of long-term effects of artificial surfing reefs are difficult to find. However, a number of commentaries on artificial surfing reef performance in the WWW and press indicate that such structures often fail to satisfy expectations that existed at the time of their construction. The Mount Maunganui artificial surfing reef in New Zealand was partly installed in October 2006 (http://www.mountreef.co.nz/MountReef/4th-Oct-2006-Mount-Reef-Delivers!_IDL=10_IDT=464_ID=2290_.html). In August 2007, it still had a "further 30% of bag filling to go." One of the large geocontainers ruptured during 2006 and was replaced in 2008. It was reported in 2009 that work was continuing to "push the reef closer to the design specs" (http://www.mountreef.co.nz/) and in July 2010, the work was ongoing: "There was also a problem with one of the containers that make up the reef – which was not properly sealed by the construction crew – and thus part of the reef is actually missing." (http://www.surfermag.com/features/rinconmtreef-opedwhenry/). Subsequently, in March 2011, another surf magazine deemed the reef "a costly mistake" and reported the local council's concerns over failure to achieve completion of the structure (http://surf.transworld.net/1000127318/news/nz-artificial-reef-a-costly-mistake/). A number of reports claimed that the artificial surfing reef was being ignored by local surfers (http://www.bayofplentytimes.co.nz/local/news/surf-reef-branded-a-dangerous-flop/3947426/).

The artificial surfing reef in Boscombe, England, was constructed as part of a scheme to regenerate the seaside resort. It was opened in November 2009, reportedly a year behind schedule and at twice the estimated cost. It was subsequently declared unsafe and was closed to the public in April 2011 (http://news.bbc.co.uk/1/hi/england/dorset/8673078.stm). This closure was blamed on substantial changes to the reef structure and fears that dangerous currents could be produced. In addition, low tides left the reef exposed out of the water. Press reports commented that the reef was being used by an average of three surfers a day. In its defense, a local councilor was quoted on the BBC as saying: "We're disappointed that the reef isn't performing better at this stage but it is innovative marine engineering. I'm not surprised that it needs some optimising."

It appears from the global experience of artificial surfing reefs that few, if any, meet their desired outcomes (Jackson and Corbett 2007). There are calls for a number to be removed as has already happened to Pratte's Reef. Problems with installation, safety issues associated with unanticipated currents, danger of entanglement in torn webbing (http://www.bayofplentytimes.co.nz/local/news/surf-reef-branded-a-dangerous-flop/3947426/), and surfing waves not living up to expectations are widely cited. Jackson et al. (2007) also point to unrealistically high expectations driven by media hype. Based on their experience of the Pratte's Reef in California, the Surfrider Foundation no longer accepts artificial surfing reefs as a form of mitigation for loss of natural surf breaks (http://www.surfline.com/surf-news/after-years-of-unspectacular-closeouts-prattes-reef-is-removed-from-el-segundo-sandbagged_19261/photos). Steinvorth (2010) argues that installation of artificial surfing reefs in areas that already have good surf breaks is inappropriate. Despite the mixed opinions of surfers regarding the outcomes of such installations and the difficulties in construction, however, new artificial surfing reefs continue to be planned and constructed worldwide.

An assessment of six artificial surfing reefs (Jackson et al. 2007) showed that only four had been completed and of these, three produced acceptable surfing conditions. An alternative assessment published on a surfing website gave a contrasting qualitative ranking of six operational (or formerly operational) ASRs on an A to F scale (Table 12.9) with no reef scoring better than a C-minus.

12.6.2 Beach Drainage Systems

Turner and Leatherman (1997) and Curtis and Davis (1998) reviewed the history of beach dewatering work up to that date. Machemehl et al. (1975) appear to have been the first to propose beach dewatering for coastal stabilization, based on flume experiments, and the first field test was conducted by Chappel et al. (1979) in Australia. The first patent of a beach drainage system was registered by the Danish Geotechnical Institute in 1985 (Vesterby 1991, 1994), and full-scale tests were conducted in Denmark between 1985 and 1991 at Thorsminde (Turner and Leatherman 1997). However, as late as 2005 Bruun concluded that beach drainage systems should still be regarded as experimental, a sentiment since echoed by

Table 12.9 Qualitative ranking of performance of artificial surfing reefs and their cost. http://oceanswavesbeaches.surfrider.org/do-artificial-reefs-work-vol-4-track-record

Name	Installation	Cost	Rating (surfability)
Burkitts Reef, Bargara, Australia	1997	AU$5000	D
Cable Station, Western Australia	1999	AU$2 million	C-
Narrowneck, Gold Coast, Australia	2000		C-
Pratte's Reef, El Segundo, California	2001	US$300,000 installation US$300,000 removal	F
Mount Maunganui, New Zealand	2005 (not yet completed)	NZ$1.5 million	D/F (but incomplete)
Opunake, Taranaki, New Zealand	2005 (not yet completed)	£935,000	Incomplete
Boscombe, England	2008 (not yet completed)	£3 million	Incomplete

Ciavola et al. (2009). Like artificial surfing reefs, beach drainage schemes are sometimes promoted as multi-functional. The system at Ravenna, Italy, for example, was primarily installed to provide water supply into reservoirs (Ciavola et al. 2009), while Curtis and Davis (1998) report on the ecological implications of such an installation in Massachusetts. Ciavola et al. (2009, p. 7317) contend that "too often this solution is presented to the coastal manager without an impartial view of the possible failure of the interventions." Nonetheless Goler (2004) reported that 33 BD (Beach Drainage) systems had been installed around the world since 1981 in Denmark, USA, UK, Japan, Spain, Sweden, France, Italy, and Malaysia, with four more under construction or approved for installation at that time.

The several monitoring studies of beach drainage systems report variable but usually inconclusive results. Chappel et al. (1979) were unable to quantify the influence of dewatering on the morphological response of a high energy beach in Australia. In a 6-year experiment at Thorsminde, Denmark, accumulation in the drained zone of 30 m^3/m of shoreline within the first year was followed by stability for a further 2 years. The beach then eroded, but possibly at a lower rate than adjacent sections of the coast. Infrequent surveys and the occurrence of a 1:100 year storm during the survey period precluded any definitive conclusions being reached on the prototype installation (Turner and Leatherman 1997). A similar conclusion was reached by Dean (1989) who found it was not possible to separate natural beach changes from those induced by a dewatering system (STABEACH) installed in Florida in 1988. Bowman et al.'s (2007) 1-year study at Alassio, Italy, concluded that the beach drainage system did not promote beach accretion. Ciavola et al. (2009) monitored a beach drainage system near Ravenna, Italy, over a 3-year period, noting a progressive accretion trend with seasonal variability that was impossible to separate from the natural behaviour of the beach. They did note that the system did not provide a definitive solution to coastal erosion. Vicinanza et al. (2010) found no positive effects of a system installed at Chiaiolella Beach,

Italy. They also reported consistent volume loss from the beach during a mild storm and damage to the system during the storm. They particularly drew attention (p753) to the "inadequacy of the dewatering system as coastal protection under high wave conditions." Curtis and Davis (1998) reported that a system installed on Florida's Gulf coast was rendered inoperable by a series of storms.

12.7 Discussion

The abundance of proposed alternative devices is an outgrowth of a number of societal issues that include:

- Intense development along ocean shorelines that are eroding.
- The reasonable expectation that erosion problems will increase as sea level rises.
- The high cost of traditional shoreline erosion response (an especially large burden for small communities).
- The demand for lower cost devices to halt erosion.
- The demand for less environmentally damaging approaches to erosion response.

Alternative shoreline stabilization device is a term applied to a category of coastal engineering structures that differ from the "standard" widely used structures. Thus, by definition, 'alternative' structures have not found wide application. These devices use distinctive materials or are emplaced in particular configurations along a shoreline that render them different from standard approaches.

Usually, however, the principles of shoreline stabilization are the same for the alternative devices as for the more widely used ones. That is, most of the devices can be said to fall into the categories of seawall, groin, or offshore breakwater, which in some fashion or other aim to trap/retain sand or reduce wave impact, the two main tasks of all shoreline erosion control devices.

Introducing any structure into a dynamic physical environment like a beach will likely promote changes in that environment. Since a beach's success and persistence is linked to its ability to adapt to changing circumstances, any structure is likely to impede that process. Beaches operate under a range of wave energy conditions to which they respond by changing shape as sediment is moved within the beach system. Such changes are commonly cyclic and beaches usually recover after storms, but not always. Those that are suffering long-term erosion are driven by either a sediment deficit or rising relative sea level, neither of which is addressed by any type of shoreline stabilization device at any meaningful timescale.

The stabilization approaches listed here have the usual impact problems, foremost of which is that engineering devices that hold the eroding shoreline in place inevitably will result in a narrowed and sometimes completely lost beach. Most of our listings, however, have either never been emplaced or have been emplaced in very few locations. Thus the experience base for these alternative structures is very sparse. In addition, monitoring, if carried out at all, has mostly been done by the manufacturers, the Florida and Puget Sound programs being major exceptions to the rule. Monitoring is typically over short time periods, and apparent successes that are promoted by the manufacturer are often undone by subsequent events such as storms. All too often, however, unexpected storms or unusual field conditions are given and accepted as a reason for failure. Storms can be expected to be the main cause of failure of all coastal engineering structures, and the alternative devices are no exception. Few literature descriptions specifically address this problem.

An interesting issue is the widespread implicit and sometimes explicit assumption in the manufacturers' descriptions of these devices that they can find use on almost any shoreline – an extension of the one-device-fits-all mentality. Of course a device that may appear to succeed on one beach may not on another. Differences in parameters, such as sand supply, sand size, wave energy and storm frequency and intensity, can be responsible for differing responses of stabilization devices.

Evaluation of success of shoreline stabilization structures is fraught with a number of hazards including over-optimistic interpretations and too-short time frames. A device should be in place for a minimum of 5 years before a reasonable evaluation of how close the device came to achieving its stated original goals can be made. At Pratte's Reef, a 10 year period was used to make such an assessment. Acknowledgement of failure is often a progressive phase in the emplacement of alternative devices. For example, the lack of success is often blamed on unexpected conditions. Most commonly this means a storm, but other commonly cited issues include poor installation. The secondary goals of the device then begin to be promoted. For example, the artificial surfing reef on Australia's Gold Coast was promoted by the press as a surfing reef. When it clearly began to fail as the sandbags shifted, the reef's purpose as an offshore breakwater was stressed by its developers. Later emphasis was placed on its function as a habitat for benthic marine organisms. Other breakwaters that failed in their original reason for being have also been labelled as important habitats. This phase of shifting goalposts often enables the purchaser and supplier to enter a prolonged period during which the public slowly forgets the device's original purpose, and the costs involved are forgotten.

Coastal managers who must make the decision as to erosion response may find it difficult to get beyond the manufacturer's claims. One approach is to seek objective evaluation of the success or failure of the same or a similar device at another location. This is problematic, however, because no two beaches are identical in terms of the processes, and success at one location does not assure a similar experience at another. We contend that the best approach if these devices are to be used is to view them as experiments and be prepared to remove them if the experiment fails. A community should have a clear statement of expectations from the contractor. Monitoring of success or failure should be done by independent parties and not by the company that installed the project.

A major issue with these devices is that they address the symptoms rather than the underlying cause of the shoreline problem. A case in point is ongoing experiments on the delta of the Chao Phraya at Bangkok. There, delta subsidence, sea level rise, damming of rivers and reduction of sediment supply to the delta, and removal of mangroves with their sediment trapping ability have contributed to annual shoreline

Fig. 12.2 Pilot emplacement of patented shoreline stabilization devices at Chao Phraya Delta involving wave baffles made of concrete

recession rates of 25 m/year for more than 40 years (Vongvisessomjai 1992). In that context a patented system of wave baffles comprising concrete pillars arranged in three lines (Fig. 12.2) has been proposed as a solution to the erosion problem and a pilot scheme has been instigated. The scheme does not address any of the underlying problems and yet, such is the need for a solution and the strength of advocacy by the developers that pilot schemes have been supported.

References

Bancroft S (1999) Performance monitoring of the Cable Station artificial surfing reef. BSc Hons Dissertation, University of Western Australia

Black KP (1998) Narrowneck reef, report 3: sediment transport modelling. Report to the Gold Coast City Council, Mar 1998

Black KP (2001) Artificial surfing reefs for erosion control and amenity: theory and application. J Coast Res Spec Issue 34:1–14

Bowman D, Ferri S, Pranzini E (2007) Efficacy of beach dewatering-Alassio, Italy. Coast Eng 54:791–800

Bruun P (2005) Beach drain. In: Schwartz ML (ed) Encyclopaedia of coastal science. Springer, Dordrecht, pp 138–140

Chappel J, Eliot IG, Bradshaw MP, Lonsdale E (1979) Experimental control of beach face dynamics by watertable pumping. Eng Geol 14:29–41

Ciavola P, Fontana E, Vicinanza D (2009) Performance of a beach drainage system at Lido Adriano (Ravenna-Italy). 33rd IAHR Congress: Water Engineering for a Sustainable Environment

Curtis WR, Davis JE (1998) Field evaluation/demonstration of a multisegmented dewatering system for accreting beach sand in a high-wave-energy environment. US Army Corps of Engineers Technical Report CPAR-CHL-98-1

Florida Department of Environmental Protection (2008) Strategic beach management plan http://www.dep.state.fl.us/beaches/publications/pdf/SBMP/Cover%20and%20Introduction.pdf

Gerstel WJ, Brown JF (2006) Alternative shoreline stabilization evaluation project. Puget Sound Action Team, Olympia, 43 pp

Goler G (2004) Numerical modelling of groundwater flow behaviour in response to beach dewatering. Masters thesis, Middle East Technical University, Turkey. http://etd.lib.metu.edu.tr/upload/12605274/index.pdf

Hutt JA, Mead ST (1998) Narrowneck reef, report 2: surfing aspects. Report to the Gold Coast City Council, Mar 1998

Hutt JA, Black KP, Jackson A, MaGrath J (1998) Narrowneck reef, report 1: Field studies. Report to the Gold Coast City Council, Jan 1998

Jackson LA, Corbett BB (2007) Review of existing multi-functional artificial reefs. In: Australasian Conference on Coasts and Ports, Melbourne, Australia, pp 1–6

Jackson LA, Tomlinson R, Turner I, Corbett B, D'Agata M, McGrath J (2005) Narrowneck artificial reef. Results of 4 years of monitoring and modifications. In: Proceedings of the 4th international surfing reef symposium, Manhattan Beach, USA

Jackson LA, Corbett BB, McGrath J, Tomlinson R, Stuart G (2007) Narrowneck reef: review of seven years of monitoring. Shore Beach 75:1–13

Machemehl JL, French TJ, Huang NE (1975) New method for beach erosion control. Engineering in the oceans, ASCE speciality conference, pp 142–160

Mead S, Black KP (1999) A multipurpose, artificial reef at Mount Maunganui Beach, New Zealand. Coast Manag 27:355–365

Mead S, Blenkinsopp C, Moores A, Borrero J (2010) Design and construction of the Boscombe multi-purpose reef. http://journals.tdl.org/ICCE/article/viewFile/1352/pdf_106

Steinvorth R (2010) Improving surfing conditions with floating wave filters. Masters thesis, Universitat Poitecnica Catalunya. http://upcommons.upc.edu/pfc/handle/2099.1/8377

Turner IL, Leatherman SP (1997) Beach dewatering as a "soft" engineering solution to coastal erosion – a history and critical review. J Coast Res 13:1050–1063

Turner I, Leyden V, Symonds G, McGrath J, Jackson A, Jancar T, Aarninkof S, Elshoff I (2000) Comparison of observed and predicted coastline changes at the Gold Coast Artificial (surfing) reef. ICCE. In: Edge BE (ed) Proceedings of the international conference on coastal engineering, Sydney. http://coastalmanagement.com.au/papers/ICCE2000-WRL.pdf

Vesterby H (1991) Coastal drain system: a new approach to coastal restoration. In: Proceedings of GEO-Coast '91, Yokohama, Japan, pp 651–654

Vesterby H (1994) Beach face dewatering – the European experience. Alternative technologies in beach preservation. In: Proceedings of the 7th national conference on beach preservation technology, Tampa, FL, pp 53–68

Vicinanza D, Guida A, Ferrante V, Ciavola P (2010) Performance of a beach dewatering system – Chiaiolella Beach, Procida Island, Italy. J Coast Res 26:753–761

Vongvisessomjai S (1992) Coastal erosion on the Gulf of Thailand. Eng Inst Thail 3:38–58

Woodruff PE (2006) Historical overview of innovative erosion control technology in Florida. Bureau of Beaches and Coastal Systems, Florida Department of Environmental Protection

Chapter 13
Bad Practice in Erosion Management: The Southern Sicily Case Study

Giorgio Anfuso, José Ángel Martínez-del-Pozo, and Nelson Rangel-Buitrago

Abstract This case study from Sicily illustrates a common sequence of events where one unwise action was countered with another, which in turn created additional problems. The situation arose through strong political interference and ignorance (or lack of concern) regarding the environmental impacts of human interventions on the shoreline and by the public perception that government has a duty to protect private property. The poor design and location of ports and harbours produced infilling problems and huge updrift accretion with concomitant downdrift erosion. The human-induced coastal retreat was counteracted by the progressive emplacement of breakwaters creating a "domino" effect. On many occasions these were constructed to protect unplanned and illegal (in the sense that they do not conform to planning regulations) beachfront summer houses. Without the presence of these structures, there would have been no need for publicly funded intervention.

Furthermore, only a narrow coastal belt close to the shoreline is used by bathers on the wide beaches formed updrift of ports and harbours and in the lee of breakwaters, most of the accreted beach being unused or partially occupied by tourist developments. Thus beach users and municipalities acquired some benefits from beach accretion at specific sites, the opposite being true in eroding areas.

G. Anfuso (✉) • N. Rangel-Buitrago
Dpto. Ciencias de la Tierra, Facultad de Ciencias del Mar y Ambientales, Universidad de Cádiz, Polígono Río San Pedro s/n, Puerto Real 11510, Spain
e-mail: giorgio.anfuso@uca.es; nelson.rangelbuitrago@mail.uca.es

J.Á. Martínez-del-Pozo
Instituto de Arqueología de Mérida – CSIC, Plaza de España, 15, Mérida (Badajoz) 06800, Spain
e-mail: jangel.martinez@iam.csic.es

J.A.G. Cooper and O.H. Pilkey (eds.), *Pitfalls of Shoreline Stabilization: Selected Case Studies*, Coastal Research Library 3, DOI 10.1007/978-94-007-4123-2_13,

13.1 Introduction

Coastal occupation has been increasing in the past few decades especially due to coastal tourism-related activities. Tourism is now one of the world's largest industries (Klein et al. 2004; Jones and Phillips 2011). In the Mediterranean region tourism is the most important activity with 298 and 400 million international and domestic tourist arrivals in 2008. The fastest rates of coastal development between 1990 and 2000 occurred in Portugal (34%), Ireland (27%), Spain (18%), followed by France, Italy and Greece (EEA 2006). In several coastal regions of Italy, France and Spain, the coverage of built-up areas in the first kilometre coastal strip exceeds 45% and, in these areas, further development is occurring in the coastal hinterland. Spain plus Italy, France, Greece and Turkey account for '*the most significant flow of tourists.... a sun, sea and sand (3S) market*' (Doods and Kelman 2008) and tourism is expected to grow at a level of 4.0% per year over the next 10 years. The European Union offers a vast market in its colder and affluent northern parts and the capacity for growth is viewed as almost unlimited.

Associated with ongoing tourism development in Europe is the high level of armouring of shorelines by coastal defences and harbours (EEA 2006), with especially high percentages of armoured shoreline in the North Sea (16%) and the Mediterranean Sea (more than 8%). The protection of built-up areas and halting of shoreline recession have largely been carried out through the construction of hard defence structures, which mainly consist of groins, jetties, breakwaters, revetments and seawalls. Beach nourishment works are common (Hanson et al. 2002; Cooper and Alonso 2006). In recent decades coastal armouring has also accompanied the construction of marinas for tourism purposes and harbours and ports for commercial activities. In this sense, Italy and the United Kingdom received the most vessel arrivals in terms of tonnage during 2003.

In many localities, arable land has also been lost to intensive agricultural activities, e.g. in the southern regions of Italy, Greece and Cyprus and the Levante in Spain, and urban development, which is especially driven by higher land prices in areas adjacent to existing settlements.

A common problem associated with coastal armouring is the "coastal squeeze" (Doody 2004), which takes place when a coastline is prevented from its landward migration by seawalls or other man-made structures. Coastal erosive processes associated with sea level rise and increasing storminess can cause the complete disappearance of the beach or salt marshes and deepening of nearshore areas fronting coastal structures (Pilkey and Dixon 1996; Doody 2004). Other, less immediately obvious examples of decreased ability to respond to changes occur when coastal defences at the base of eroding cliffs cut off sediment supply to beaches alongshore (Runyan and Griggs 2003). Environmental problems are also sometimes related to beach nourishment works; nourished material usually increases water turbidity and sedimentation processes affect seagrass meadows and sediment quality in nearby beaches. As a result human activity on beaches generally produces loss of ecological value, decreases in biodiversity and diminution of landscape value (Williams and Micallef 2009).

The case study presented here deals with the pitfalls of coastal armouring along 90 km of coastline including sandy beaches and cliffed sectors within the administration of the Ragusa Province (South Sicily, Italy). As observed in many Mediterranean countries (EEA 2006), coastal occupation in South Sicily has essentially been due to the tourism-related urban developments. Summer houses, apartments, hotels, restaurants, promenades and seafront roads have been constructed on the wide backbeach and dunes. In addition, ports, harbours and breakwaters have interrupted longshore drift and caused erosion of downdrift beaches (Cooper et al. 2009).

In this study, the coastline, defined as the instantaneous water line position (Boak and Turner 2005), as well as coastal defence structures, ports and harbours have been mapped using different cartographical series: a georeferenced topographic map (1967, scale 1:25,000), two registered and geometrically corrected photogrammetric flights (1977 B&W, scale 1:17,000 and 1987 colour, scale 1:12,500), the 1999 orthophotographs series and, finally, the 2008 geographic data provided by the STIR Web Mapping Service (http://www.sitr.regione.sicilia.it).

Coastal evolution was calculated following the "end point rate" method (Jiménez et al. 1997) between successive photographs and the 1977–2008 interval, the investigated time spans corresponding to short and medium-term evolution according to Crowell et al. (1993). The total shoreline mapping error due to the geometrical accuracy and spatial resolution limitations of the photographs, and the uncertainty in water line position, has been assumed to be 10 m. Consequently, coastal variations lower than 10 m and beach surface variations less than 2,000 m^2 (calculated along longshore sectors 200 m in length), have been discarded and are not displayed in the final cartography (Martínez and Anfuso 2008).

For the quantitative assessment of human-made structures' impacts on the study area, coastal structures were mapped as line segments and the coefficient of infrastructural impact K (Aybulatov and Artyukhin 1993) was obtained by dividing the total length of all maritime structures (groins, moles, seawalls, revetments, breakwaters, etc.) by the entire length of the coastal section under consideration (in this case coastal sectors were 500 m in length). According to this methodology the extent of infrastructural impact could be then classified as *minimal* at $K = 0.0001–0.1$; *average* when $K = 0.11–0.5$; *maximal* at $K = 0.51–1.0$ and *extreme*, when $K > 1.0$.

13.2 Study Area

The study area is in Ragusa Province, South Sicily (Italy, Fig. 13.1). The littoral zone is composed of sandy beaches rich in quartz (≅65%), carbonates (≅30%), feldspars and heavy minerals. The beach face and nearshore areas have a low gradient and consist of fine sand with bathymetric contours running parallel to the coastline. One or more longshore bars are frequently observed and they

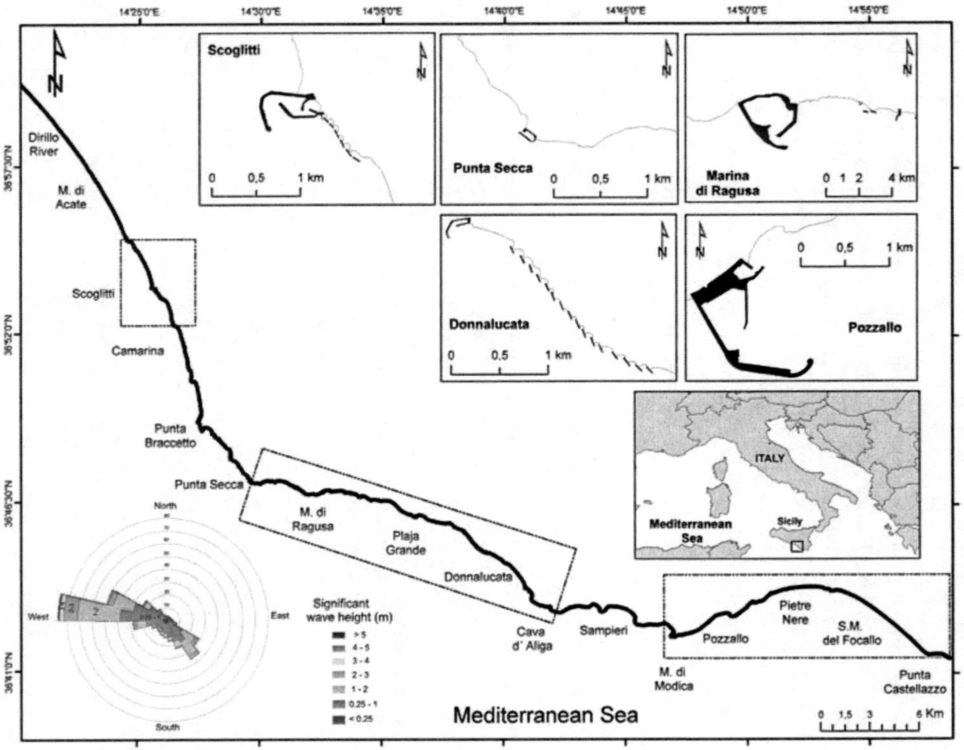

Fig. 13.1 The Ragusa Province littoral with main ports, harbours and defence structures and location of the four investigated sectors. Significant wave height data characteristics recorded over the 1998–2008 period obtained from the offshore directional wave buoy located at Mazzara del Vallo in the Sicilian Channel over a depth of 20 m (Italian Wave Climate Network)

control breaking wave processes which usually take place far from the shoreline, giving rise to large surf zones with spilling breakers that are typical of dissipative beach states. Beaches are backed by dune ridges and cliffs while promontories divide the littoral into morphological cells. In most cases, sediments produced by cliff retreat are very fine and are rapidly winnowed by waves and there is no significant sedimentary input to the littoral budget from fluvial sources (Martínez and Anfuso 2008).

The study area is microtidal (astronomical spring tidal range c. 20 cm), with prevailing winds from the W, SW and SSW. According to the data from the nearest offshore buoy, most frequent and severe storms approach from the W and WNW with significant wave heights greater than 5 m. Less severe storms approach from the SE, with maximum significant wave heights of 3 m (Fig. 13.1).

Because of its coastal orientation, the sector between the Dirillo River mouth and the Punta Secca promontory is essentially affected by storms from a westerly direction and the sector extending from the Punta Secca to the Punta Castellazzo promontories is influenced by storms approaching from the east and west directions, which give rise to important longshore currents. The main longshore transport along the Ragusa Province littoral is southeast directed but an opposite transport is also recorded, especially during spring and autumn. As a result of the coastline orientation, the westward directed transport is especially significant in the sector between Cava d'Aliga and P. Castellazzo (Fig. 13.1).

Along the investigated littoral, ports, harbours, coastal protection structures and headlands acquire a great importance in coastal compartmentalization because they work as convergent, divergent or transit limits (terminology following Carter 1988) dividing the littoral into morphological cells.

Since the 1960s ports and harbours have been constructed at Scoglitti, P. Secca, Marina di Ragusa, Donnalucata and Pozzallo. Rope haulage zones have been constructed at Casuzze, Sampieri, Marina di Modica and Pozzallo, and small, pre-existing agricultural and/or fishing villages have seen great expansion linked to the construction of summer houses and local tourism. The coastal population peak is recorded in July and August, when people move to the coast from the hinterland, generating a significant environmental pressure, i.e., increased demand for water supplies, sewage disposal, etc.

In terms of land occupation, increased human pressure on the coast has also occurred through the development of intensive agricultural activities. The lack of a management policy and the huge and rapid increase of human occupation resulted in urban sprawl and considerable coastal stress. In fact, most human activities and construction have been developed within the "protection" and "influence" zones defined by the Italian Coastal Act (2004). Within the protection zone, extending 150 m landward, any kind of construction is supposedly prohibited, and within the influence zone, which is 150 m landward extended, any construction needs to be approved by the local Planning Office. A management response is now required to protect coastal and, specifically, beach resources upon which the local economy is based.

13.3 Coastal Evolution and Armouring

The study area can be regarded as a single large physiographic unit divided into several cells of different dimensions, limited by fixed, natural and artificial boundaries. In the study area, sediment broadly moves from west to east because of longshore transport which impinges on natural and human-made coastal structures, giving rise to a series of accreting/eroding areas. Hence, over the analysed intervals, significant areas of accretion have been observed close to harbours and ports, usually on the western sides of these structures, with erosion processes prevailing downdrift (Martínez and Anfuso 2008; Anfuso and Martinez 2009). This situation is evident at Scoglitti, Marina di Ragusa and Pozzallo ports and Donnalucata harbour.

All of these ports and harbours have experienced sediment infilling because their entrances are located within the surf zone. In an effort to solve the infilling problems, their jetties have been extended several times. This has simply exacerbated downdrift erosion, which in turn has been countered by the progressive construction of offshore breakwaters. These have locally halted coastal retreat giving rise to tombolo formation and the creation of a series of swash-aligned shoreline cells but they have also shifted erosion processes downdrift. The progressive addition of engineering structures to counter the problems caused by others has been termed the "domino" effect (Cooper et al. 2009). Coastal defences have been progressively emplaced over the investigated period with an associated increase in coastal armouring. In this paper, the impact of coastal structures and coastal armouring are assessed particularly in four coastal sectors: P. Zafaglione–Scoglitti (Fig. 13.2); P. Secca-Plaja Grande (Fig. 13.4); Playa Grande-Cava d'Aliga (Fig. 13.5); and Pozzallo-P. Castellazzo (Fig. 13.7).

In the P. Zafaglione-Scoglitti sector (Fig. 13.2), Scoglitti is a tourist coastal town with 3,000 and 66,000 inhabitants during winter and summer months, respectively. It is served by a port, essentially devoted to fishing activities. The port jetties have been modified several times since the 1960s (Fig. 13.3a) and the northern dock was enlarged again in 2008, at a cost of ten million Euros, converting the structure into an absolute cell limit that totally impedes sediment bypass.

Between 1987 and 1999 the port jetties created huge updrift accretion with c. 42,000 m^2 of new formed beach surface (corresponding to shoreline advance of c. 10 m/year). A further increase of c. 15,000 m^2 (c. 1.8 m/year) occurred in the 1999–2008 period and a total accretion of c. 55,000 m^2 was recorded over the 1977–2008 period (c. 4 m/year), (Fig. 13.2).

Accumulated sediments are not removed by waves from the second and third quadrants, thus increasing the sedimentary deficit along the area between P. Zafaglione and Scoglitti port. This sector shows coastal squeeze-related problems, in fact it is constituted by a narrow sand beach backed by a cliff that is unable to migrate landward because it is artificially stabilised in order to protect the promenade and coastal road. In recent years, erosion processes have caused beach loss and rock falls in the backing cliff that was then additionally protected by rip-rap revetments and seawalls (Fig. 13.3b).

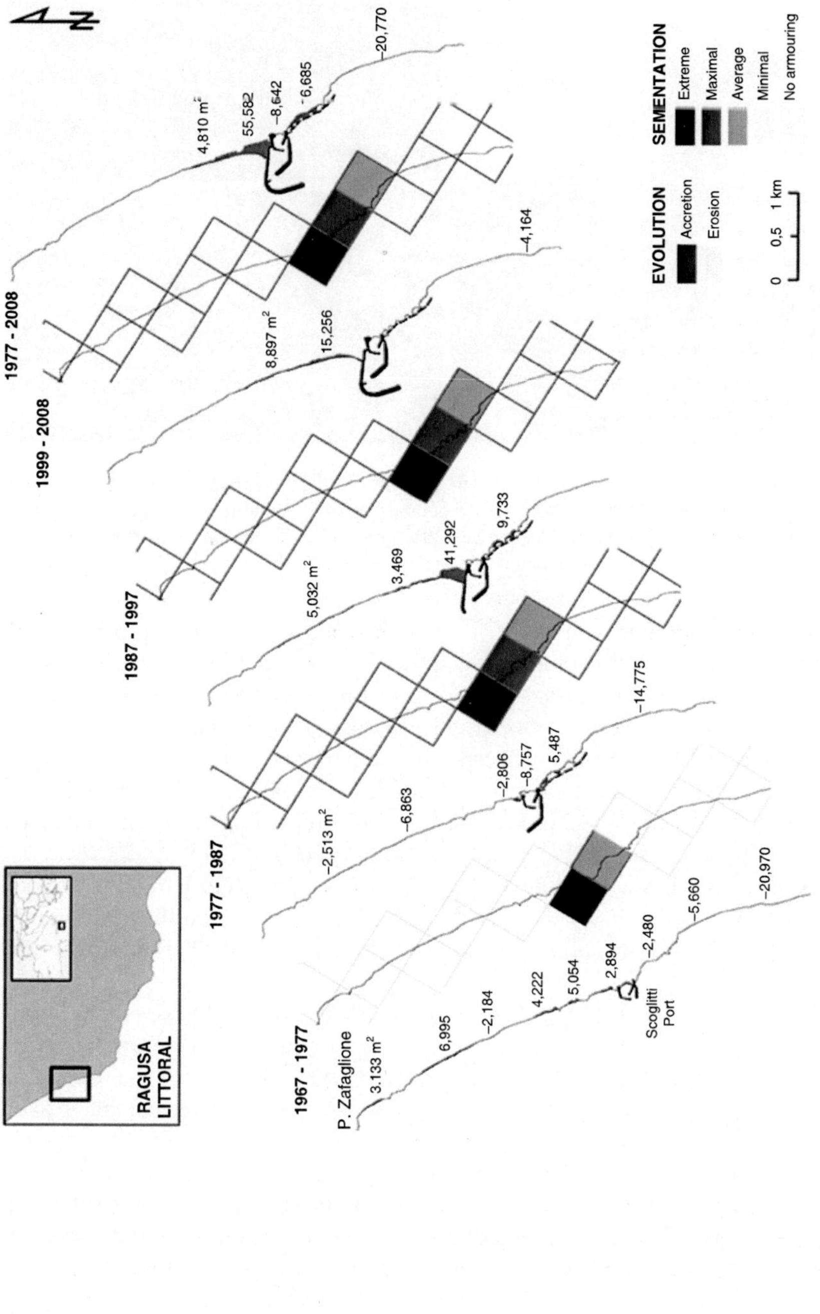

Fig. 13.2 Coastal evolution and armouring along the P. Zafaglione – Scoglitti sector

Fig. 13.3 Phases of Scoglitti port enlargement (**a**); littoral sector between P. Zafaglione and Scoglitti port, view from north to south (**b**, December 2010); southern breakwaters emplaced downdrift (south) of Scoglitti port (**c**, December 2010) and beach accretion updrift (west) of the dock and haulage zone at Marina di Ragusa (**d**, April 2006)

South of Scoglitti port (downdrift), six breakwaters have been progressively constructed to 'solve' erosion problems that then migrated further downdrift (southward) and affected the coastal road which is now protected by a rip-rap revetment.

Coastal armouring evolution reflected the progressive emplacement of structures; armouring was "extreme" in 1977, because of the port, and "extreme" to "average" in 1987 because of the construction of six offshore breakwaters. In the following decade, the northern breakwaters favoured tombolo formation while the southern ones suffered great erosion and did not form tombolos (Fig. 13.3c).

In the P. Secca-Plaja Grande sector (Fig. 13.4), the small harbour at Punta Secca produced accretion on the eastern side and recorded infilling problems. Approximately 12,000 m^3 of sediments were trapped in the harbour over the 1985–1995 period.

At Marina di Ragusa, one of the most important tourist coastal towns in the study area, a dock and a haulage zone for small recreational boats were emplaced in the 1970s. The dock produced accretion updrift (west, Fig. 13.3d) and erosion downdrift (east) which was counteracted by the successive emplacement – over the 1987–1999 period – of two 70 m large breakwaters and few small groins which gave rise to a heavy swash aligned coastline.

At M. di Ragusa the new port, inaugurated in July 2009, is already suffering infilling problems and producing accretion at both sides of the structure, working as a fixed, convergent limit (Fig. 13.4). Erosion areas have been observed between

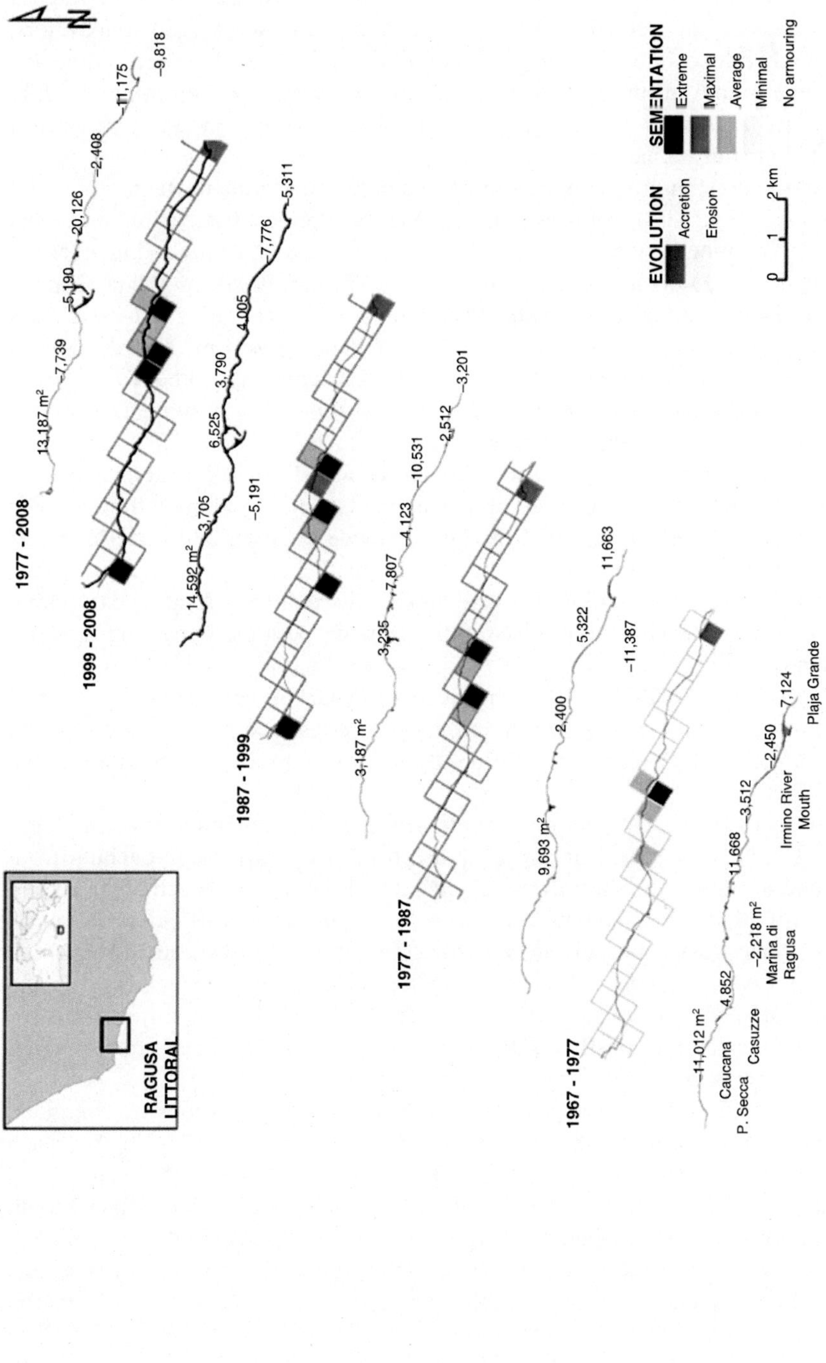

Fig. 13.4 Coastal evolution and armouring along the P. Secca – Plaja Grande sector

Punta Secca harbour and M. di Ragusa port, namely at Casuzze and Caucana; in the latter locality, 70,000 m^3 of gravel was nourished in 2004 along an 800 m-long coastal sector to protect Byzantine archaeological remains and the coastal road. In order to increase the stability of the artificial beach four submerged groins of calcareous rock blocks were emplaced. On the eastern side of M. di Ragusa port, a large artificial beach (c. 6,000 m^2), protected by two submerged breakwaters, was created with sediments dredged during port construction. Dredged sediments, which came from Caucana and Casuzze areas, were taken from their original cell in this way and transferred to the adjacent morphological cell, between M. di Ragusa port and Donnalucata harbour.

Increasing erosion problems were observed between Marina di Ragusa port and Donnalucata (Fig. 13.4), especially in the Natural Protected Area located at the Irminio River mouth, where sand was winnowed and the underlying deposits outcropped (c. 12,000 m^3 was lost between 1977 and 2008). At Plaja Grande, where a 240 m long breakwater was emplaced over the 1967–1977 and 1977–1987 periods c. 60,000 m^2 (c. 3.5 m/year) and 16,000 m^2 (c. 1.8 m/year) of beach accretion was recorded, respectively. Presently, the breakwaters form a swash-aligned shoreline impounding large quantities of sediments that are definitively lost to the longshore transport system.

Within the Plaja Grande-Cava d'Aliga sector (Fig. 13.5), Donnalucata is a small coastal village with c. 10,000 inhabitants during summer time, served by a small jetty in 1967. In 1977 this was extended and an additional one was constructed beside it to create a small harbour for fishing and recreational activities. The western and eastern beaches were linked by shallow offshore bars along which longshore transport took place, thus allowing periodic bypassing between the beaches.

Between 1967 and 1977 c. 19,000 m^2 (c. 4.1 m/year) of new beach were formed and erosion was counteracted by the progressive construction of 16 breakwaters that again produced downdrift migration of the erosion problems according to the "domino" effect.

Between 1977 and 1987, in response to sediment accumulation in the harbour, principally linked to sediment and algae deposition by waves approaching from the second quadrant, the docks were extended and the updrift beach accreted and rotated as it adjusted to the new configuration. On the downdrift side, substantial erosion threatened the beachfront houses east of the main town and 16 offshore breakwaters were constructed. Great quantities of sand then accumulated in the lee of the offshore breakwaters forming a series of tombolos; as a result, east of the last breakwater, the coastline suffered erosion at spatially variable rates ranging between 1 and 5 m/year.

Between 1987 and 1999 the groins were extended again in response to continued sediment accumulation in the harbour. This led to further accretion in the beach to the west (c. 30,000 m^2, c. 6.6 m/year), the gain of sediment is likely to be derived from longshore drift coming from the east, especially the Irminio River mouth area. The same trends continued during the 1999–2008 period and, when considering the 1977–2008 period, the sediment gain around the offshore breakwaters (c. 90,000 m^2) equalled that lost from the area to the east (c. 92,000 m^2). Accretion

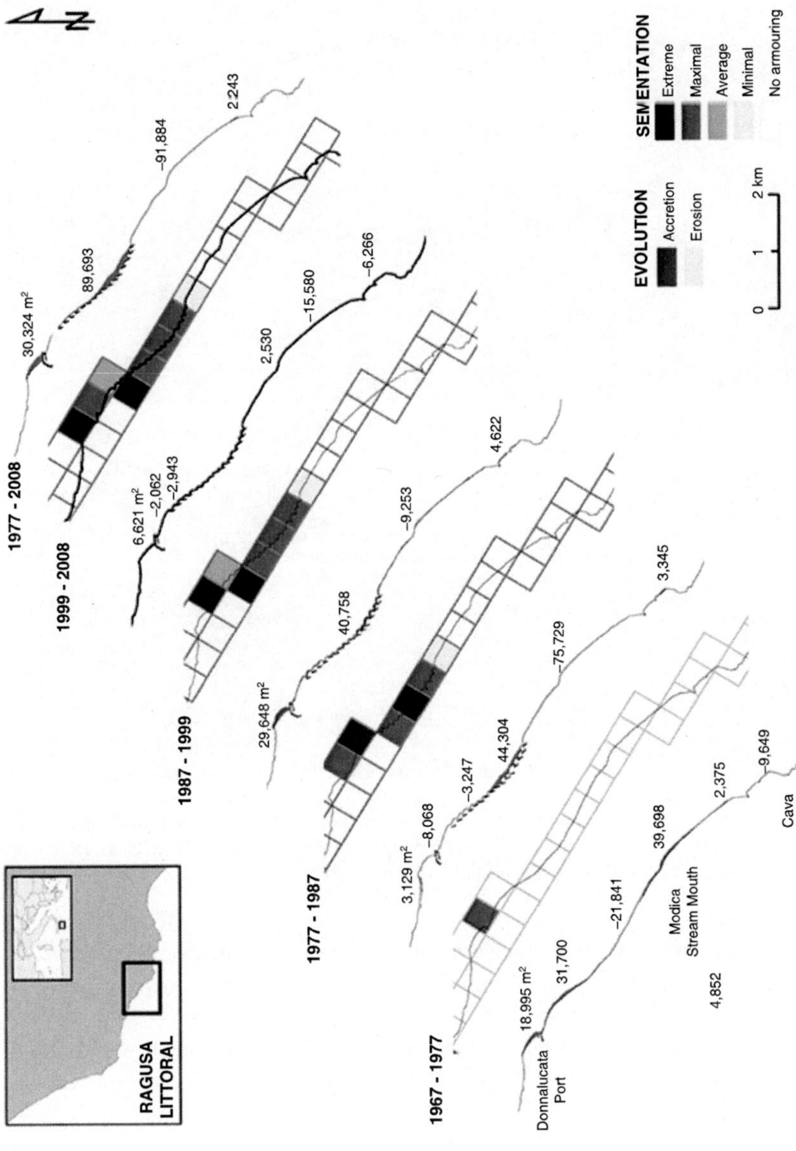

Fig. 13.5 Coastal evolution and armouring along the Plaja Grande – Cava d'Aliga sector

Fig. 13.6 Migrating dunes at the northern breakwaters emplaced downdrift (east) of Donnalucata harbour (**a**, April 2011); summer houses protected by rip-rap revetments downdrift (east) of easternmost breakwater at Modica Stream mouth (**b**, April 2011)

in the lee of offshore western breakwaters produced an important surplus of sediments which gave rise to the formation of landward migrating foredunes at places threatening summer houses and invading agricultural areas (Fig. 13.6e). East of the breakwaters extensive erosion processes reduced the beach width and a series of seawalls and rip-rap revetments were constructed to protect beach front properties (Fig. 13.6f).

At Donnalucata, the extent of coastal armouring was greatly increased and extended eastward; the amount of armouring slightly decreased over the 1987–1999 period because of the progressive formation of tombolos in the lee of the breakwaters.

In the Pozzallo-P. Castellazzo sector (Fig. 13.7), human impacts were related to the modification of Pozzallo pier and to the breakwaters constructed at Pietre Nere

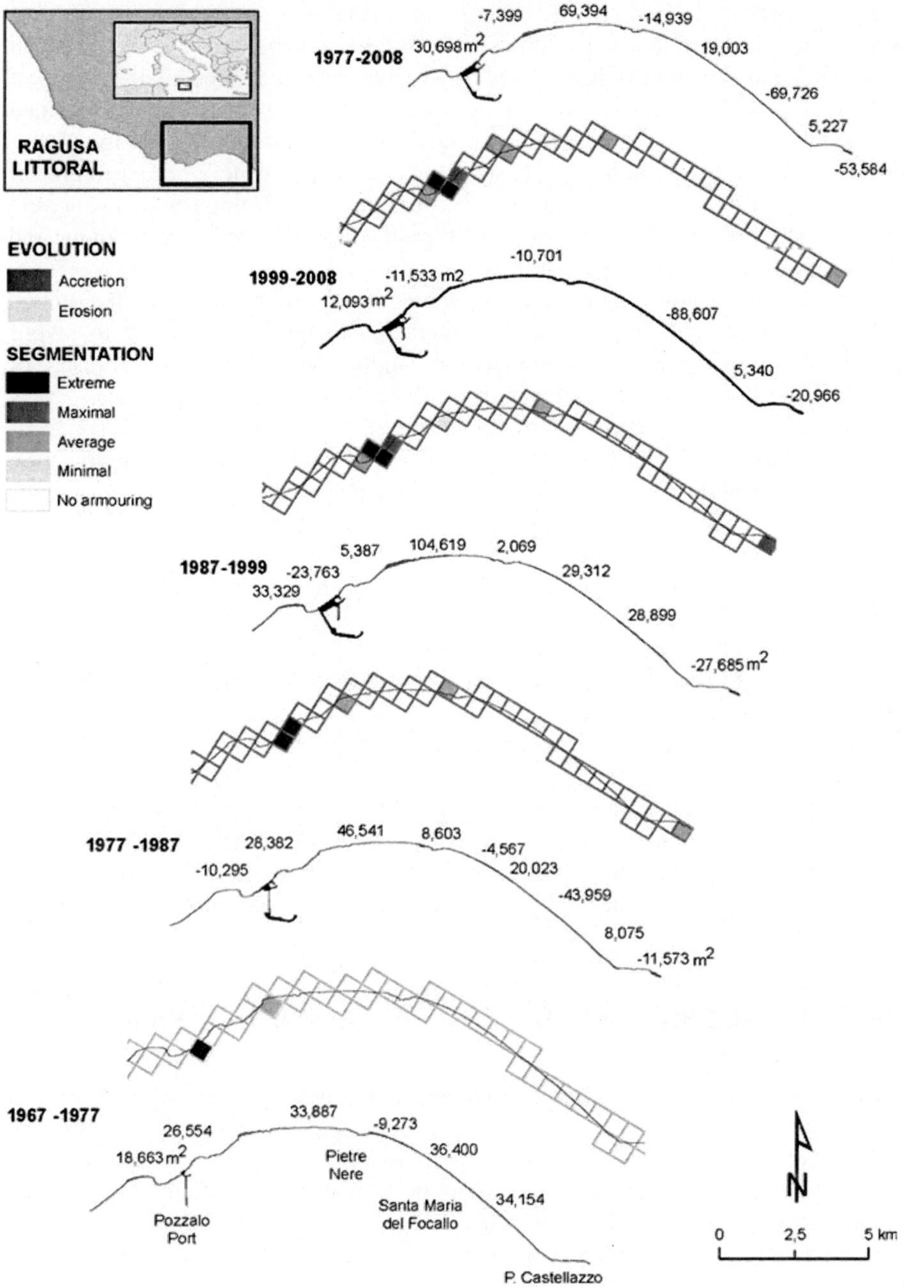

Fig. 13.7 Coastal evolution and armouring along the Pozzallo – P. Castellazzo sector

and P. Castellazzo. Pozzallo is the most important coastal town in the Ragusa Province with a permanent population of c. 17,500 inhabitants. The town, which has a long maritime tradition, is served by a port constructed in the 1970s to support the developing industrial activities. It originally comprised an offshore dock joined to the mainland by a pier which did not greatly affect littoral drift. In the 1980s, however, it was modified by the construction of a new east pier in order to protect the structure from waves coming from the third quadrant. Currently, this structure, which is still underused, is divided into different sectors devoted to recreational and fishing boats and cargo ships.

Because of the coastline orientation, wave fronts approaching from the second quadrant are of great importance and interact with the Pozzallo port to favour progressive and increasing sedimentation on the eastern side. As a result of this infilling, formerly deep water in front of the cliffed coastal sector is now very shallow and the urban beach of Pozzallo recorded an accretion of about 100,000 m^2 of beach surface (c. 7 m/year) over the 1987–1999 period after the modification and enlargement of the port structure. Tourism developments, i.e. bars, restaurants, beach facilities and a promenade, have been constructed on the recently-formed beach.

Sediments impounded east of the Pozzallo port are not sheltered from western approaching waves because on the lee side of the port erosion processes were observed at Pietre Nere (where a 150 m long breakwater was emplaced) and at Santa Maria del Focallo, where coastal retreat affected the littoral road that was protected by concrete blocks. Over the 1999–2008 period, about 88,000 m^2 (c. 3.2 m/year) of beach surface reduction took place at S. Maria del Focallo (Fig. 13.5).

Lastly, four breakwaters were emplaced at the eroding cliffs forming the P. Castellazzo headland in order to preserve archaeological sites. The breakwaters caused an increase in beach surface area and the formation of tombolos that in subsequent decades experienced successive cyclic faces of erosion and accretion.

13.4 Coastal Structure Impacts and Management Policies

The observations show that beach behaviour is not primarily impacted by coastal dynamics but by ill-placed developments constructed since the 1960s. Coastal infrastructure includes beachfront summer houses, restaurants, coastal roads and greenhouses and all of these greatly affect beaches and dunes (Fig. 13.8a, b). The extension of ports and harbours initially designed for recreational and small-scale fishing activities (at P. Secca, Marina di Ragusa and Donnalucata) or fishing and commercial activities (at Scoglitti and Pozzallo) have also had significant impacts.

The ports have suffered from an unsuitable location in terms of hydrodynamics and poor design and they have experienced problems through accumulation of sand and *Posidonia Oceanica* debris and cannot be used for long periods. For instance, at Donnalucata, during 2007, the harbour was totally filled by algae and sand and

Fig. 13.8 Beach erosion and damages caused by run off processes at M. di Acate (**a**, December 2010); environmental degradation associated with greenhouses north of M. di Acate (**b**, December 2010); white gravel sediments coming from the nourished sector at Caucana (**c**, April 2006) and the mechanical machinery used in the attempt to remove gravel sediments (**d**, July 2010)

was completely useless; a similar situation presently exists at the Pozzallo port sector for recreational boats.

In order to solve infilling problems, port docks were elongated several times which exacerbated downdrift erosion while periodically dredged sediments – that could have been used for beach nourishment or just bypassed downdrift – were accumulated in city dumps or deposited offshore (Anfuso and Martinez 2005).

All of the ports and harbours and the breakwater at Plaja Grande gave rise to wide beaches but only a c. 50 m wide strip close to the shoreline is used by bathers, most of the beach surface being unused or only partially occupied by tourist developments, especially restaurants and beach facilities. Hence, beach users and Municipalities acquired some minor benefits from beach surface accretion at specific sites. The opposite is true in eroding areas. Erosion processes at Santa Maria del Focallo (Ispica Municipality) reduced the beach surface area, diminished the beach attractiveness and reduced tourism demand, and created additional expense to repair the damaged coastal road and construct protective structures. The eroded sediments accumulated at Pozzallo town where the newly enlarged beach presently hosts a number of economically beneficial activities for the local residents and the Municipality.

The term 'coastal protection' means different things to different people (Cooper and McKenna 2008; McKenna et al. 2008). To environmentalists and ecologists it means letting nature take its course to protect the ecosystem while to coastal landowners and engineers it means constructing something to protect property. The latter interpretation is prevalent along the Ragusa littoral. But the protection is successful only if its impacts are disregarded. These are, however, usually very high in socio-economic, demographic, ecological, physical and climatic terms (Fabbri 1998).

The use of science in environmental decision-making has not increased over past decades, thus resulting in a huge knowledge gap between scientists and decision makers (O'Connor et al. 2009). The main problems which prevent the integration of science into the environmental decision-making are uncertainty surrounding scientific information, unusable scientific information and lack of active and continuous communication between different actors (O'Connor et al. 2010). This has been true through recent decades in Sicily, where no relation at all existed between scientists and managers, and a general management plan was never implemented. Nowadays, some improvement is taking place in this sense because of the development of environmental studies, collaborations with research entities and the formulation of general guidelines for ICZM (Integrated Coastal Zone Management) at a regional level.

Despite the existence of the aforementioned guidelines, beach protection and erosion management were transferred in 2010 from the Provinces to the Municipalities that only have a very local view of erosion problems and quite often lack the technical capacity for proper coastal management decision-making.

The response of policy-makers and managers in Sicily to coastal erosion suggests a concern for short-term human interests over long-term strategic goals despite the fact that strategic goals in ICZM should prevail over local ones. McKenna et al. (2008) show that there is still sufficient flexibility in European ICZM recommendations to justify advancing the self-interest of individuals or small groups at the expense of the public good or long-term sustainability. Generally, the local principles, based on small spatial and/or temporal perspectives, greatly reflect the bottom-up participatory, consensus-based approaches in ICZM at the expense of generic, strategic principles. In Ragusa decision-making is not based on bottom-up community participation. Rather, decisions are taken at a personal level, because of very specific political or economic interests and often breakwaters were emplaced under the guise of "urgent" conditions that allowed Administrators to order their construction without following the normal legal procedures, e.g. public procurement.

This case study shows a common sequence of events where one unwise action is countered by another, which in turn creates more problems. This was because of a weak system of regulation, strong political interference and ignorance (or indifference) of the implications of the environmental impacts of human interventions. The analysed situation was exacerbated by a public perception that government had a duty to protect private property, a situation that has been noted elsewhere in Europe (Cooper and McKenna 2008).

As a result, the massive engineering interventions at Donnalucata were precipitated by unplanned and illegal constructions (illegal in the sense that they do not conform to planning regulations) of beachfront summer houses and greenhouses for intensive agriculture. Without the presence of these illegally built structures, there would have been no need for publicly funded intervention. The degradation caused by efforts to defend the interest of a small number of stakeholders is not only suffered by society in general, through loss of environmental quality and amenity and public funding of defence structures, but also by nearby stakeholders whose (downdrift) properties became at risk after the emplacement of structures (Cooper et al. 2009). Similar situations were observed at Scoglitti, Marina di Ragusa, Plaja Grande and Pietre Nere, where breakwaters were successively constructed to enlarge beach width provoking downdrift erosion.

Hence, the shoreline stabilization 'solutions' did not solve the problem but just treated the symptoms and, in many cases, produced side effects which created additional problems, the most important being downdrift erosion. Furthermore, at the medium-term scale, coastal structures prevent landward migration and inhibit beach-dune and beach-shoreface interactions which are especially important during extreme events. They increase coastal squeeze, which requires more hard structures and/or greater nourished sand volumes for beach maintenance under sea level rise conditions (Cooper and Alonso 2006). This is a common problem at many places in the study area as well as in many other Mediterranean coastal sectors where beaches have been transformed into "sandy solariums" and are experiencing continuous and constant deterioration, under the perception that a beach is similar to a golf camp or a playground which need continuous maintenance. In such situations, humans are the primary geomorphic agent and beach construction and maintenance depend on human interventions. This creates a dramatic rise in expense for maintenance as even more infrastructure is built at the coast.

Additional problems associated with the emplacement of coastal protection structures are the increased risk of drowning and the negative impact of structures on coastal scenery (*sensu* Williams and Micallef 2009), especially evident at Donnalucata where breakwaters create a major visual impact.

Beach nourishment works at Caucana in 2004 may produce severe consequences on seagrass beds, which have a great ecological value and reduce the energy of incoming waves. The seagrass areas form the natural source of calcareous skeletal debris that at places comprises up to 80% of the natural beach sediment (Garcia and Servera 2003). At Caucana, nourished sediments essentially consisted of coarse sand and gravel obtained by crushing the local Miocene limestone (the Ragusa Formation) with a cost of 12€/m^3 and a total cost for the nourishment works of two million Euros. High water turbidity was observed during the nourishment works and for a few months afterwards particularly during energetic wave conditions. Under these conditions, sediments are deeply remobilised and the white silt-sized fraction is suspended. The silt-sized fraction is linked to the crushing process and also to the abrasion process, which progressively rounded and reduced the size of the injected sediments.

At present, gravel sediments are partially covered with sand and the problems of turbidity previously described did not impact the sea-meadows fronting the nourished beach sector. On the other hand, in 2005 nourished sediments started to bypass the eastward artificial submerged groin and arrive at the adjacent beach (Fig. 13.8c). This produced great discontent among beach users and pushed the local Municipality into a ridiculous attempt, carried out in July 2010, to remove the gravel which by that time was deeply mixed with the natural sand along the whole beach length (Fig. 13.8d).

Other beach nourishment works are projected for several sectors in the study area. Most important problems linked to the execution of such projects are the availability of suitable borrow material in land deposits, because no geophysical surveys have been carried out to identify marine sources. Furthermore, the idea of using sediments trapped updrift, or within ports and harbours, for nourishment projects as well as the idea of placing permanent bypassing systems, have not been considered.

Lastly, several economic considerations emerge when analysing the solutions adopted to counteract coastal erosion problems over the past decades. On one side, the huge costs associated with beach nourishment and construction and maintenance of hard structures were mainly supported by the Regional Government through European Regional Development Funds. On the other side, income from property taxes and other revenue generated by coast-dependent tourism was gained by local Municipalities. Furthermore, the cost of the protective structures is quite high when compared with the economic returns from beaches, essentially consisting of activities, such as the renting of summer houses, which do not contribute taxes.

According to the previous assumptions, hard structures and nourishment operations at many places along the Ragusa littoral are probably only partially economically justified. An exhaustive cost-benefit analysis should be carried out because such analyses give a reliable basis for decision-making processes in coastal erosion management.

Acknowledgements Thanks go to Mike Scott (UBC Vancouver, Canada) for improving the quality of English and to Morris Floridia for field assistance. This work is a contribution to the Andalucía PAI Research Group RNM 328.

References

Anfuso G, Martinez JA (2005) Towards management of coastal erosion problems and human structure impacts using GIS tools: case study in Ragusa Province, Southern Sicily, Italy. Environ Geol 48:646–659

Anfuso G, Martinez JA (2009) Assessment of coastal vulnerability through the use of GIS tools in South Sicily (Italy). Environ Manage 43:533–545

Aybulatov NA, Artyukhin YV (1993) Geo-ecology of the World Ocean's Shelf and Coasts. Hydrometeo Publishing, Leningrad

Boak E, Turner I (2005) Shoreline definition and detection: a review. J Coast Res 21(4):688–703

Carter RWG (1988) Coastal environments. Academic, London

Cooper JAG, Alonso N (2006) Natural and anthropic coasts: challenges for coastal management in Spain. J Coast Res SI 48:1–7

Cooper JAG, McKenna J (2008) Working with natural processes: the challenge for coastal protection strategies. Geogr J 174(4):315–331

Cooper JAG, Anfuso G, Del Río L (2009) Bad beach management: European perspectives, vol 460, Geol Soc Am, Special Paper. Geological Society of America, Boulder, pp 167–179

Crowell M, Leatherman SP, Buckley M (1993) Shore-line change rate analysis: long term versus short term data. Shore Beach 61(2):13–20
Doods R, Kelman I (2008) How climate change is considered in sustainable tourism policies: a case of the Mediterranean islands of Malta and Mallorca. Tour Rev Int 12:57–70
Doody JP (2004) 'Coastal squeeze' – an historical perspective. J Coast Conserv 10(1–2):129–138
EEA (2006) The changing face of Europe's coastal areas. In: Breton F, Meiner A (eds) EEA report no 6, Luxembourg office for official publications of the European Communities, Luxembourg
Fabbri KP (1998) A methodology for supporting decision making in integrated coastal zone management. Ocean Coast Manag 39:51–62
Garcia C, Servera J (2003) Impacts of tourism development and water demand and beach degradation on the island of Mallorca (Spain). Geogr Ann 85:287–300
Hanson H, Brampton A, Capobianco M, Dette HH, Hamm L, Laustrup C, Lechuga A, Spanhoff R (2002) Beach nourishment projects, practices, and objectives – a European overview. Coast Eng 47:81–111
Jiménez J, Sánchez-Arcilla A, Bou J, Ortiz M (1997) Analysing short-term shoreline changes along the Ebro delta (Spain) using aerial photographs. J Coast Res 13(4):1256–1266
Jones A, Phillips M (2011) Disappearing destinations. CABI, Wallingford
Klein YL, Osleeb JP, Viola MR (2004) Tourism generated earnings in the coastal zone: a regional analysis. J Coast Res 20(4):1080–1088
Martínez JA, Anfuso G (2008) Spatial approach to medium-term coastal evolution in south Sicily (Italy): implications for coastal erosion management. J Coast Res 24(1):33–42
McKenna J, Cooper JAG, O'Hagan AM (2008) Managing by principle: a critical analysis of the European principles of Integrated Coastal Zone Management (ICZM). Mar Policy 32:941–955
O'Connor MC, Cooper JAG, McKenna J (2009) Integrating science into shoreline management practice and policy: an Irish perspective. J Coast Res Spec Iss 56:1267–1270
O'Connor MC, Cooper JAG, McKenna J, Jackson DWT (2010) Shoreline management in a policy vacuum: a local authority perspective. Ocean Coast Manag 53:769–778
Pilkey O, Dixon K (1996) The corps and the shore. Island Press, Washington
Runyan K, Griggs G (2003) The effects of armoring seacliffs on the natural sand supply to the beaches of California. J Coast Res 19(2):336–347
Williams A, Micallef A (2009) Beach management, principles & practice. Earthscan, London

Chapter 14
The History of Shoreline Stabilization on the Spanish Costa del Sol

Gonzalo C. Malvárez

Abstract The assessment of failure or success in coastal stabilization depends on various factors. Most of the arguments for the implementation of coastal protection and/or stabilization rest on the need perceived by the local population, users and managers of a coastal stretch to stop shoreline recession or modify beach behaviour in some way. However, it is critical to establish at the onset of a process of beach stabilization what is the long-term trend of the beach system and, consider the long-term implications of maintaining any intervention in the coastal behaviour.

Costa del Sol in southern Spain is one of the most heavily developed coastal stretches in Europe (and the World) and is internationally renowned for attracting both mass- and high-income tourism. Annually more than nine million tourists visit the 100 km steep coastal segment, which supports a permanent population of over 1.2 million people. A significant proportion of the national economy is based (directly or indirectly) on tourism. Faced with threats to infrastructure from shoreline mobility, investing in coastal stabilization is the obvious response from the Public Administration, which is, by Law, responsible for the safe keeping of the shoreline and the Public maritime-terrestrial land around it.

The question considered here is: how healthy was the morphosedimentary system at the onset of mass tourism and how healthy is it now that up to 96% of some coastal council areas comprise consolidated urban land? When beaches began to suffer severe erosion in the 1970s, the question arose of whether to develop an alternative tourism economy based on golf courses and marinas or whether to try to stabilize and maintain a recreational beach system?

G.C. Malvárez (✉)
Area of Physical Geography, Universidad Pablo de Olavide, de Sevilla, Sevilla, Spain
e-mail: gcmalgar@upo.es

J.A.G. Cooper and O.H. Pilkey (eds.), *Pitfalls of Shoreline Stabilization: Selected Case Studies*, Coastal Research Library 3, DOI 10.1007/978-94-007-4123-2_14,

This paper provides an analysis of the environmental and socio-economic context and the sequence of shoreline stabilization approaches that have been applied but failed. Coastal stabilization techniques essentially attempt to stop nature from finding its equilibrium, but in Costa del Sol despite several major changes in strategy no approach has been successful.

14.1 Introduction

14.1.1 Population Dynamics and Socioeconomic Context

The Costa del Sol as a global tourist destination is a coastal stretch of some 100 km long by (effectively) 5 km wide (Fig. 14.1). In 2005 alone it received over nine million tourists of 50 nationalities. This places pressure on the resources and infrastructure, whose resident population has increased from 380,873 in 1950 to 1,136,712 in 2006. In some towns, like Marbella, population increased from 8,982 in 1940 to 12,069 in 1960, to 100,036 in 2001.

The Costa del Sol tourism infrastructure developed without rational spatial planning and growth was simply determined by demand (Malvarez-Garcia et al. 2000, 2005). The result has been a transformation of the coastal landscape with now irreversible environmental impacts (loss of sand from beaches, large scale alteration of river basins, pollution of aquifers, etc.). Some of the associated economic and social impacts have been interpreted as positive trends; between 1991 and 2007 for example there was an increase of 248% in tourist places and 60% in the resident population. The rapid growth in coastal tourism has been the main driver of this change. Additional to littoral resources, the rural hinterland provides cultural resources, adventure, golf, etc. with significant demand from residential tourism.

The response from the Government to tourism growth has been to increase the area of land available for development. For instance, from 1998 to 2005 the consolidated urban land in the eastern Costa del Sol extended by 23%, and in 2001 the land zoned for development already accounted for 16,771 ha. Some councils are proposing in their new planning documents to urbanize 30% of the municipality, both as developed and development-zoned land (Navarro-Jurado 2005).

Some stretches of Costa del Sol currently experience considerable population pressure. The average density in some municipalities exceeds 5,000 inhabitants per km^2 which is among the highest in coastal Europe (EEA 2006). In Europe the highest increase in artificial surfaces (20–35%) between 1991 and 2001 took place in the coastal zones of Portugal (34%), Ireland (27%), and Spain (18%) within the 10 km-wide coastal zone. In several coastal regions of Italy, France and Spain the coverage of built-up areas in the first kilometer coastal strip exceeds 45%. Costa del Sol has been undergoing this process of human modification since 1956 (Fig. 14.2).

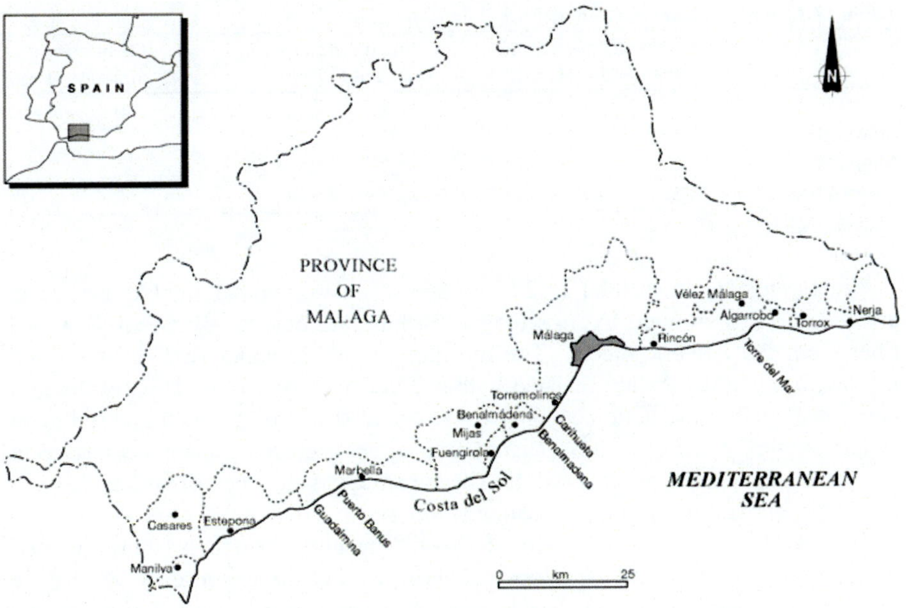

Fig. 14.1 Location map of Costa del Sol, showing places referred to in the text

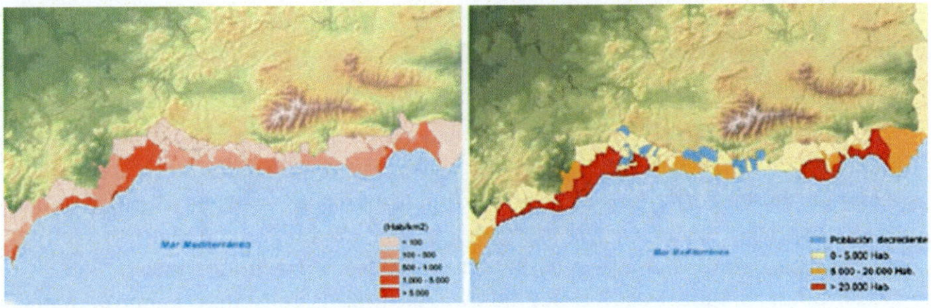

Fig. 14.2 Population density and trends in Costa del Sol. After CMA, 2009

14.1.2 Environmental Setting

In the study area (Fig. 14.1) the main hydrodynamic influence on the coast is wave action; tidal range is negligible. The wave regime in this section of the Mediterranean Sea can be classified as low energy but this is punctuated by storms that bring water level surges and greater waves (e.g. Backstrom et al. 2008). The mean significant wave height is 1.0 m and mean zero crossing period is 5.0 s with an almost equal bimodal directional spectrum dominated by westerlies and easterlies. The beach morphology is adjusted to high frequency waves in the form of steep-narrow foreshores (Table 14.1).

Table 14.1 Wave climate in Costa del Sol

	Deepwater significant wave height (Hs m)	Mean zero crossing period (Tz secs)	Mean wave direction (Deg)
Modal East	1.0	5.0	80
Storm East	2.5	7.0	80
Modal West	1.0	4.5	255
Storm West	2.5	7.0	255

Malvarez-Garcia et al. 2000

Tidal range is microtidal (<20 cm average astronomical tidal range). The effective fetch is limited to an average 500 km. and only rarely do swell waves filter from the Atlantic Ocean. The morphology of the inner shelf is steep and narrow; oceanic depths are reached within 2 km from the coast. This results in a concentration of wave action on a narrow fringe of steep coastal shelf and sediment supply is mainly reworked fluvial sands and supply is episodic and concentrated in time around seasonal heavy rainfall. These characteristics mean that beaches are highly dependent on short-term sediment supply.

Energy dissipation takes place on steep foreshores where offshore-directed sediment transport rapidly develops off-shore bars at the expense of beach face erosion. In these reflective beach environments, during high-energy events (i.e. storms) higher rates of energy supply translate into violent plunging or collapsing breaker types that produce shoreline retreat and coarsening of beach material due to enhanced cross-shore sediment transport.

One significant characteristic of morphodynamic evolution in Costa del Sol is the very limited development of cellular circulation. Littoral cell boundaries are extremely variable given the energy regimes of waves and tides combined with the continental shelf morphology. This provides the basis for:

- Short crested, high frequency waves, caused by limited fetch and low energy achieved by local wind waves. These waves are significantly steep and cannot generate bottom friction to initiate sediment transport in water depths beyond 3 m generally. The wave geometry (generally trochoidal and short crested) is highly reactive to changes in bathymetry and thus refraction is intense.
- Tidal range is almost negligible, thus wave action is constant on approximately the same water level. Despite the statistical presence of 70% calm wave fields from offshore records, this is generally not observed on the coast as sea breeze is very active during the spring, summer and autumn periods.
- Steep surf zones generate semi reflective scenarios and intense sediment transport in narrow mid surf zones. This implies strong short-lived longshore drift periods and potential for rip currents.

14.2 Shoreline Stabilization

The Spanish Coastal Act, the highest level guidance for land planning along the coast, establishes that all coastlines are considered public land and thus, it is mandatory to allow access along and onto the public land. This concept is the

Fig. 14.3 Promenade on sea wall. The 1960s wall now supports major park-like use for the parks and gardens municipal system

base for the convergence of separate issues, for example, the recreational and protective functions of promenades constructed on sea walls. Once the beach is backed by urban land, competing interests originate largely because of the past failure of the planning process to deal effectively with the early growth of the urban communities and the key role that shoreline stabilization represents in the defence of the tourism infrastructure.

14.2.1 Sea Walls: First Try

Physical protection of the coast followed the rise in value of shoreline real estate and the first engineering works affecting the villages of the Costa del Sol were undertaken in the 1960s as promenades were constructed on dunes and the back-beach area (Fig. 14.3). These were often subject to inundation during storms and were followed by demands for further protection or strengthening.

Promenades enabled access to the beach and they became important, both for their protective and recreational roles. Sea-walls fronting the promenades were considered the most feasible and appropriate shoreline stabilization method. However, the Costa del Sol began to experience an increasing erosion problem in the

Fig. 14.4 Malagueta beach in Malaga after effects of sea-wall. The b is gone. Photograph from Archivo Histórico Provincial de Málaga

1970s after the construction of most of the sea-walled promenades because of erosion induced by their structural characteristics.

As is common with sea walls, the effects of undertow erosion did not take long to appear and narrowing of beaches was widespread along the coast (Fig. 14.4). This effect was accentuated along the Costa del Sol given the acute angle of approach of waves which commonly approach at a 45° angle to the shoreline and create significant littoral drift in the narrow surf zone. The sand in this context is transported to deep water and recovery by fairweather waves is impossible since the sand is beyond wave base. In addition, sea wall end effects translated erosion downdrift, to which the response was to extend the sea walls.

Promenades, built on sea walls also incorporated significant functions for accommodation of urban infrastructure such as water pipes, electricity etc. Thus they were never intended to be modified. Some of the requirement for modification resulted from frequent flooding on the land side because the sea walls were too high to allow the seaward discharge from ephemeral streams. Thus flooding came from both sides of sea walls when storms hit the Costa. With these issues the authorities sought a different approach to shoreline stabilization.

14.2.2 Groynes: Second Try

The next phase of efforts to stabilize the coast came in the form of groynes. Straight structures, normal to the beach, were deployed in an attempt to protect the promenades, maintain the beaches and stabilize the eroding shoreline. However, the process causing the erosion of the beach was not tackled because, although the groynes were efficient in controlling longshore drift, they failed to stop off-shore-directed sediment transport. This led to the subsequent use of hammer-head groynes designed to cope with orthogonal as well as longshore movement.

Fig. 14.5 A large groyne field at Marbella in the 1970s. The hammerhead design coped with onshore and longshore erosion but hampered water circulation at a time when sewerage plants were still not fully operational and discharges took place directly into the artificial embayments

These groynes became a common feature all along the Costa del Sol (Fig. 14.5) in the 1970s. Groyne fields were also deployed in an attempt to minimize erosion and control deposition around new marinas, like Puerto Banús, a luxury tourist resort with high investments in real estate.

The issue of littoral drift along the Costa del Sol is one of the main issues in coastal stabilization. As elsewhere, (Cooper and Pilkey 2004) drift is not well understood at a regional level and the bidirectional wave climate severely influences the functioning of groyne fields and jetties at marinas. The port of Fuengirola, constructed in the 1970s on an existing mooring jetty provides a good illustration. Port engineers chose opposite directions for net longshore drift in the 1950s and 1970s designs, but none was fully successful and sedimentation in the marina has been a constant issue for navigation and shoreline erosion downdrift (both east and west).

14.2.3 *Beach Nourishment: Third Try*

A new approach to protection was introduced with the implementation of the Coastal Act of 1988; this was aimed at starting a new era in soft coastal protection that superseded the hard solutions of the past. It involved the widespread application of beach nourishment (Fig. 14.6). Beach nourishment, however, turned out to be unsuccessful in terms of the longevity of the beach fill and the associated economic implications. Projection of outcomes of nourishment projects are very

Fig. 14.6 Beach nourishment is a world wide shoreline stabilization method. Local hydrodynamics should aim at equilibrium profile. Malagueta beach in Malaga 1990. Photo from Demarcación de Costas de Málaga. Ministerio de Medioambiente

much dependent upon understanding local morphodynamic conditions, and particularly the processes operating when modal conditions are exceeded (during storms).

Experiences on the Costa del Sol show that the application of beach fill has not been entirely satisfactory partly because of difficulties in solving the fundamental problem of sediment starvation in the coastal system. The former riverine sources of sediment have been much reduced by dam building and the sediment stored in coastal dunes has been lost under urban development. Offshore sediment deposits were used in beach nourishment schemes and reasonably resilient beaches were built, but only because the material that was pumped onto the beach was so coarse or had such a high component of limestone that it became highly consolidated and partially cemented. In other cases, off-shore material was excessively shelly, making it unpleasant for recreational purposes. The first attempts to nourish Marbella's beaches failed partly through use of sands drawn from the mouth of the Guadalhorce River. The upper surface of these sands rapidly cemented, creating an abrasive surface that was unsuitable for recreational use.

In contrast, sediments utilized for nourishment of Málaga's beach had ideal characteristics, although it was still prone to sand loss during storms. The failure to achieve a reasonable life span is strongly influenced by the distortion that beach nourishment introduces into the littoral system. The sudden addition of large volumes of beach material creates a significant shift in the hydrodynamics of shoaling waves, well beyond the expectations of the design phase.

This seems to have been the case on beaches such as Marbella and Málaga, where replenishment had to be repeated soon after the first fill. The reason for this could be the entrainment of large volumes of sediment in a littoral cell whose dynamics have been distorted into a more effective transport machine. Over seven million m^3 of sand were pumped onto 27 km of beaches along the Costa del Sol in 1992 alone. Continued annual expenditure on beach nourishment is necessary to maintain the beach resource and stabilize the shoreline

14.3 Phases of Shoreline Stabilization: An Example

An interactive viewer (Ojeda-Zujar and Cabrera-Tordera 2006) allows direct comparison of various orthophotos of the Costa del Sol. Figure 14.7 shows a sequence of changes at the Playa de El Fuerte de Marbella (The Fort). Figure 14.7a shows the beach in 1956 on which the sea wall was subsequently to be constructed on mid-beach. The small stream opposite the old Fort was channelized and built upon. The beach disappeared within 10 years of seawall construction and groynes were built in the 1970s (Fig. 14.7b). Note the significant gain in beach surface and shoreline length. However, water quality in the embayment soon became a problem. Water treatment plants were established only in the early 1990s with the implementation of the newly enacted Coastal Act of 1988. The beach was nourished after demolition of the groynes in 1992 and built to a width of 60 m. Figure 14.7c shows that in 1998, despite significant readjustment in the early years of the nourishment scheme, the beach was wider than originally. This standard width, which was applied to all Spanish beaches when possible, was based on a cost effective analysis performed to estimate whether coastal protection could be supported by tax increases given the additional surface provided for recreational purposes. This beach was fringed by two terminal groynes, but beach re-fills were necessary throughout the 2000s and to the present. Figure 14.7d shows the beach in 2009. The scars of all former stabilization attempts can be seen, starting from the tree-lined old sea wall that is now a major promenade lined with exclusive businesses and El Fuerte, the old Fort, converted to a 5 Star Spa-Hotel.

14.4 Discussion

The rapid development of tourism and the associated efforts at shoreline stabilization recorded in the Costa del Sol are unparalleled in Europe. The situation prompts a reflection on why did this all happen and what was its influence on the status of the coastal sedimentary system?

Many old industrial buildings in Costa del Sol, (*Ingenios* or sugar cane factories, Fig. 14.8), which were abundant in the 1870s, can still be found in ruins scattered along the coast. The floodplains of the rivers were too wet for other agricultural

Fig. 14.7 Sequence of shoreline stabilization programmes at Playa de El Fuerte, Marbella –(**a**)1956-(**b**)1976-(**c**)1998-(**d**)2009

Fig. 14.8 An old sugarcane refinery (Ingenio) in the 1800s from eastern Costa del Sol. From archives of the Casa de los Lario de Malaga

practices and sugar became a prosperous bussiness. *Trapiches* (sugar mills) were often associated with the sites of previous industrial activities (chiefly mining) that had dominated earlier.

The blast furnace of La Concepción at Rio Verde, a roofless ruin today in wasteland in the hinterland of Puerto Banus for instance, employed some 1,084 men in 1828 and by 1844 was producing 72% of all the iron produced in Spain (Gil-Delgado and Jiménez-Barrientos 2007). This significant industry was based on three main elements: proximity to the iron and limestone mines of Sierra Blanca, hydro-power from the Rio Verde and the abundance of woods in the surrounding area to provide fuel. Forests were depleted fast by the successful industry and soil erosion in the steep Sierra would have accelerated.

However, the abandonment of industrial activity (mining, furnaces, sugar cane) during the early 1900s led to economic decline. In Marbella, like all along the coast in the 1940s, the most frequent cause of death was reported to be malaria. The large rivers of the west of the area, the Rio Verde (Fig. 14.9), Rio Real, Guadaiza and Guadalmina provided high water discharges during the winter months but remained stagnant for long warm periods in the summer creating ideal conditions for mosquitoes. In the high Sierra de Grazalema (approx. 1,100 m.), rainfall exceeds 2,000 mm year^{-1}. The many rivers and streams running southeast supported high sediment yields to the coast and created ephemeral fresh water lagoons in the

Fig. 14.9 River mouth of the Rio Verde in 1956 (**a**) and 2009 (**b**). Note the marked drift to the SW and blockage creating lagoons in (**a**)

summer. These lagoons were breached by fluvial discharge each winter, providing an influx of sediment to the coast.

In this microtidal environment sediment yield from rivers is redistributed by wave action (particularly during storms) in a narrow coastal fringe and the short-term equilibrium (i.e. stability of shoreline position) is highly dependent on high-energy events in both fluvial and marine environments

In southeastern Spain rivers are typically ephemeral and characterised by relatively short and steep catchments. This leads to rapid concentration and significant water run-off. These ephemeral rivers discharge onto a steep and narrow coastal shelf. The steep shelf provides an unstable environment for large accumulations of sediments.

The natural sedimentary system at the onset of the tourism era is evident from examination of historical air photos. A comparison between 1956 and 2009 orthophotography showed abundant sand on beaches with development of wide supratidal beaches and aeolian dunes (Fig. 4.10). Parabolic dunes were present at Los Monteros and vegetated dunes east of the Guadalmina River for example (Fig. 4.10). Under these conditions it appears that the riverine system yielded a high sediment load that was reworked by a vigorous littoral drift. The contemporary situations shown in Fig. 4.10 show the extent to which urban development has covered the former littoral sediments and how narrow the contemporary beaches are.

Fig. 4.10 The abundance of marine sediments on beaches of the 1950s generated significant dune fields; some parabolic dunes in "El Saladillo" (*left set*)

14.5 Conclusion

The continuous attempts to control natural processes along the highly developed Costa del Sol failed (and continue to fail) for various reasons. The main reason is that the project-based approaches to stabilization did not include sufficient acknowledgement of regional sedimentary processes.

Among the variety of issues that were not accounted for in coastal protection and shoreline stabilization was the then established trend that the sustainability of the measure had to be guaranteed. Hard engineering was the preferred option because of the demand to protect and extend the newly developed economic resource (tourism) and getting rid of the erosion threat to infrastructure was a priority. Meanwhile promenades, build upon sea walls, provided additional valuable land for beachfront development. In many cases, promenades and infrastructure for development (i.e. roads, water treatment networks, etc.) were established upon the backshore, and most times on significant sand bodies. This is where the natural and human systems collide: the sand bodies indicated a well-established sedimentary system that could fluctuate according to sediment supply and wave conditions but sea walls focussed on establishing the landward limit for shoreline fluctuation. Obviously this created beach narrowing and destruction which prompted a new approach.

When promenades have been established upon seawalls, even if the position is encroaching onto the sea rather than defending a retreating position, and beaches are needed as the main economic resource, a groyned shoreline seemed the best option. If the system is still presumed to have sediment availability, it would be reasonable to assume that littoral drift would have provided the necessary material to support beach creation within the artificial embayments. The actual situation was quite different because sediment yield was reduced by river damming (to provide water supply) and changes in landuse practices (that reduced soil erosion). Groynes also require detailed and precise calculation of spacing and in the 1970s wave data was not available for long-term diagnosis of wave-induced processes. In addition, the highly bi-directional wind/wave regime is particularly problematic in the design of groyne fields.

An additional element that is perhaps critical in the chronic failure of shoreline stabilization is that Costa del Sol, like many Mediterranean shores, is dominated in summer time by a significant sea breeze. This element may have been critical in the lack of success or sustainability of the final and third attempt in shoreline stabilization: nourishment. Along with the severe lack of sediment available to the beach system (which by this stage was clearly identified) the forcing factors were related to low energy environments common in Mediterranean conditions. This in the 1990s was corroborated by incipient but reliable wave records. These wave records, available in some cases since 1985, measured offshore waves. These records suggest that wave climate in Costa del Sol is dominated by calm conditions (over 77%), whereas sea breezes are very important in the coastal dynamics. Therefore nourishment projects based on these wave data may have underestimated energy conditions. In addition to this, the lack of tide also enhances the role of surf zone dynamics operating at a consistent sea level during the sea breezes.

Pitfalls in shoreline stabilization in Costa del Sol, and many other Mediterranean shores, may therefore be linked to:

- Lack of understanding of sediment dynamics in historical times
- Use of standard shoreline stabilization methods under the assumption of an inflated sediment budget
- Design problems related to groyne spacing. The shoreline, already having encroached seawards of its original position, was fronted with groynes that were incapable of controlling highly oblique and bidirectional wave fields.
- Severe water quality issues related to very enclosed bays generated by hammerhead groyne fields
- Beach nourishment failure is related to lack of available sediment in the system as a whole
- Nourished beaches work in unpredictable ways if cellular cells are not understood. Wind fields and indented coastline generate a complex cellular circulation frequently dominated by opposite current fields
- Underestimation of wave energy and absolute miscalculation of sea breeze effects on morphodynamics.

The continued efforts from engineers, scientists and managers to contain shoreline change in Costa del Sol may have failed for multiple reasons. Failures in shoreline stabilization efforts have been the constant theme in the experience; no matter how hard it has been attempted, Costa del Sol beaches have never achieved anything close to their original (pre-development) state and the system is now an utterly anthropic coast (Cooper and Alonso 2006).

References

Backstrom JT, Jackson DWT, Cooper JAG, Malvárez G (2008) Storm-driven shoreface morphodynamics on a low-wave-energy delta: the role of nearshore topography and shoreline orientation. J Coast Res 24:1379–1387

Cooper JAG, Alonso I (2006) Natural and anthropic coasts: challenges for coastal management in Spain. J Coast Res Spec Iss 48:1–7

Cooper JAG, Pilkey OH (2004) Longshore drift: trapped in an expected universe. J Sediment Res 74:599–606

European Environment Agency (EEA) (2006) The changing faces of Europe's coastal areas. EEA report no 6/2006, Copenhagen, p 107

Gil-Delgado Ó, Jimenez-Barrientos JC (2007) Altos hornos "La Concepción" de Marbella (Málaga), *Bienes Culturales*. Revista del Instituto del Patrimonio Histórico Español 7:137–141

Malvarez-Garcia G, Pollard J, Dominguez R (2000) Origins, and measurement of stress on the coast of Southern Spain. Coast Manag 28:215–234

Malvarez-Garcia G, Pollard J, Dominguez R (2005) The planning and practice of coastal zone management in Southern Spain. J Sustain Tour 11:204–223

Navarro-Jurado E (2005) Indicadores para la evaluación de la capacidad de carga turística. Ann Tour Res en Español 7(2):397–422

Ojeda-Zujar J, Cabrera-Tordera A (2006) Utilidades y Funcionalidades de un Visor Tridimensional Interactivo en la Gestión del Litoral (Sigla: Sistema de Información Geográfico del Litoral de Andalucía). In: *El Acceso a la Información Espacial y Nuevas Tecnologías Geográficas. Granada.* Ed de la Universidad de Granada, pp 1707–1717

Chapter 15
Coastal Defense in NW Portugal: The Improbable Victory

Helena Granja and José Luís Pinho

Abstract Coastal defense in Portugal has a long history. The first attempts to resist coastal change date from early in the twentieth century, when the town of Espinho saw its most seaward streets damaged by wave impact. A coastal defense work (a "muralha", or wall) was built. The 1980s saw the building of several groins at different locations in Portugal. The fight against the sea continued during subsequent years in a tentative effort to achieve an "artificially stable" coastal zone. Coastal retreat and thinning of beaches continued, especially downdrift of structures. The pattern has progressively revealed the inefficiency of efforts to stabilize the coastline. The underlying causes have not been addressed and indeed are poorly understood. Coastal defense is simply a temporary and palliative means of addressing the impacts of coastal erosion and, sometimes gives a false sense of security to coastal populations. In this chapter an overview of the NW Portuguese coastal defense structures is presented, their efficiency is discussed and some representative examples are described. Some recommendations are presented to create a more efficient coastal zone management policy.

15.1 Introduction

Defenses against the sea in NW Portugal began at the beginning of the twentieth century. At that time, the density of population in the coastal zone was still low though some coastal towns had already suffered some infrastructural damage from

H. Granja (✉)
CIIMAR/Earth Sciences Department, University of Minho, Campus de Gualtar, Braga 4710-057, Portugal
e-mail: hgranja@dct.uminho.pt

J.L. Pinho
Department of Civil Engineering, University of Minho, Campus de Gualtar, Braga 4710-057, Portugal

J.A.G. Cooper and O.H. Pilkey (eds.), *Pitfalls of Shoreline Stabilization: Selected Case Studies*, Coastal Research Library 3, DOI 10.1007/978-94-007-4123-2_15,

their seaside location. The town of Espinho was confronted with serious erosion by the end of the nineteenth century and in 1909 the first seawall (the "muralha") was constructed, only to be destroyed by the sea during a storm in 1910.

But real problems arising from coastal erosion happened when people moved to the coastal zone for tourism and leisure, at the beginning of the second half of the twentieth century in the first phase and for permanent living in the last few decades. The number of houses and buildings grew exponentially without much concern for planning, though land use policies from CZM plans were implemented and an older law had existed for a long time (the "Law of the maritime public domain" dates from 1864). With these constructions, coastal risks and the need for defense appeared. The cycle of building followed by defense works was followed by a worsening of erosion impacts. Since the 1990s, in the NW coastal zone, the sedimentary deficit has become more pronounced leading to sand loss from beaches (Granja and Loureiro 2007; Loureiro 2007), steepening of beachfaces, and increased cliff retreat rates.

The NW zone of Portugal is mesotidal with a highly energetic wave climate. Main wave crest orientation is from the northwest, inducing a drift current from north to south. However, this current is, in some areas, reversed due to the presence of some natural (bars, ebb tidal deltas, rocky outcrops) and artificial (breakwaters, jetties, groins) obstacles that promote local wave diffraction. This is the case at the Leixões Harbour breakwaters (Fig. 15.1) which promote a south to north drift that has caused updrift accretion of Matosinhos beach to the North.

The Douro estuary spit points to the north (as do other rivers such as the Lima, Cávado and Ave) by a south to north drift produced by wave diffraction around the ebb-tidal delta and some rock outcrops. South of it, the main drift is again North–south, promoting the accumulation of sediments updrift transverse obstacles (*e.g.* Espinho groins and Madalena submerged outfall) and increasing erosion downdrift.

The causes of erosion on the Portuguese coast are: (i) lack of sediment supply due to dam construction, dredging in harbours, extraction of sediments and depletion of sand deposits on the inner shelf; (ii) influence of coastal defense works, (iii) increase in the mean sea level and in the frequency and intensity of storm events. Until now, it was not possible to clearly demonstrate the relative importance of each cause for erosion processes. However, recent monitoring programs on some coastal stretches have revealed evidence that should be further investigated in order to quantify the rates and causes of erosion.

Many dunes have been almost entirely eroded along the north and central Portuguese coast. In the northern segments sandy beaches have been replaced by gravel beaches. Field observations showed that coarse sediments and reworked shells are preferentially deposited in the beachface, meaning a lack of finer sediments in the nearshore. Transverse structures located south of occupied coastal areas are becoming increasingly inefficient. This evidence shows that in order to keep the sandy beaches on the Portuguese coast, sediment sources need to identified and preserved. This means that erosion must be allowed to happen freely on some coastal stretches in order to temporarily maintain others, in locations where cliff retreat is the major sand source for transport by littoral drift. A compromise between coastline retreat and retaining sandy beaches must be achieved.

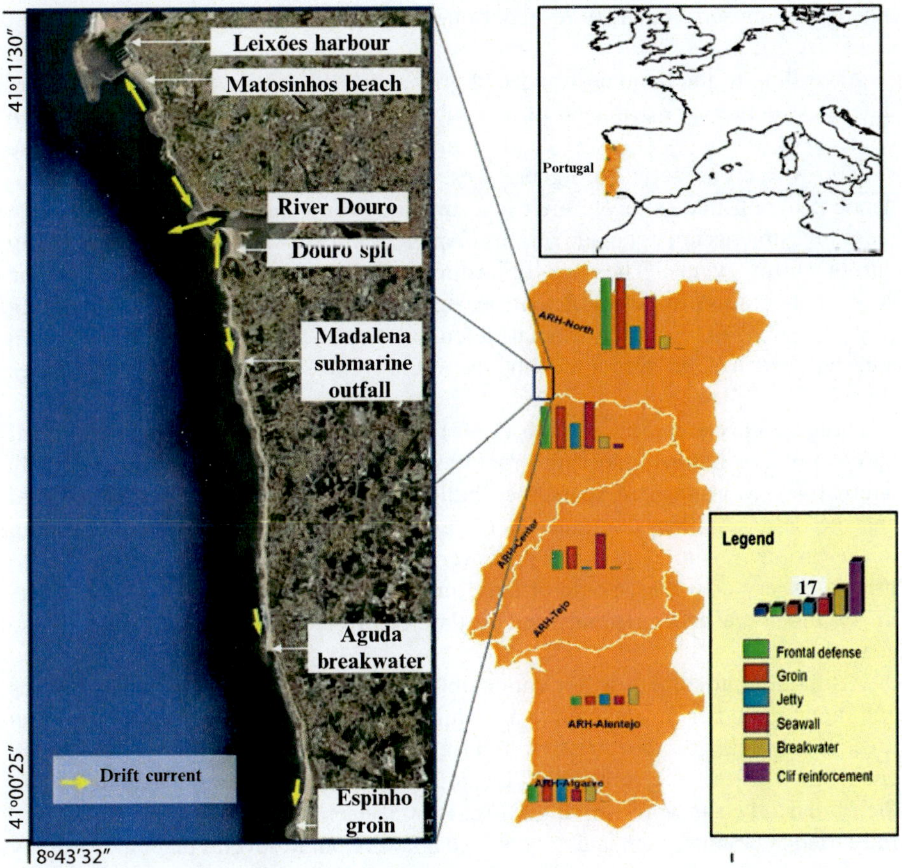

Fig. 15.1 Location of coastal defense structures and main direction of drift currents in the West coast of Portugal

This chapter presents an overview of the Portuguese erosion defense policy, the number and type of defense structures along the coast, their efficiency and some illustrative examples of the nature and impacts of coastal structures. It concludes with some recommendations regarding the erosion problem in Portugal and future measures that must be taken in order to adapt to this reality.

15.2 Defense Policy

Several different types of coastal defense structures have been built in Portugal: frontal defenses, groins, jetties, seawalls, breakwaters and cliff reinforcements (Fig. 15.1). These structures are mainly concentrated in the northern region, where they present a density of about one coastal defense structure per 10 km of coast (IA 2011).

According to the national water authority (IA 2011) almost 70% of the coastal defense structures are damaged (this is subdivided into low, intermediate and high degrees of damage) (Fig. 15.2). Beyond the initial costs of coastal defense structures, the energetic Portuguese west coast wave climate requires a maintenance program that involves high investments throughout their life cycle. When these investments are not made serious damage occurs to the structures, particularly during storm events. The lack of sediment in the coastal zone causes some structures (groins) to be inefficient because they cannot promote beach building (Fig. 15.3). In many cases, frontal defenses have increased sediment loss from adjacent beaches, in turn reducing the effectiveness of the defense structures (Fig. 15.2).

Granja and Soares de Carvalho (1995) questioned the effectiveness of coastal "protection" by hard engineering structures and pointed out the need for less costly alternatives. More than 15 years later, building in risk-prone areas has not stopped and the defense policy continues to favor these kinds of hard structures. Adding to the problem, some small harbours were recently constructed in an attempt to protect fishery activities. Some of these infrastructures have caused local impacts in the erosion/accretion trends and further contributed to disequilibrium of the coastal zone.

Artificial nourishment is not compatible with Portuguese west coast dynamics and nourished beach sediments are rapidly lost. But neither do groins and revetments stabilize the coast or reduce erosion. On the contrary, they have contributed to the acceleration of erosion rates in several coastal stretches and their financial costs were very high (Fig. 15.2). Beach nourishment was unsuccessfully tried at Granja beach and geotextile bags were put in several places, especially at the base of cliffs, but they have short lives and do not prevent retreating by terrestrial processes like gullying (as also happens with revetments).

Detached breakwaters which have the potential to promote tombolo building, seem to be a less harmful hard solution. While they might create local 'solutions', however, they contribute to downdrift erosion. Though costly on exposed energetic coasts like the Portuguese west coast, increased benefits from their use with beach fills may encourage their future construction (Nordstrom 2000). Coastal stretches where natural beaches have mostly "survived" are those that present rocky platforms or outcrops that promote dissipation of wave energy and tombolo genesis, like detached breakwaters do.

15.3 Examples of Defense

In this section some examples of coastal defense structures are presented and their effects on local morphodynamics are characterized.

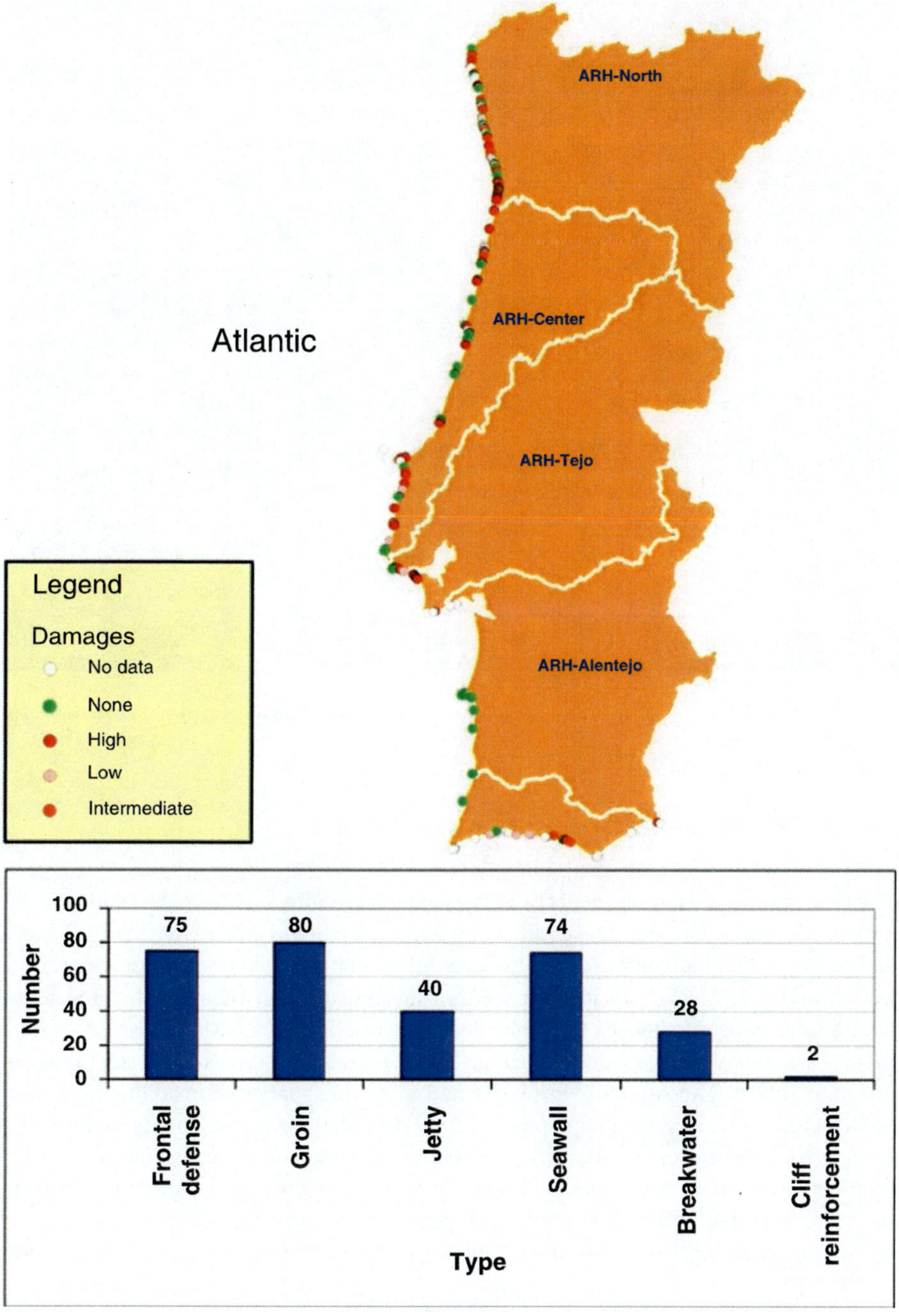

Fig. 15.2 Spatial distribution of coastal defense structures by hydrographic regions and total number of coastal erosion defense structures by type, in Portugal (IA 2011)

Fig. 15.3 Examples of coastal erosion defense structures: (**a**) Espinho groins and (**b**) Cortegaça groins and revetment

15.3.1 Ofir-Apúlia Groins and Revetments

The Ofir-Apúlia coastal segment (Fig. 15.4) extends southwards from the sandy spit in Esposende and is one of the cases where construction should never have been allowed and consequently hard defense should never have been initiated (Granja and Soares de Carvalho 1991, 1995). During the 1970s, the construction of a hotel, three condominiums and some houses on the backshore was allowed. Erosion was very active in the following years, especially during storms, threatening dunes on which building had taken place. The situation was worsened by gullies that developed due to poorly constructed rainwater drainage pipes. Hotel foundations were damaged in 1983 leading to stabilization efforts, first by planting *Carpobrutus edulis* and afterwards with construction of a revetment. Soon the revetment presented signs of failure at the same time that cliff retreat partially exposed the foundations of condominiums.

A groin was built in 1986 at Pedrinhas, 1,700 m south of the hotel. It did not create a protective fillet beach but promoted serious erosion processes southwards in an area that had been relatively stable until then. A new groin immediately south of the hotel was built in 1987. Meanwhile downdrift of Pedrinhas groin, erosion cut an active cliff and precipitated the construction of a revetment in front of the houses. The groin was successively elongated due to progressive cliff retreat and loss of beach sands.

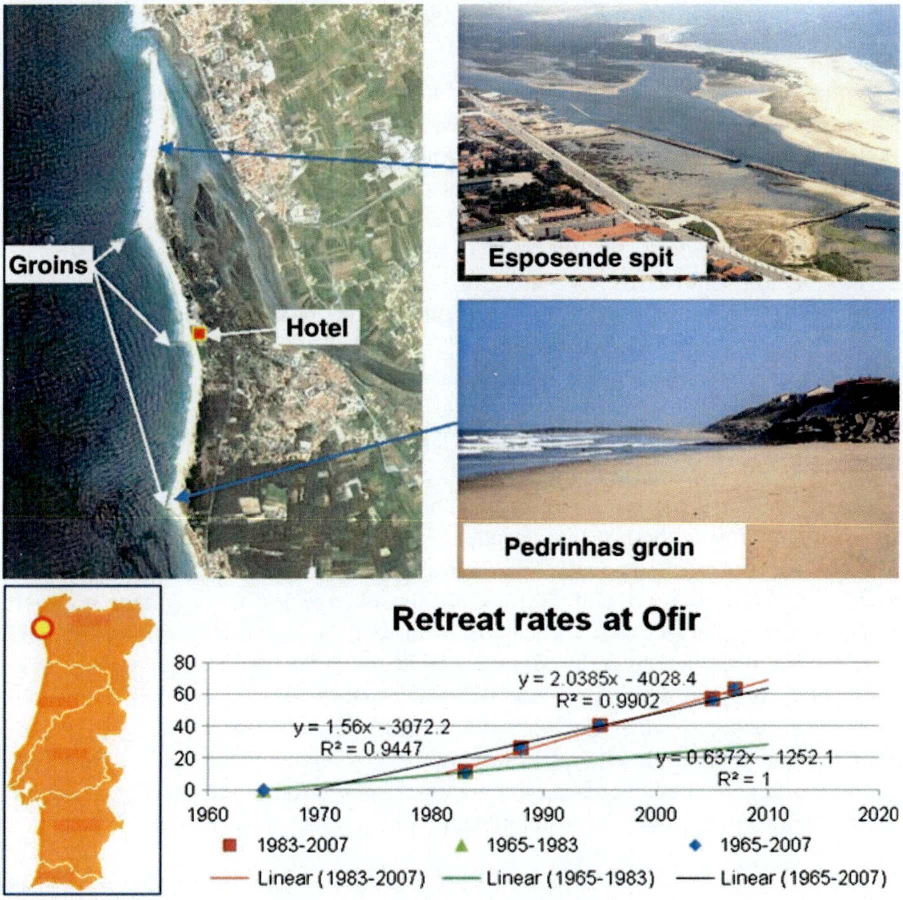

Fig. 15.4 Examples of coastal erosion defense structures: Ofir-Apúlia case study

Almost simultaneously, another groin was built north of the seaside houses of Ofir spit but soon was removed (in 1991) because it was found to be inefficient. Another groin, with a southward curving tip, was built north of the residences, causing an erosion focus at the north end of the revetement in front of the houses; more defensive works at the base of damaged houses were subsequently built. At Apúlia, in 1988, another groin was built, favoring the enlargement of the updrift beach and the rapid retreat of the cliffs southwards.

The defense option taken in this coastal segment increased erosion, leading to the construction of successive hard defense structures and consequent morphodynamic disequilibrium in an area that was relatively stable until then (Pedrinhas-Cedobem). Today, this groin is shorter (in 1988 it was 240 m long) and inefficient but the retreat rates downdrift continue to be very high (for example, 1.69 m/year at Pedrinhas-Cedobem), as happens also with the groin at Ofir, which is also inefficient, with downdrift high cliff retreat rates (1995–2010: 2.3 m/year; 1983–2007: 2.04 m/year; 1965–2007: 1.56 m/year; 1965–1983: 0.64 m/year).

15.3.2 Esmoriz-Cortegaça-Maceda Groins and Revetments

Espinho (Fig. 15.3a), due to its long history of erosion, experienced several successive solutions of hard defense. The two long groins spaced 120 m apart, built in 1988, and the frontal seawall, did not prevent sand loss from beach. Later on, new interventions, including artificial nourishment between the groins, were implemented.

The enlargement of the breakwaters at Espinho promoted the beginning of severe erosion southwards. Defense structures were built and enlarged during the 1980s and 1990s, at Esmoriz, Cortegaça and Maceda beaches (Fig. 15.5).

The groin at Cortegaça beach (Fig. 15.3b) was built just north of a camping site. With a northerly curved tip, it never created an updrift fillet beach. The front of Cortegaça was defended using revetments, resulting in the loss of the beach. Soon, Esmoriz beach suffered from the same high retreat, causing another frontal defense. The groin at Maceda beach did not defend any building or infrastructure updrift and enhanced a severe erosion downdrift, with the retreat (loss) of tens of meters of forested land during the last two decades.

Based on comparative analyses of aerial photographs from 1958, 1967, 1990, 1995, 1996, 2001, 2003 and 2004, coastline change for the segment between Cortegaça and Maceda beaches was determined (Fig. 15.5). Concerning the cliff retreat rates, if the whole period between 1958 and 2004 is considered using linear regression analysis, the retreat rate is 3.5 ± 0.4 m/year. But, in fact, two main periods can be assessed: (i) between 1958 and 1990, when the rate is 2.4 ± 1.0 m/year; (ii) and post-1990, when the rate is 6.2 ± 0.5 m/year (Henriques 2006). This means an accentuated increase of retreating rates after 1990, which cannot be dissociated from the defense structures built during the end of the 1980s.

15.3.3 Mira Beach Groins

Mira Beach is located on the Portuguese coast of the centre region. In the mid 1980s, two groins were built about 800 m apart. The northern one is located about 500 m south of Mira beach village. These two coastal groins were severely damage by storms and in 1990 the southern groin was almost destroyed. Both groins were repaired in 2005. They are located in a fine sand beach that has been monitored since 2008.

Referring to aerial photographs of 1990 and 2005, it is possible to evaluate general trends in coastline evolution (Fig. 15.5). Despite the fact that some previous studies reported that this sector was temporarily stable, this monitoring project showed a general trend of shoreline retreat. This is a textbook example of a local intervention attempt that included destruction of the initial groins due to storms and the failure of shoreline stabilization due to a natural transgressive trend.

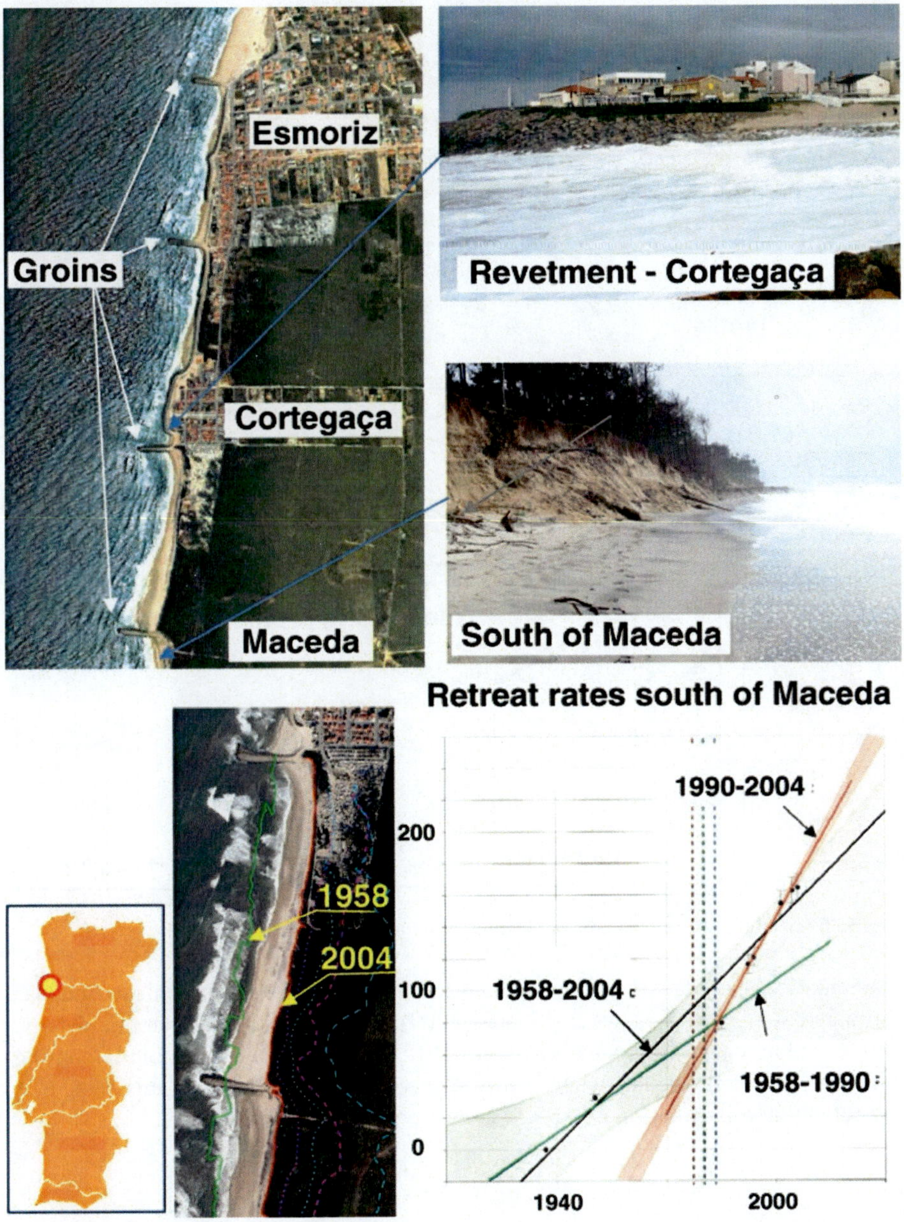

Fig. 15.5 Examples of coastal erosion defense structures: Esmoriz-Cortegaça-Maceda case study (Adapted from Henriques 2006)

By the end of 2005, the coastline already reflected the influence of the reconstructed groins, evidenced by two discontinuities observed in Fig. 15.6. They reflect loss of sediments downdrift of the groins and accretion updrift. Referring to the simulation of the coastline retreat between 1990 and 2005 and using

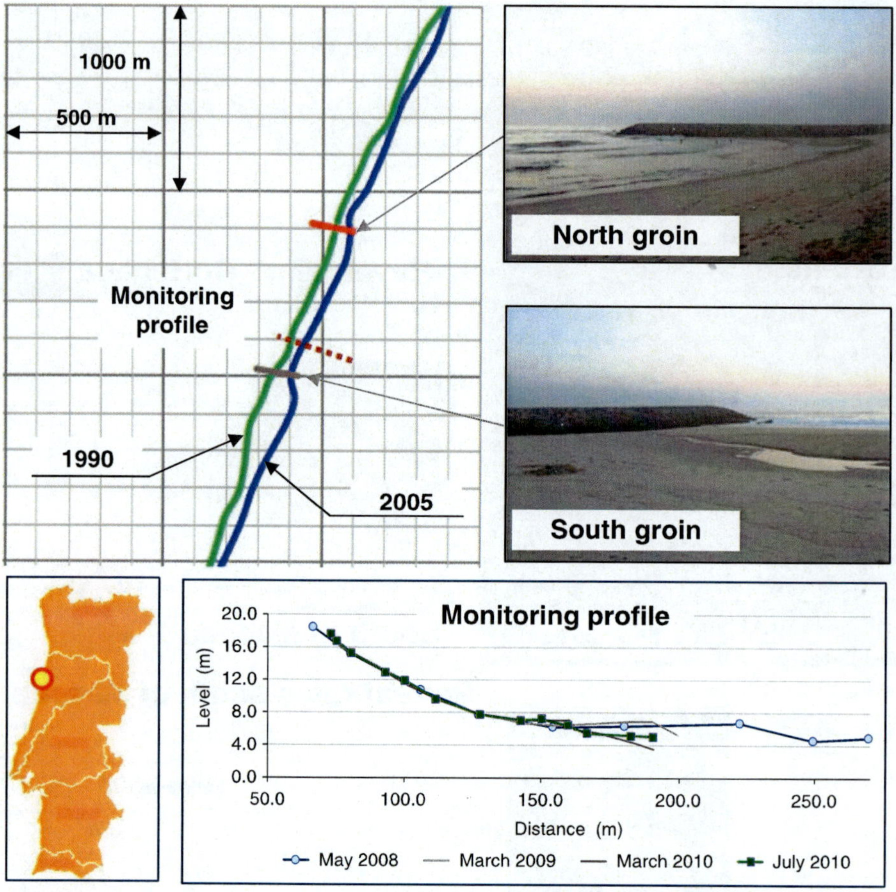

Fig. 15.6 Examples of coastal erosion defense structures: Mira beach case study

mathematical models, the average annual sediment transport during this period was estimated at 1.3×10^6 m^3/year (Mendes 2009). It must be stressed that this volume is in circulation due to input from sandy cliff high retreat rates (probably coming from the northern stretch, where erosion is very intense). Due to this fact, it is expected that this situation will became worse in the near future.

15.3.4 Aguda Beach Breakwater

Aguda is located on the coastal stretch south of the Douro Estuary (Fig. 15.1), where the coastal border is narrow, with a subtidal rocky platform and many outcrops creating small tombolos between pocket beaches. Fishery communities, though presently very small, fight against adverse wave conditions when they want to go out or return with their boats (Granja et al. 2010).

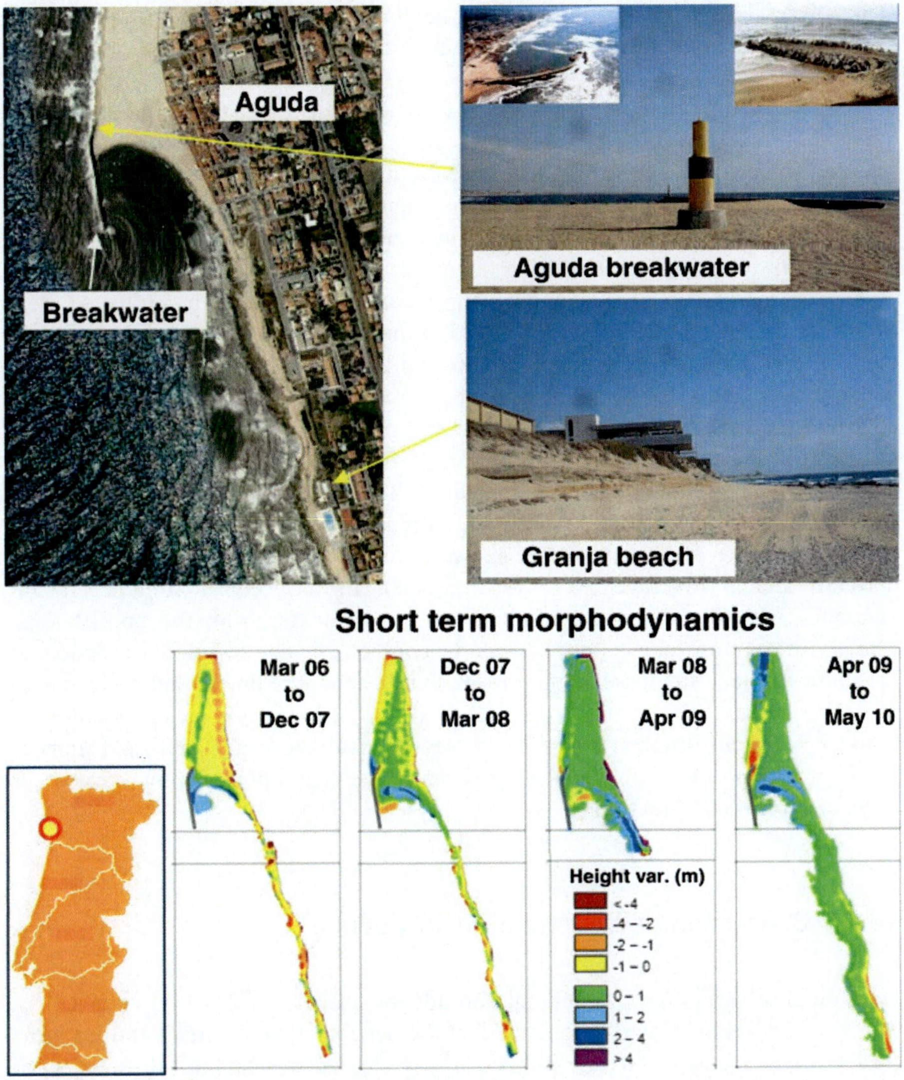

Fig. 15.7 Examples of coastal erosion defense structures: Aguda beach case study

The Portinho da Aguda (Fig. 15.7) was built between October 2001 and July 2002, at Aguda village. This coastal stretch presents a rocky platform that protected the coastal border against the direct wave impact, in an area where the erosion impact knowledge is well known. The harbour was created by a detached curved breakwater anchored on rocky outcrops. A temporary groin, starting on the beach, was made to allow the access and transport of material for the breakwater construction. After some months and still during the construction of the breakwater, a fillet beach formed against the temporary groin, promoting the enlargement of the updrift beach and the erosion of the downdrift coastal segment (Granja Beach). When construction was completed, the temporary groin was not removed (Rosa-Santos et al. 2009). This fact did not allow the re-establishment of the littoral drift, promoting sand retention against the groin and creating a deficit of sediment in the downdrift segment.

To remedy this situation, sand transfer from the updrift area to Granja Beach was attempted, in an effort to compensate for the accentuated loss of sediments at the famous old seaside resort. But, due to the very high energy conditions, sands do not remain very long on the beach, making recurrent spring or summer recharges necessary to maintain at least minimal conditions for beach users.

Sediment retention started during the construction phase against the temporary groin, occurring either alongside the groin/breakwater or inside the protected area. The most frequent wave directions are from the west (60% of the buoy records in the period of May 2001 to September 2009) followed by the northwest (34% of the buoy records) and southwest (5% of the buoy records). For the most frequent wave directions, diffraction tends to interrupt the previously installed north to south coastal drift in this segment. During storms from the south and southwest, this coastal defense structure is inefficient and can be significantly affected by sediment deposition inside the artificial bay, forcing dredging to take place.

The annual net sediment balance shows a tendency towards accumulation at the northern stretch (average 50×10^3 m^3/year). In the central stretch, erosion alternates with accretion that results from sand transfer from the updrift area. This sediment transfer was not enough to compensate the sedimentary deficit of the south segment. In this segment, the maximum annual sediment deficit was about 20×10^3 m^3 and the average was 1.3×10^3 m^3 (this result includes sediment transfer volumes). Erosion inside the protected area can result from sand transfer works or transport by wave induced currents (wave diffraction near the breakwater or waves from SW direction).

15.4 Discussion of Benefits and Impacts

Hard defense works have beneficial and adverse effects (Pilkey and Neal 1991). The case studies presented here show that the benefits are ephemeral and lead to a false sense of security. Adverse impacts involve high costs and the enhancement of local erosion problems. Trying to make something static when it is naturally dynamic is not a sound solution because it promotes an unsustainable situation.

For all the analyzed situations, it is apparent that the defense of the NW Portuguese coast was not achieved. Over several years, large sections of the coast have been lost to erosion and others have been severely damaged. Monitoring of the coast before and after construction of hard defense structures clearly shows an increase in cliff retreat rates and shoreward migration (when possible due to an open accommodation space at the backshore) of beaches, southwards of their location, after construction. In some situations, the shoreward beach migration is halted by the presence of infrastructure (e.g. seaside roads, seawalls, houses); in these cases beaches thin and steepen or even disappear.

As observed in the case studies, retreat rates increased downdrift following the construction of groins. Besides interrupting longshore drift, the groins interfere with wave refraction and breaking that in turn affects circulation in the surf zone. When groins fail at their root, sand bypassing resumes and cliff retreat rates decrease downdrift. This is the case of Ofir, Pedrinhas and Maceda groins. However, due to the high and steep cliffs caused by previous beach erosion, the natural recovery of the beaches is not possible. It is too late.

Revetments, built as emergency solutions after severe storm damage, promote the disappearance of beaches due to wave reflection, turbulence and increased backwash velocity that remove sediments. These engineering structures do not solve the cliff retreat problem which is largely due to gravity slumps. In addition, revetments create unstable areas at cliff ends, especially on the downdrift side, that result in rapid retreat (Granja and Soares de Carvalho 1991).

Breakwaters have been built for small harbor protection. The case studies (and similar research at Pedra Alta; Soares de Carvalho et al. 2005) show that these structures can reduce wave energy, enhance deposition, and contribute to the establishment of a tombolo where there is sufficient sediment supply. But downdrift, the erosion problem is enhanced.

Concerning hard defense structures, particularly groins, adverse impacts outweigh benefits. As reported here, it is irrefutable that after the construction of groins, rates of downdrift retreat increase. Being maintained, and without any other preventive measure, these structures will contribute to huge coastal retreat scenarios.

In addition, revetments function as headlands, reinforcing the erosive end effect, shaping a re-entrant coastline in retreat between artificial promontories (Carvalho and Granja 1997), as already happens at present at *e.g.* Pedrinhas, Esmoriz and Cortegaça. The fight against waves of the Atlantic Ocean in a sediment-starved situation using local hard defense structures seems to be an inappropriate solution to the erosion problem.

15.5 Conclusions and Recommendations

It is evident that Portuguese shorelines are retreating. It is also true that people don't want shoreline retreat. It is also true that coastal defense efforts are often viewed as an age-old conflict of men against the sea. It is not easy to provide solutions. Most reports converge on four options: retreat, defend, adapt, do nothing. But where, how, when and according to what criteria?

It seems obvious that for different coastal stretches, different solutions are required: (1) retreat: first priority for ICZM purposes; (2) defend: only when a town or a high interest utility justify it; (3) adapt: *work with nature* finding less harmful solutions; and (4) do nothing: allow natural processes to freely evolve. These are considered below.

15.5.1 Defend

This option should only be applied in coastal segments where social and economical interests of the population and the region do not allow any alternative. It should be applied after careful analysis of cost-benefit analyses, including the costs of maintenance and repair of the structures and the restoration of affected downdrift areas. Research and monitoring programs are fundamental skills that are required to support decision making.

This option is costly when all factors are taken into consideration. Costs should be shared by local municipalities and residents (such as coastal defense tax that is based on the user pay principle) and not just public funding. Sand dredging from harbours and from saturated areas updrift of transverse structures should be compulsorily relocated downdrift in such a way that its inclusion in longshore drift be effective. This option should not be applied in cases where a seaside expansion of a pre-existent center is proposed. Espinho is an example of this option.

15.5.2 Retreat

The analysis of each case must be done according to the legislation and financial support for demolitions, relocations and compensations. This option should also consider harmful defense structures. Seaside expansion of population centers and creation of new leisure and touristic places should be perpendicular and not parallel to the coast, as prescribed by the law. Examples include several places along the coast such as Mar, and Ofir-Pedrinhas.

15.5.3 Adapt

To adapt to reality means permitting sufficient space for shoreline processes to operate (shoreline retreat). To *work with nature* is a conservative solution that is not harmful to the environment, although it is defined in different ways by different workers (Cooper and McKenna 2008).

Clearly, in some cases, the defense options do not fit well with existing conditions. In those cases, new experiments should be undertaken to better understand the local situation. For example, there is a need to "help" the aeolian component to reinforce the dune-beach system with global negative sedimentary budgets. Beach nourishment must adapt to the overall coastal marine dynamics and beach equilibria while using sediments that are in harmony with the system dynamics. This observation applies to the Gaia coast and Granja Beach in particular.

15.5.4 Do Nothing

This approach allows the natural evolution of some coastal segments where development has not yet occurred. In this way, the whole system can freely migrate according to the natural dynamics where there is accommodation space for change. Belinho Beach is a good example of this approach.

Finally, it should not be forgotten that there are factors that contribute to the starvation of sand sources, phenomena that began at the most northern sector of the Portuguese NW coastal zone and that have progressively extended southwards. In beaches, coarse sediments are dominant as are extensive areas of rocky outcrop. Some sediment sources are local, mainly derived from retreating cliffs. All of these facts must be taken into consideration when a defense policy is envisaged.

Acknowledgments The authors would like to thank all organisations and collaborators that have participated in several monitoring programs developed in the Portuguese West coast. We thank C. Finkl for a helpful review of the manuscript.

References

Carvalho GS, Granja HM (1997) Realismo e pragmatismo: uma necessidade para o aproveitamento dos recursos naturais da zona costeira (O exemplo da zona costeira do noroeste de Portugal). In: Colectânea de ideias sobre a zona costeira de Portugal, Publicação da Associação EUROCOAST-Portugal, pp 25–66

Cooper JAG, McKenna J (2008) Working with natural processes: the challenge for coastal protection strategies. Geogr J 174:315–331

Granja H, de Carvalho S (1991) The impact of "protection" structures on the Ofir-Apúlia coastal zone (NW Portugal). Quat Int 9:81–85

Granja H, de Carvalho S (1995) Is the coastline "protection" of Portugal by hard engineering structures effective? J Coast Res 11(4):1229–1241

Granja H, Loureiro E (2007) Why are shingle beaches replacing sandy beaches? (Coastal zone of NW Portugal). In: Coastal sediments' 07 Proceedings, New Orleans, LA, pp 545–560

Granja H, Bastos L, Pinho JLS, Gonçalves J, Henriques R, Bio A, Magalhães A (2010) Small harbours risks: lowering for fishery and increasing erosion. The Case of Portinho da Aguda (NW Portugal). In: Proceedings of Littoral 2010, London, Sept 2010

Henriques R (2006) Monitorização da zona costeira tendo em vista a sua vulnerabilidade - apli-cação à zona costeira do noroeste de Portugal. PhD thesis, University of Minho, Braga, p 525

IA (2011) Portuguese water authority. url: http://www.inag.pt

Loureiro E (2007) Indicadores geomorfológicos e sedimentológicos na avaliação da tendência evolutiva da zona costeira. Aplicação ao concelho de Esposende. PhD thesis, University of Minho, Braga, p 329

Mendes J (2009) Application of mathematical models for morphodynamics and sediments transport predictions in coastal zones. MSc thesis, University of Minho, Oct 2009 (in Portuguese)

Nordstrom K (2000) Beaches and dunes of developed coasts. Cambridge University Press, Cambridge, 338 p

Pilkey OH, Neal WJ (1984) Living with the Shore. Duke University Press, Durham, USA

Rosa-Santos P, Veloso-Gomes F, Taveira-Pinto F, Silva R, Pais-Barbosa J (2009) Evolution of coastal works in Portugal and their interference with local morphodynamics. J Coast Res Spec Iss 56: 757–761

Soares de Carvalho G, Granja H, Henriques R (2005) Recent coastal defense by engineering structures and their consequences (problems in the coastal zone of NW Portugal). In: Viggosson G (ed) Proceedings of the second international coastal symposium ICS 2005, Hofn, Iceland, Book of abstracts, pp 188–192

Chapter 16
Stabilizing the Forgotten Shore: Case Study from the Delaware Bay

Hilary Stevens and Arthur Trembanis

Abstract The western shore of the Delaware Bay estuary is lined with low-profile fetch-limited barrier beaches. Prime Hook National Wildlife Refuge was established in 1963 to provide habitat to migratory waterfowl. Two study sites are adjacent to the refuge, Prime Hook Beach with a community of houses along the water and Fowler Beach, which is undeveloped. Construction in the region has impacted coastal hydrology and longshore sediment transport throughout the last century. After the 1962 nor'easter, soft stabilization and beach scraping became common. Scraping artificial dunes has also become a method of choice to isolate freshwater marshes from the bay. Both methods require frequent repetition as well as permits and lengthy planning processes. Benefits from this type of stabilization include an increase in recreational appeal, preservation of freshwater habitat and sustained property values for private landholders. Negative impacts of stabilization include overhead costs to taxpayers, degradation of natural habitat, loss of aesthetics of natural beaches, and continued public expectation of government intervention. As the threats of rising sea level and more frequent storms increase, beach overtopping and marsh inundation will become more frequent. Coastal development and infrastructure is at continued risk of flooding. Future management must take a long-term perspective on these realities and plan for the changing coastline.

Abbreviations

DelDOT	Delaware Department of Transportation
DNREC	Delaware Department of Natural Resource and Environmental Control

H. Stevens
Coastal Resources Center, University of Rhode Island, Narragansett, RI 02882, USA
e-mail: hilary@crc.uri.edu

A. Trembanis (✉)
Department of Geological Sciences, University of Delaware, Newark, DE 19716, USA
e-mail: art@udel.edu

J.A.G. Cooper and O.H. Pilkey (eds.), *Pitfalls of Shoreline Stabilization: Selected Case Studies*, Coastal Research Library 3, DOI 10.1007/978-94-007-4123-2_16,

NOAA United States National Oceanographic and Atmospheric Administration
PHNWR Prime Hook National Wildlife Refuge
USFWS United Stated Fish and Wildlife Service

16.1 Introduction

Delaware Bay is a wide, shallow estuary located in the mid-Atlantic region of the east coast of North America. Like most developed coastal regions, it has a history of engineering the coast to promote shipping, infrastructure, development, and recreation areas. Early projects such as dredging and breakwaters were primarily oriented to support shipping. Into the twentieth century more emphasis was put on engineering the coast to stabilize the shoreline and barrier inlets. More recently, soft stabilization such as nourishment and construction of artificial dunes are the favored methods of combating erosion and barrier evolution.

Shoreline retreat has been occurring on most of the western shore of Delaware Bay for decades (Maurmeyer 1978; French 1990). Shoreline retreat rates of between 0.4 and 2.9 m/year have been reported in the area between Slaughter Beach and Broadkill Beach, from 1843 to the present. Sea level is rising in this area (Nikitina et al. 2000), and the barriers are retreating in response to that trend. The low relief of fetch-limited barriers makes them particularly susceptible to storm overtopping (Lewis et al. 2007). Many recent efforts at shoreline stabilization such as recurrent dune scraping have been implemented in direct response to storm damage (M. Stroeh, pers. corr. January 2011).

Small coastal communities along the western shore of Delaware Bay include a mixture of year-round residents and seasonal vacationers. They have become increasingly concerned with problems of erosion, flooding, and loss of property, and expect governmental support in dealing with these problems (Drew and Kraft 1980; M. Stroeh, pers corr. January 2011). Prime Hook National Wildlife Refuge was established in 1963 in recognition of the valuable habitat offered by the area's marshes for migratory birds (USFWS 2010). As a result, management for preservation of those marshes has become a high priority.

This location offers an excellent example of the variety of issues facing many coastal communities struggling to balance numerous stakeholders, budget concerns, and political realities in the face of rising sea level and increasing coastal populations.

16.2 Background

16.2.1 Site Description

16.2.1.1 The Delaware Bay Region

Delaware Bay covers 2,100 km^2 and is relatively shallow, averaging 10 m depth (Lewis et al. 2005) with a maximum of 45 m at the mouth (NOAA 2010). Its outlet

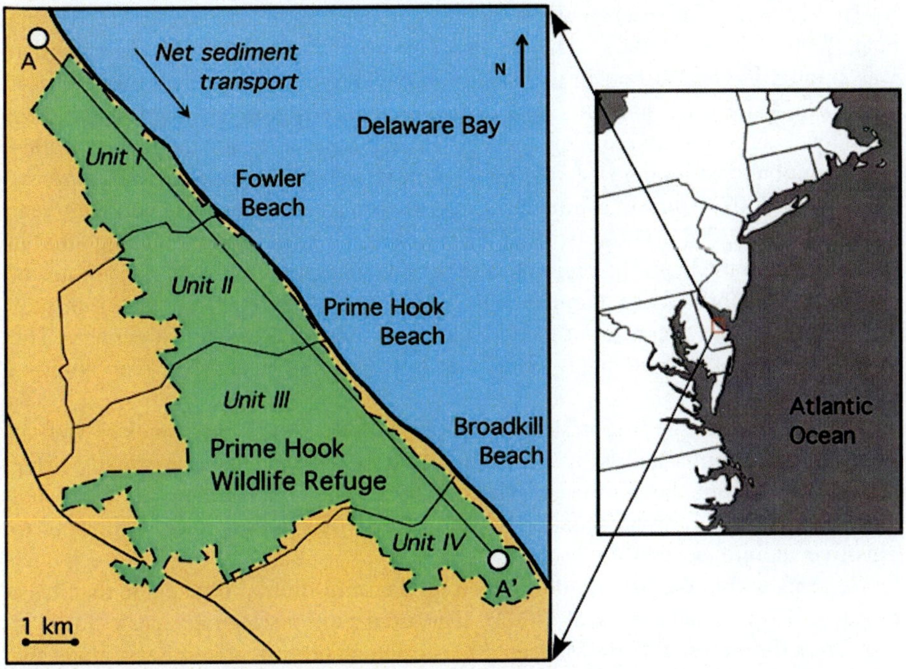

Fig. 16.1 Prime Hook National Wildlife Refuge includes 4,100 ha of coastal wetlands with some inholdings of private land along the coast. It is divided into four units for hydrologic control. Net sediment transport in this area is to the southeast. Transect A-A' indicates the profile detailed in Fig. 16.2

to the Atlantic Ocean is confined by Cape Henlopen and Cape May. The 17.5 km wide opening is relatively narrow compared with the 45 km maximum width of the Bay and this narrow opening therefore limits the influence of ocean waves on the bay wave climate (Jackson 1995). This study focuses on Fowler Beach and Prime Hook Beach, which are located on the western shore of Delaware Bay, approximately 16 km from the mouth of the bay (Fig. 16.1).

The Atlantic coast of North America is a tectonically passive trailing margin (Nummedal 1983). As a result, the entire region has long been free of significant volcanic or seismic activity. Delaware Bay is a drowned river valley that was formed through repeated cycles of sea level transgression and regression (Knebel et al. 1988). The most recent cycle began with the last glacial maximum, the Wisconsin, approximately 18,000 years ago. Sedimentation of the estuary ensued from about 8,000 years ago continuing up to the present (Fletcher et al. 1990). Transgressive sediments have been laid down over the eroded surface of the pre-Holocene basal material. The topography of that unconformity has a strong effect on the present-day morphology through its influence on the behavior of the overlying sediments and depth-limited physical processes (e.g. tidal currents and waves) (Kraft et al. 1981; Khalequzzman 1997).

The Delaware Bay is currently undergoing local relative sea level rise (Nikitina et al. 2000; Sella et al. 2007). The Delaware Bay area was part of the forebulge that was created by the Wisconsin ice sheet that covered much of the northern part of the North American continent. As a result, the entire region has been undergoing subsidence since the time of deglaciation. Recent measurements taken with highly precise global positioning systems (GPS) indicate subsidence rates of 1–3 mm/year (Sella et al. 2007). Current estimates of eustatic sea level rise are around 2 mm/year (Douglas et al. 2001). These two factors result in the local sea level change that is now evident in the bay. The region has experienced relative sea level rise at rates of 1–3 mm/year throughout the Holocene. The rate slowed to about 1 mm/year 1,250 years before present, but has dramatically increased in the last century. The bay is currently experiencing 3–4 mm/year of local relative sea level rise (Nikitina et al. 2000).

The mid-Atlantic region is subject to occasional seasonal hurricanes as well as extra-tropical storms known as nor'easters. These storms characteristically bring heavy winds from the northeast. The combination of elevated water level and increased wave energy associated with storms often causes severe impacts to sensitive shorelines (Jackson et al. 2002).

Currents within the bay are dominated by the semi-diurnal tidal cycle that has a range of 1.6 m (Jackson et al. 2002). The tidal currents average peak flows of 1–1.5 m/s (Sommerfield and Madsen 2003) oriented northwest-southeast along the primary axis of the bay.

Lewis et al. (2005) described the barrier beaches of Delaware Bay and demarcated them as occurring nearly continuously along the western (Delaware) shore. These barriers are typically narrow (<25 m) with minimal (1–2 m) natural dune development. Because of their low profile, fetch-limited barriers are particularly sensitive to sea level change (Pilkey et al. 2009). 2005 LIDAR data indicate that dunes on the Fowler and Prime Beaches were between 0.5 and 1 m in elevation (Delaware DataMIL 2009).

The region is home to a variety of important ecological resources. The beaches of Delaware Bay are nesting sites for the horseshoe crab *Limulus polyphemus*, a protected species that is closely monitored by the Delaware Division of Fish and Wildlife (DNREC 2010b; USFWS 2010). Numerous species of migratory waterfowl utilize the freshwater marshes of Prime Hook Wildlife Refuge (DNREC 2010b; USFWS 2010). Additionally, bald eagle populations in Delaware have boomed in recent years and many have been sited near this area (DNREC 2010a).

16.2.1.2 The Study Sites

Prime Hook and Fowler Beaches are good examples of fetch-limited barrier beaches. Both locations are undergoing rapid changes in response to rising sea level, which pose challenges for planning and management.

The underlying geology of Fowler Beach comprises a thin layer of sand underlain by marsh material (Fig. 16.2). Between the headland beaches of Fowler

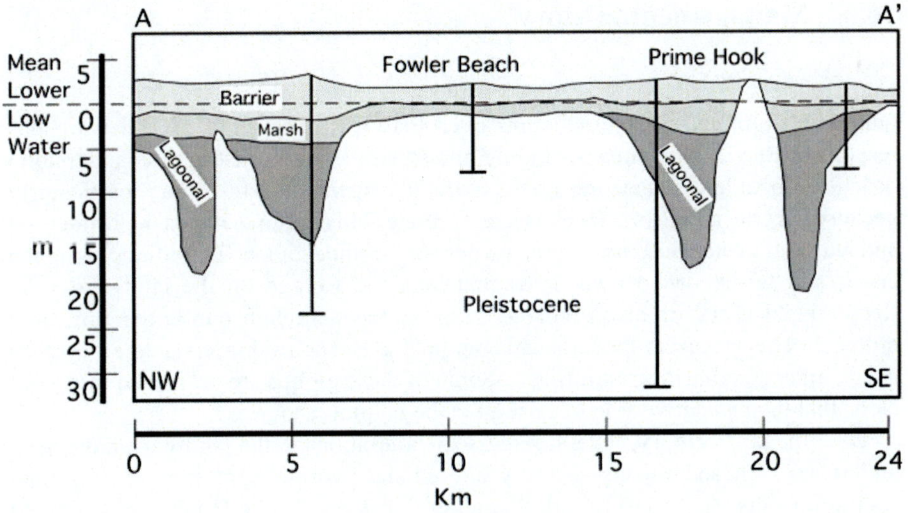

Fig. 16.2 The underlying geology of this area reflects 18,000 years of marine transgression into a river valley, as demonstrated by the indicated cores. Pleistocene rock with deeply incised stream channels is overlain by transgressive lagoonal and marsh sediments. Barrier sands form the surface layer. A headland protrudes at Prime Hook (After Kraft et al. 1981)

and Prime Hook there is an incised paleo-channel where a tributary to the ancestral Delaware River flowed. This paleo-channel is infilled with Holocene lagoonal sediments. South of Prime Hook Beach is a narrow exposed Pleistocene headland and another deep valley (Kraft et al. 1981). This headland is a source of sediment to the surrounding area (Maurmeyer 1978).

Fowler Beach is undeveloped, with only a narrow road and small parking lot. There are some undeveloped private inholdings along the shore (USFWS 2010). Prime Hook Beach has a small community within two blocks of the water that, as of 2004, consisted of 206 home- or landowners. Of those, 43 are year-round residents and the remainder are seasonal. The beach itself is private. Prime Hook Road is the only access route to this community. Part of the road lies below the mean higher high water line and is increasingly flooded. Landward of this area is low-lying agricultural land that is also frequently subject to coastal flooding (USFWS 2010).

Prime Hook National Wildlife Refuge (PHNWR) covers about 4,100 ha of coastal fresh- and salt-water marshes (Fig. 16.1). This area was established in 1963 as federal land to provide habitat to freshwater marsh plant species and migratory birds that pass through the region (USFWS 2010). PHNWR is divided into four units based on water impoundments (Fig. 16.1). The Delaware Department of Natural Resources and Environmental Control (DNREC) handles issues of coastal management, including planning for future stabilization efforts. Prime Hook Refuge is a popular site for birding, hunting, and fishing. Recreational users enjoy access to the wildlife and natural areas through maintained roads and facilities.

16.3 Management History

Numerous stakeholders are involved with management decisions in this area. The federal government has jurisdiction over PHNWR. The state of Delaware has responsibilities in the area including maintenance of infrastructure and public beaches. Local landowners are vested in the management of this shore, particularly because the beach in Prime Hook is itself private. This group includes the residential and seasonal owners in Prime Hook, owners of the undeveloped inholdings along the coast, and farmers who own fields that abut the Refuge on the landward side. Recreational users are also involved in the process through public meetings and outreach. This group comprises a wide range of users such as birders, hunters, beach-goers, hikers, and fishing enthusiasts. Some of these groups are organized into local clubs that take an active role in caring for the natural resources.

Over the past century, several large-scale alterations to the landscape have been undertaken. These projects severely altered the hydrology of the area and had lasting impacts. As of 1911, both Slaughter Creek and Prime Hook Creek drained directly into Delaware Bay. Both stream channels were cut off and diverted by a storm. Since that time they have been directed through the use of canals to control fresh and tidal flow (M. Stroeh, pers. corr. January 2011). The creation of mosquito ditches and drainage of low-lying uplands during 1930s and 1940s brought about a slow conversion of salt marsh to freshwater marsh (USFWS 2010). In 1988, the PHNWR further altered the area hydrology by constructing impoundments totaling 1,620 ha of freshwater marsh (USFWS 2010).

16.3.1 Shoreline Stablization History

Various forms of shoreline stabilization have been utilized in this region. Hard stabilization such as breakwaters and inlet stabilization were constructed near the mouth of the bay to make the area more safe and functional for shipping traffic. More recently, soft stabilization such as nourishment and dune scraping has been used to reinforce barrier beaches inside the bay.

Breakwaters constructed near the mouth of the bay in 1828 (Del Sordo 1988) led to a decrease in sediment supply to lower bay beaches by altering the littoral drift and sedimentation at Cape Henlopen, and thus cutting off supply from the ocean beaches (French 1990).

Roosevelt Inlet at Lewes was cut and stabilized with two jetties in 1937. In 1950, three groins were constructed at Broadkill Beach. Erosion at the north end of the beach remained a problem, so two additional groins were added in 1954 and concrete blocks were emplaced on the beach in the following years and dune vegetation was planted (Drew and Kraft 1980).

Following the extremely severe 1962 nor'easter known as the Ash Wednesday Storm, soft stabilization became common practice in this region. In 1962, following the storm, the Army Corps of Engineers emplaced 15,400 m^3 of sand along 1,190 m of Prime Hook Beach, at an average density of 12.9 m^3/m length of beach (DNREC 2010a, b; French 1990). Broadkill Beach was nourished six times between 1957 and 1976, totally 643,000 m^3 of sand were deposited (Drew and Kraft 1980). In 1999, another 27,680 m^3 of sand was deposited on Broadkill Beach by state and local government agencies (PSDS 2002). In 2008, the 13 northernmost lots in the Prime Hook community were filled with 900 m^3 of sand (DNREC 2010a, b).

Dune scraping is also a common form of shoreline stabilization in this region. This method involves bulldozing sediment from the landward side of the barrier, typically overwash fans, into an artificially enlarged dune. Dune scraping has been used along the shorelines of PHNWR Units I and II six times between 1988 and 2006 in response to storm events that overtopped the barriers. In the winter of 1991–1992, geotubes were emplaced in addition to dune scraping as a further erosion control measure. The tubes quickly broke up and had to be removed from the marsh (Michael Stroeh, pers. corr. January 2011).

As of 2008, the FWS ceased scraping along Unit I, north of Fowler Beach Road, and allowed the area to flood through a newly formed inlet with salt water from the Bay (USFWS 2010). The FWS has proposed additional beach scraping along the Prime Hook shoreline in response to the 2009 storms and the inlets and overwash caused by the storms (USFWS 2010).

PHNWR submitted a proposal for scraping dunes in November 2010, in response to 2009 storm damage. This is the first time the Refuge has gone through the Environmental Assessment process for this type of work. The proposal presented several management options, including no action and two levels of dune scraping (USFWS 2010). The final assessment supports the option of scraping dunes along 915 m of Fowler Beach and to fill recently opened inlets. The assessment acknowledges local subsidence as well as sea level rise. The assessment also describes that Unit II, which was fresh until the 2006 storm that opened an inlet into it, is expected to remain partially saline. Additionally, Unit III will also likely receive some salt water. This scraping project is described as a means to delay further management action and provide better understanding of the area's needs in preparation for the forthcoming Comprehensive Conservation Plan (USFWS 2010). The Comprehensive Conservation Plan (CCP) for the Refuge was first announced in 2005 and is underway to plan for 15-year management of the Refuge (USFWS 2010).

Public opinion about the dune scraping is mixed and the project has generated controversy and media attention, including a documentary by the local news station (WBOC 2011). Local landowners support the scraping to reduce coastal flooding. Beachfront homeowners are very concerned about increased erosion and risks to their property. Some local farmers seek to reduce the saltwater intrusion that impacts adjacent fields (M. Stroeh, pers. corr. January 2011). However, other groups such as Delaware Audubon seek to protect habitat for shore birds and oppose the scraping (M. Stroeh, pers. corr. January 2011; Delaware Audubon 2011).

The scraping project was initially intended to go forward during the winter of 2011 but has been delayed by lawsuits filed by Delaware Audubon. The scraping cannot take place during the spring or summer to avoid disrupting nesting seasons. It is unclear if enough sediment would remain in the system to make scraping feasible by 2012, so the future of the project is uncertain.

The state of Delaware contracted with PBS&J, a private consulting firm, to create a management plan for the Delaware Bay coastline. This management plan (DNREC Mar 2010) recommends nourishment to protect the bay coastline from erosion. It outlines three potential nourishment strategies, ranging from 18,300 to 134,600 m^3 in initial emplacement with additional emplacements every 4–5 years recommended for maintenance. It additionally recommends planting 30,000–70,000 units of beach grass on the newly built up dunes to stabilize them.

16.3.2 Shoreline Change

Shoreline change in this area has been tracked by numerous studies (Fig. 16.3). Maurmeyer (1978) compiled records from the Army Corps of Engineers and other researchers regarding the adjacent beaches. The proximal locations on Slaughter Beach to the north were retreating at 2.2 m/year between 1910 and 1956. The Broadkill Beach northern terminus was retreating at a rate of 1.3 m/year from 1843 to 1964. French (1990) measured shoreline change between 1882 and 1977 based on historic T sheets and aerial photographs. Those results indicate net accretion in Prime Hook at a rate of 0.4 m/year. During that same time period, Fowler Beach eroded in a sporadic manner, averaging 1.5 m/year. More recently, the USFWS 2010 environmental assessment cites 45–60 m of shoreline retreat along Unit II from 1986 to 2010, yielding an average of 1.9–2.5 m/year.

Between 2000 and 2010, a series of storms caused rapid evolution of the coast along Fowler Beach (Figs. 16.3 and 16.4). A spring nor'easter in May 2006 created large washover fans and opened an inlet into Unit I, north of Fowler Beach Road. This inlet led to salt water intrusion into Unit I. Despite scraping to create the artificial dune along Unit II after that storm, another large storm in November 2009 overtopped that area, wiping out the artificial dune, and opened two inlets into Unit II. Units I and II now have active inlets to the bay and have undergone changes in vegetation and wildlife as well as wave energy in the marsh (M. Stroeh, pers. corr. January 2010 and January 2011).

Maurmeyer (1978) used field observations to map nearshore currents, and found currents in this area to vary seasonally, with northward movement in the summer and southward in the winter. Winter currents prevail, so the net sediment transport is southward. The 2010 Coastal Management Plan included a planning-level hydrologic model of the bay. Under this model, longshore current is southward in this area. During storm conditions, Fowler and Prime Hook beaches are inundated and flow is seaward, and then southward.

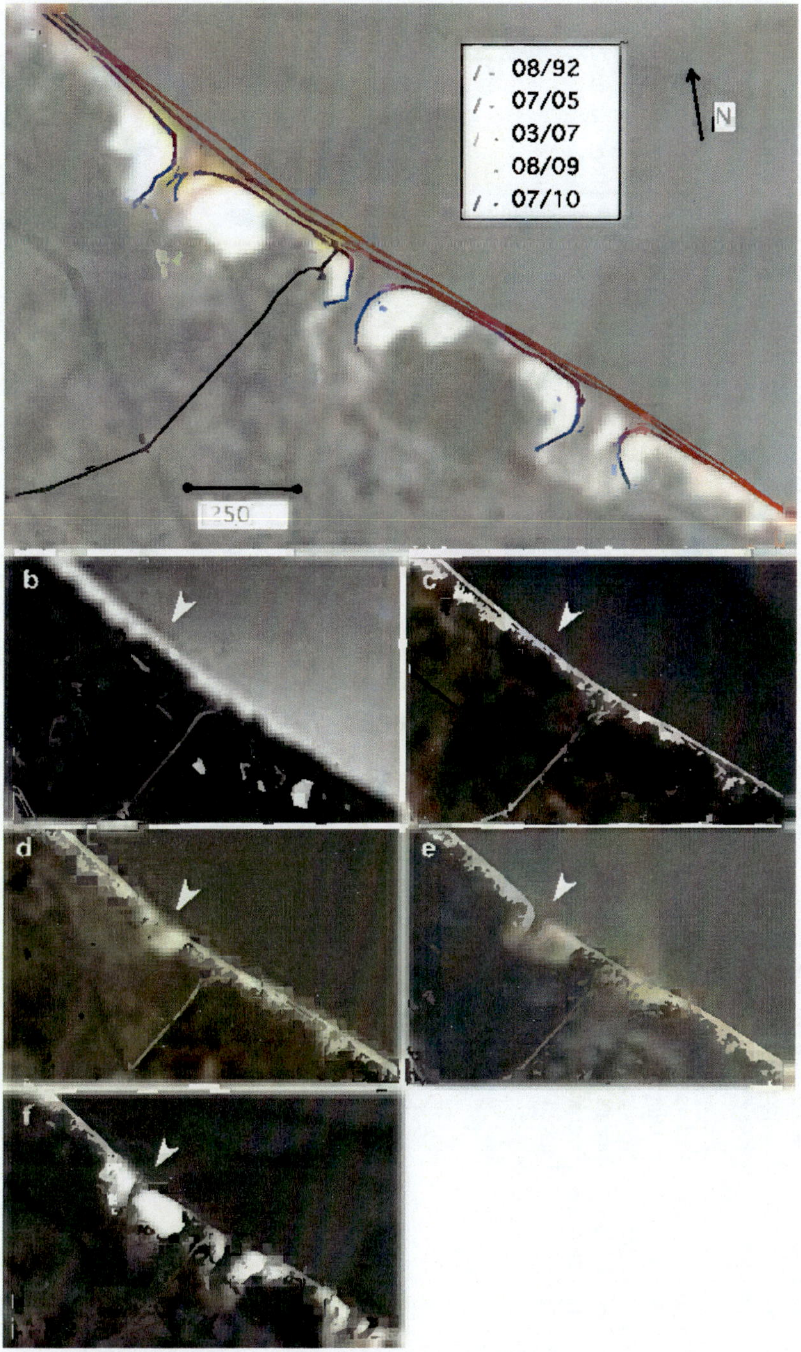

Fig. 16.3 (**a**) Historic shorelines laid over a July 2010 satellite image show retreat and the formation of inlets near at Fowler Beach. (**b**) Image dated March 1992; (**c**) July 2005; (**d**) March 2007; (**e**) August 2009; (**f**) July 2010 (Imagery source: Google Earth accessed January 20, 2011)

1910 No major landscape alterations yet	1930s -40s Ditches dug to control water flow	1962 Ash Wednesday Storm and subsequent nourishment	1988 Impoundment	1999 Soft Stabilization	2006 Hurricane Ernesto	2008 Mother's Day Storm	2009 Veteran's Day Storm
Both Prime Hook and Slaughter creek drain into Delaware Bay	Begins slow conversion from salt marsh to freshwater march	Erosion and coastal flooding; Prime Hook received m^3 of sand	FWS impounded 1620 hectares of fresh-water marsh	Broadkill nourished with 28,000 m^3 of sand; artificial dunes scraped on Fowler	Inlet into Unit I opened; artificial dunes scraped back into place	New inlets into Units I and II; Unit II is scraped; Prime Hook received 900 m^3 of sand	Inlet into Unit II widened after coastal flooding; further scraped proposed

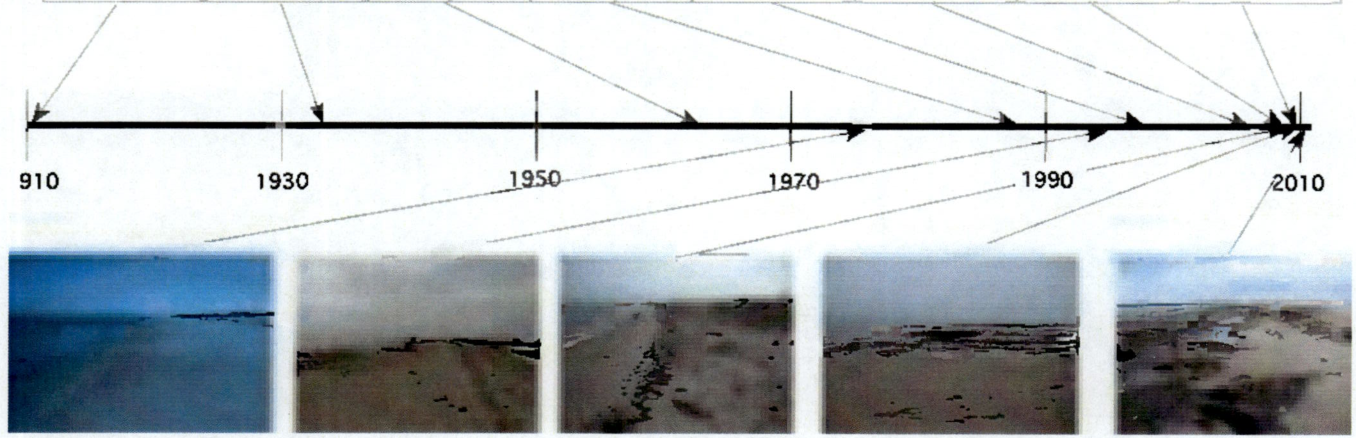

Fig. 16.4 The region has a long history of storm damage and stabilization efforts in response. This series of photographs show the recent changes at Fowler beach, dated: (**a**) March 1977; (**b**) July 1998; (**c**) February 2009; (**d**) November 2009; (**e**) January 2010. *Photos* (**a–b**)*: Evelyn Maurmeyer. Photos* (**c–e**)*: Hilary Stevens*

16.4 Impacts of Stabilization

The maintenance of shorelines is a complicated and costly process. There are direct costs and benefits associated with hard or soft stabilizations, but also many indirect impacts on coastal communities. In many cases, the public feels that maintenance of beaches, including protecting private property, should be a government responsibility. Public opinion in Delaware coastal towns has favored nourishment as a form of shoreline stabilization (Drew and Kraft 1980).

16.4.1 Direct Impacts

The primary benefits of these projects are the reduced risk of flooding and preservation of desired habitat. The cessation of overwash inlet formation leading to salt water intrusion into the marsh is the primary goal of shoreline stabilization in the PHNWR. The protection of private property in the Prime Hook community is also an important goal of this project. Public opinion favors action to maintain public beaches, which in this case is only Fowler Beach, and private landowners can be very vocal about the need to protect their property.

Preservation of freshwater marsh to provide habitat for vegetation and waterfowl is one goal of the PHNWR. When inlets open and salt water intrudes into the marsh, the nature of the habitat changes. The inlets provide new habitat for some shore bird species. However the loss of freshwater marsh impacts vegetation and migratory bird species. The newly flooded marsh creates a salt marsh habitat, although there is also some concern that the area is deep enough to be lagoonal in nature (M. Stroeh, pers. corr. January 2011). Calculating the value of any given habitat depends on which species are considered.

Residents and landowners adjacent to the PHNWR, like residents in many coastal areas, are very concerned with the accelerating rate of coastal erosion and the risk to property it poses. They have put pressure on the state and the Refuge management to take action to reduce the risk of flooding in the marsh and to stabilize the beach. Scraping dunes is widely perceived to be an effective, comparatively inexpensive and aesthetically tolerable method to preserve homes and private property. However, the 2006 and 2009 storms quickly and completely wiped out the artificial dunes and did nothing to prevent the formation of overwash fans and new inlets.

Stabilization of shorelines has cost state and federal taxpayers for decades already, and will continue to do so under the management plans currently being put forward. The costs incurred under the Delaware state management plan including nourishment of bay beaches and planting vegetation are estimated at $984,924–1,775,589 over 10 years (DNREC 2010a, b). The dune scraping plan recommended by the Refuge is estimated to cost $13,000–19,000 for each scraping episode (USFWS 2010).

16.4.2 Indirect Impacts

In addition to these direct costs for nourishing the beaches, maintenance of infrastructure such as roads and impoundments incurs costs to the federal and state governments. For example, the FWS installed and maintains the impoundments and water control systems while the Delaware Department of Transportation (DelDOT) is responsible for maintaining the roads throughout this area.

The altered aesthetics of the stabilized shoreline can be considered a cost or benefit, depending on the success of the project. Construction projects such as groins and jetties are generally considered unattractive and detract from the natural beauty of a beach landscape. Aesthetic detraction is especially true for projects that fall into disrepair and become eyesores that are dangerous to the public, such as the failed geotubes at Prime Hook. Dune scraping at Fowler, where the sediment layer is thin, creates a disproportionately high dune and leaves the adjacent exposed peat denuded of sediment and vegetation. Nourishing beaches usually appeals to the public as it creates a high and wide beach that is suitable for recreation. However, finding good quality source material is the critical element. If the sediment used is too coarse or poorly sorted, the resulting beach is unappealing for recreational use, too fine, and the material will be rapidly removed from the system.

Any stabilization of the shoreline impacts sediment transport along the coast. The nourishment recommended in the state management plan would increase the sediment available for transport along the western coast. Nourishment introduces more sediment into the system, although in this case the source is not specified. Because it is generally of a different character than the existing sediment, it will shift in response to wave energy and temporarily increase sediment transport. However, this effect is short-lived and further nourishment will eventually be required, as outlined in the management plan. Additionally, adding sediment to the system may have deleterious impacts on organic build-up deposits such as marsh material and intertidal hard bottoms like the sand builder worm (*Sabellaria vulgaris*).

16.5 Planning for the Future

The residents and community leaders along the Delaware Bay have a number of challenges to consider when planning shoreline management going forward. Protection of private property and property values, habitat preservation, public safety, and shoreline change are all factors to be considered. Long term planning depends on estimates for sea level rise, storm intensity, and development. Strategies for managing the shoreline in this area include hard stabilization, soft stabilization, or no engineering action (Table 16.1).

Table 16.1 Shoreline management strategy options

Option	Pros	Cons
No action	No initial cost to taxpayer; improved habitat for horseshoe crabs and shore birds	Politically unpopular; potential loss of private property or infrastructure such as flooded roads
Scraping	Comparatively less costly; does not require import of sediment	Unnatural appearance; required after each storm event; requires base sediment supply in situ; loss of shorebird inlet habitat
Nourishing	Politically popular; wider beach boosts recreational value; adds sediment to the system	Dredging is costly and requires permits; trucking in sand requires upgrade of road; diminishes habitat value by altering sediment characteristics; requires periodic repetition
Jetties and seawalls	Traps sediment in area immediately updrift	Unaesthetic; increases erosion in adjacent areas; heavy initial construction costs; requires maintenance
Abandonment/ removal of Fowler Beach Road	Reduced overhead cost to refuge; more natural appearance	Reduced access to refuge for recreation or maintenance; reduced access to private inholdings

Climatic predictions call for continued and potentially rapid rise in sea level (Overpeck et al. 2006) as well as increased frequency and/or severity of storms (Knutson et al. 2008) in the North Atlantic. These factors will both contribute to continued retreat of the shoreline with more frequent over-topping of dune and barriers. Coastal wetlands will be inundated and the remaining salt marshes are likely to become more saline. Any long-term planning must account for these changes. Recent efforts by state and local agencies (e.g. DNREC) to develop long-term planning with particular consideration of inundation effects of continued relative sea level rise are an important part of the process. More important still is strong enforcement and regulation to adhere to scientifically informed decisions about management alternatives. Scraping and nourishment will not be an adequate panacea and will only delay the inevitable and could inadvertently promote further imprudent development which will then also be in harm's way increasing the risk of loss later on.

Coastal populations are increasing and that is expected to continue (Bookman et al. 1999). In this region, that trend is manifest in increased demand for seasonal and retirement property, and a higher percentage of permanent residents. The watershed is becoming more densely populated. This change in land use from agrarian to residential has wider impacts on the hydrology and water quality, as well as the recreational demand on the Refuge and beaches.

Acknowledgements This work is a result of student research funded by the University of Delaware. We would also like to thank Evelyn Maurmeyer, Michael Stroeh, and Rob Young for their help and advice.

References

Bookman CA, Culliton TJ, Warren MA (1999) Trends in U.S. coastal regions 1970–1988. Report, NOAA, Silver Spring, MD

Del Sordo SG (1988) National Register of Historic Places Registration Form: National Harbor of Refuge and Delaware Breakwater Harbor Historic District. National Park Service. http://pdfhost.focus.nps.gov/docs/NRHP/Text/89000289.pdf. Accessed 12 Jan 2011

Delaware Audobon web site (2011). http://www.delawareaudubon.org/conservation/scrapingsuit.html. Accessed 28 Jan 2011

Delaware DataMIL web site (2009). http://datamil.delaware.gov/geonetwork/srv/en/main.home. Accessed 24 Apr 2009

Delaware Department of Natural Resources and Environmental Control (DNREC) (2010a) DNREC's annual mid-winter survey finds record number of bald eagles. Press release. 40 (16), 2 Jan 2010

Delaware Department of Natural Resources and Environmental Control (DNREC), (Mar 2010b) Management plan for the Delaware Bay Beaches. Dover, Delaware, pp 327

Douglas BC, Kearney MS, Leatherman SP (2001) Sea level rise history and consequences. Academic, New York

Drew K, Kraft JC (1980) Geology of Delaware barriers: environmental planning. In: Proceedings of the conference coastal zone '80, Hollywood, FL, pp 830–849

Fletcher CH, Knebel HJ, Kraft JC (1990) Holocene evolution of an estuarine coast and tidal wetlands. Geol Soc Am Bull 102(3):283–297

French GT (1990) Historical shoreline changes in response to environmental conditions in West Delaware Bay. Masters Thesis, University of Maryland, pp 240

Jackson NL (1995) Wind and waves: influence of local and non-local waves on mesoscale beach behavior in estuarine environments. Ann Assoc Am Geogr 85(1):21–37

Jackson NL, Nordstrom KF, Eliot I, Masselink G (2002) "Low Energy" sandy beaches in marine and estuarine environments: a review. Geomorphology 48:147–162

Khalequzzman M (1997) Conformability of the Holocene transgressive barrier-nearshore facies along the estuarine and oceanic coasts of Delaware. PhD dissertation, University of Delaware, p 350

Knebel HJ, Fletcher CH, Kraft JC (1988) Late Wisconsinan-Holocene paleogeography of Delaware Bay; a large coastal plain estuary. Mar Geol 83(1–4):115–133

Knutson TR, Sirutis JJ, Garner Vecchi GA, Held IM (2008) Simulated reduction in Atlantic hurricane frequency under twenty-first-century warming conditions. Nat Geosci 1:359–364

Kraft JC, John CJ, Marx PR (1981) Clastic depositional strata in a transgressive coastal environment: Holocene Epoch. Northeast Geol 3(3–4):268–277

Lewis DA, Cooper JAG, Pilkey OH (2005) Fetch limited barrier islands of Chesapeake Bay and Delaware Bay. Southeast Geol 44(1):1–17

Maurmeyer EM (June 1978) Geomorphology and evolution of transgressive estuarine Washover barriers along the western shore of Delaware Bay. PhD thesis, University of Delaware

Nikitina DL, Pizzuto JE, Schwimmer RA, Ramsey KW (2000) An updated Holocene sea-level curve for the Delaware coast. Mar Geol 171(1–4):7–20

NOAA Coastal Services Center Digital Coast web site (2010). www.csc.noaa.gov/digitalcoast/data/coastallidar/index.html. Accessed 16 Apr 2010

Nummedal D (1983) Barrier Islands. In: Komar PD, Moore JR (eds) CRC handbook of coastal processes and erosion. CRC Press, Boca Raton, pp 77–121

Overpeck JT, Otto-Bliesner BL, Miller GH, Muhs DR, Alley RB, Kiehl JT (2006) Paleoclimatic evidence for future ice-sheet instability and rapid sea-level rise. Science 311:1747–1750

Pilkey OH, Cooper JAG, Lewis D (2009) Global distribution and geomorphology of fetch-limited barrier islands. Jour Coastal Res 25:819–837

Program for the Study of Developed Shorelines (PSDS) (2002). Beach nourishment. Retrieved Jan 2010 from http://www.wcu.edu/1038.asp

Sella GF, Stein S, Dixon TH, Craymer M, James TS, Mazzotti S, Dokka RK (2007) Observation of glacial isostatic adjustment in "stable" North America with GPS. Geophys Res Lett 34, L02306, doi:10.1029/2006GL027081

Sommerfield CK, Madsen JA (Oct 2003) Sedimentological and geophysical survey of the upper Delaware estuary. Report to the Delaware River Basin Commission

United States Fish and Wildlife Service (Nov 2010) Final Environmental Assessment for Dune Work at Prime Hook National Wildlife Refuge. Milton, Delaware

WBOC television station website (2011). http://www.wboc.com/Global/story.asp?S=13728199. Accessed 28 Jan 2011

Chapter 17
Shoreline Stabilisation: Lessons from South Wales

Michael R. Phillips

Abstract Shoreline stabilisation in South Wales was driven by historical industrial development. Coal from the hinterland was brought to the shore, especially during the nineteenth century for use in metal industries that were located on the coast, and for export. At that time coastline impacts were not understood and development legacies have proven to be environmentally and financially expensive. Even when early in the twentieth century seawall and groyne impacts were recognised, coastal defence strategies still followed previous patterns. With industrial decline, economic regeneration has been supported by coastal location but unfortunately, lessons from the past have not been learned. Case studies show damage to new developments resulted from inappropriate infrastructure location, while undermined seawalls are now being protected by rock armour. This is becoming increasingly serious as there is little available funding to undertake effective remedial measures. It is concluded that new strategies which consider prevailing coastal processes are needed for long-term management; otherwise paving the shoreline may become more common on the Welsh coastline.

17.1 Introduction

The coastal zone has evolved in response to many natural and anthropogenic factors and processes. Furthermore, climate change and sea level rise now pose significant threats to coastal regions. In the United Kingdom, accepted strategies to manage coastal areas under threat are: hold the line; advance the line; retreat from the line; and no active intervention. However, each of these options comes with its own problems and consequences, especially when human activities are

---

M.R. Phillips (✉)
School of Built and Natural Environment, Faculty of Applied Design and Engineering, Swansea Metropolitan University, Mount Pleasant, Swansea, Wales SA1 6ED, UK
e-mail: mike.phillips@smu.ac.uk

J.A.G. Cooper and O.H. Pilkey (eds.), *Pitfalls of Shoreline Stabilization: Selected Case Studies*, Coastal Research Library 3, DOI 10.1007/978-94-007-4123-2_17,

threatened. Ketchum (1972) classified human activity in the coastal zone under the following groupings: residency; industrial and commercial; agriculture, aquaculture and fishing; tourism and recreation; waste disposal; conservation; and military and strategic. While all these can be currently identified on the South Wales coast, traditional industrial and commercial activities are in decline and accommodating growing development pressures is an issue. The industrial revolution, and especially industrial growth in the nineteenth and twentieth centuries, caused many present-day coastal problems (Bullen 1993). This inheritance has further complicated current protection and management of Welsh coastal assets. An evaluation of the present day status of a coastline is fundamental in deciding whether to actively manage or refrain from intervention (Simm 1996). In many vulnerable coastal areas beaches lie in front of constructed defences and it is important to differentiate between the general health of a beach and its function in coastal defence. A beach may be at a very low level and its volume may be reducing, but this does not necessarily mean that the beach needs managing to improve its defensive capability (Simm 1996). However, implicit in understanding any temporal shoreline behaviour is data acquisition. Managing the coastal zone involves policies at various geographical scales but economic drivers often lead to inappropriate local decisions. These often ignore fundamental coastal processes and lessons learned elsewhere. Consequently via case studies at Llantwit Major, Porthcawl, Port Talbot and Llanelli (Fig. 17.1), this chapter will assess consequences of historic and contemporary Welsh shoreline stabilisation. Cause and effect will be examined at each location with lessons learned and policy implications subsequently reviewed.

17.2 Case Studies

17.2.1 Llantwit Major – Colhuw Beach (OS Ref: SS955673)

In 1972 the Glamorgan Heritage Coast became the first coastline designated as such in the UK. Approximately 23 km long, this protected coastline has spectacular cliffs, interrupted only by the occasional bay such as Col-Huw beach, approximately 1 km south of the historic town of Llantwit Major (Fig. 17.1). The geology is Lias limestone cliffs backing a beach consisting of limited sand, Lias mudstones, shales and limestone pebbles, which range from a few millimetres to a couple of metres in diameter (Bullen 1993; SBCEG 1999). The spring tidal range is 9.6 m and the beach is subject to direct wave attack from the prevailing southwest and refracted wave attack from the southeast (SBCEG 1999). Consequently, sediment drift is strongly eastwards, driven by the oblique dominant wave approach. Due to beach shape, topography and coastal processes, good surfing waves are generated.

Fig. 17.1 Case study locations

One of the few accessible bays along this section of coastline, monitoring by the shoreline management partnership estimates cliff erosion at *circa* 0.08 m year^{-1} (SBCEG 1999).

Col-Huw is a mixture of private and local authority ownership; the local authority running the car park and the Surf Life Savers club, with the Glamorgan Wildlife Trust Reserve adjacent to the beach. It is popular for swimming and surfing with a café located on the shoreline (SBCEG 1999). Williams and Davies (1990) described it as a 'honey-pot' recreational beach, mainly a local visitor attraction with some tourist interest from other parts of Wales and elsewhere (SBCEG 1999). The area is prone to significant storms and high waves, and has been the subject of

Fig. 17.2 Storm damage to Surf Life Saver Club

many coastal defence measures. In the early 1990s, storms severely damaged the Surf Lifesaver building, cafe and car park (Fig. 17.2). Several options in response to coastal retreat were considered, including doing nothing except rebuild the Lifesaver club, but the chosen solution was partial setback and upgrade of the building with substantial hard engineering protection. This consisted of a curved, convex Carboniferous Limestone block revetment, approximately 80 m in length, 20 m in width and 6 m in height (Fig. 17.3). It was constructed with 2,900 tonnes of hardcore and 6,000 tonnes of armour stone, made up of *circa* 7 tonne blocks. It incorporated a concrete beach access ramp and is the largest engineering structure along this coast (Williams et al. 2002). However, in severe south-westerly storms, the revetment is frequently overtopped and outflanked (Fig. 17.3).

Williams et al. (1998) questioned whether the revetment should have been constructed in a coastal area where emphasis is on natural beauty, arguing that anthropogenic intervention had reduced this part of the Glamorgan Heritage Coast's aesthetic quality. Williams et al. (2002) correctly argued that any attempt to hold the line along this particular section of coastline, which is subject to beach erosion and rapid longshore drift, would always be difficult. Logically the buildings should have been relocated inland from this position of natural instability and ironically, following construction of the revetment, shoreline management adopted a policy of managed retreat for this coastline (SBCEG 1999).

Fig. 17.3 Reconstructed Life Saver Club fronted by new revetment

17.2.2 Porthcawl (OS Ref: SS825765)

Porthcawl (Fig. 17.1) became important in the mid nineteenth century because of extensive coal exporting from its harbour. It has a spring tidal range of 8.9 m (SBCEG 1999) and under storm conditions, surges can add 2 m to tide heights (Williams 2002). Following the South Wales coal industry decline and consequent lack of trade for smaller ports, Porthcawl became residential, and the town centre focused on a developing a tourism industry. Shoreline stabilisation started in 1887 with the construction of Brogden's seawall (Fig. 17.4). A combination of spring tides and south-westerly storms caused parts of this wall to fail and in 1906 it was superseded by a vertical seawall. In 1934, exposure of its base and timber piles resulted in the seaward construction of a new 450 m long vertical wall (Williams 2002; Carter 1988, Fig. 17.4). Between the two walls, a low level promenade was created. Hard engineering structures such as seawalls, groynes, piers, etc. are traditional responses for coastal infrastructure protection, but unfortunately they are expensive and tend to promote erosion. Local beach levels continued to fall and in 1965 a concrete buttress was added, which by the late 1970s was in disrepair. Various responses were considered and in 1984/5 the Coastal Protection Authority (CPA) decided to construct a bitumen-grouted revetment (Fig. 17.5). It was made up of a 0.150 m thick asphaltic concrete layer on top of sand fill, with angled bitumen-grouted cobbled slopes, the height

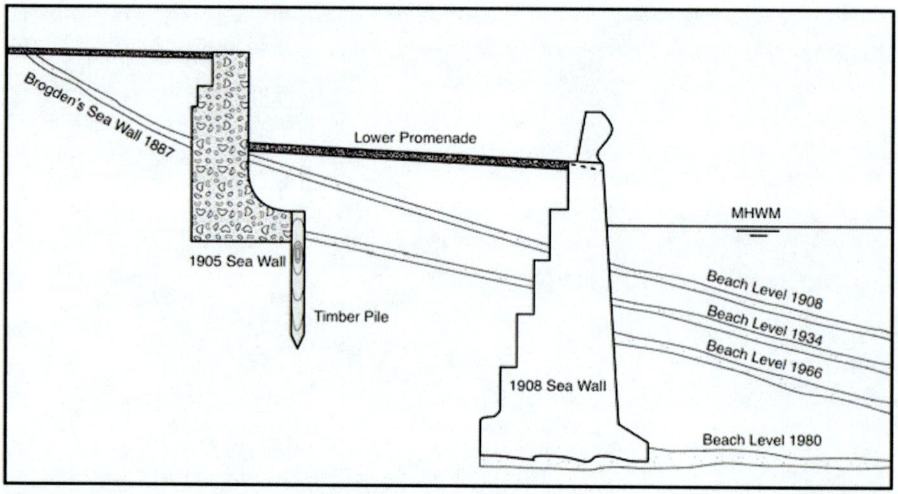

Fig. 17.4 Temporal variation in beach level (after Carter 1988)

Fig. 17.5 Bitumen revetment covering upper beach

from bedrock to seawall base being 3 m (Williams 2002). Although not aesthetically pleasing, the project was successful in preventing further erosion and damage to hinterland infrastructure. Beach tourism is supported by the nearby Coney Beach and Rest Bay to the east and west respectively. The CPA undertakes maintenance work on the bituminous 'town' beach (Fig. 17.5) although the cost is minimal compared with traditional coastal defence measures. However, Porthcawl's main breakwater receives significant wave impacts during storms and requires regular repairs, its structural integrity being of concern to the CPA (SBCEG 1999). This case study shows how shoreline stabilisation locks local authorities into a cycle of expenditure, justified by the protection of high value infrastructure.

17.2.3 Port Talbot (OS Ref: SS755875)

Port Talbot (Fig. 17.6) is located on the eastern flank of Swansea Bay and since 1100AD there has been evidence of shoreline volatility along the whole of this coastline (Bullen 1993). The town grew up around an iron and steel industry which expanded during the nineteenth and early twentieth centuries. Coal mined from the South Wales valleys was brought by rail to Port Talbot for producing steel, and this was subsequently exported via the old docks (Fig. 17.6). Workers' homes were constructed around the steelworks, most being low-cost terraced properties and little thought was given to issues such as tourism. The coastal environment was originally sand dunes with wide beaches typical of a macro-tidal environment. Dunes were removed for the steelworks and associated development, but systems still exist to the northwest and southeast. The shoreline is highly exposed to extreme southwesterly storms and the spring tidal range is 8.4 m. Coal waste is often found on the beach, brought ashore by wave action (SBCEG 1999). The macro-tidal range is a disadvantage for all South Wales ports because ship access and egress is limited over a narrow window during the tidal cycle. Furthermore, ship size is governed by lock dimensions. With design changes at the end of the Second World War resulting in many new ships being too large to negotiate Port Talbot docks' locks, it was the end for many previously commercially viable trading routes. Therefore, unless something could be done to address this situation, the steelworks would have inevitably gone into decline. It was decided that the large cargoes of imported iron ore and coal necessary for economic steel production required a sheltered deep water harbour and consequently, British Steel (now Tata Steel) and the British Transport Docks Board (now Associated British Ports) entered into a commercial venture to construct the Port Talbot Tidal Harbour (Fig. 17.6: OS ref: SS750880). During its construction, 11.2×10^6 tonnes of gravel, sand and silt were removed (Bullen 1993) and with an extensive industrial hinterland, mostly occupied by Tata Steel and related industries, it now dominates the area (SBCEG 1999, Fig. 17.6). However, constructing seaward of the natural shoreline has impacted on natural processes and negatively affected the north and south adjacent coastlines, including the beaches at Aberavon and Margam (Fig. 17.6). Between 1960 and 1976 maintenance dredging for the harbour equalled the capital spoil removal (Bullen 1993) and this on-going process removes sand to a spoil area in Swansea Bay where there is a weak shoreward recirculation.

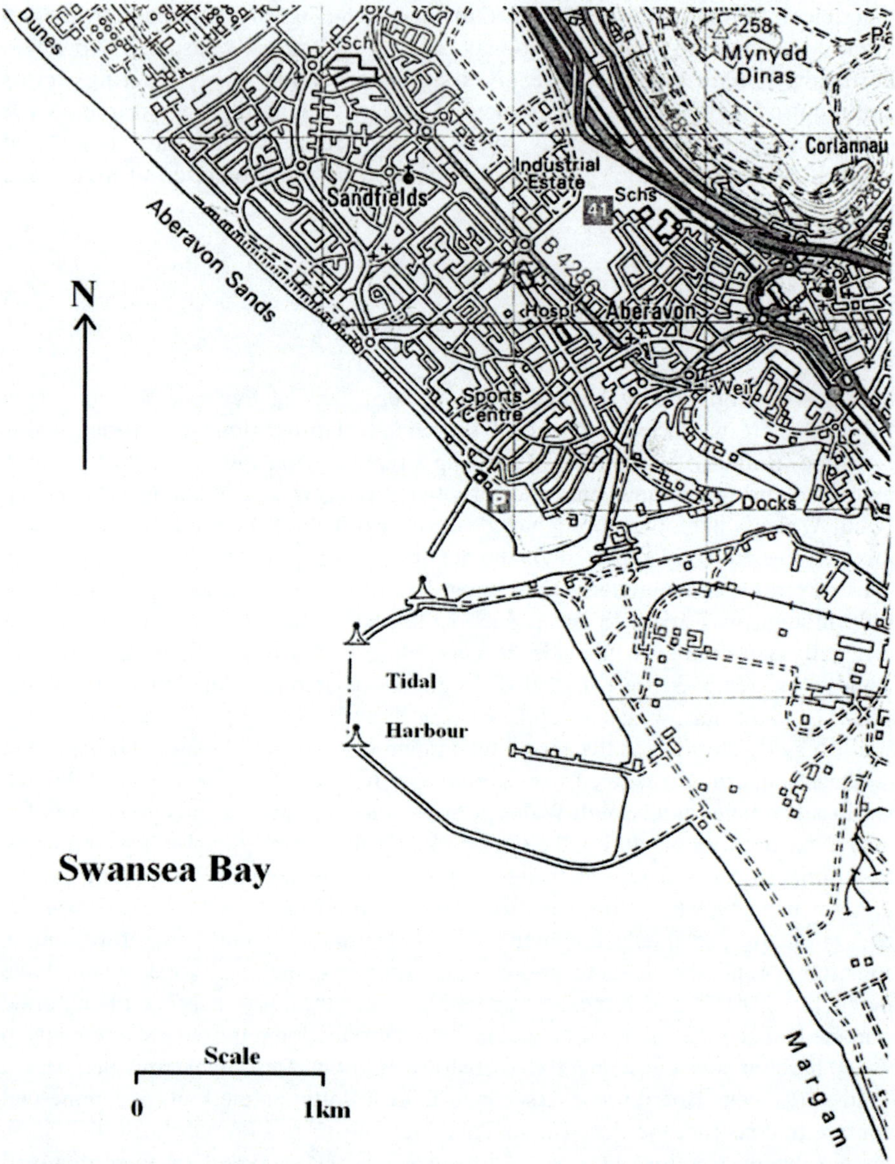

Fig. 17.6 Port Talbot coastline (Reproduced by kind permission of Ordnance Survey, Crown Copyright. All rights reserved)

Aberavon Sands (OS ref: SS735904) is a recreation beach (Fig. 17.6) with a long seawall/revetment and *circa* 2.5 km long promenade. Following completion of the tidal harbour, local authority records between 1958 and 1975 showed an average fall in beach level of 0.8 m. Analysis of a 1991 aerial survey showed that between 1975 and 1991 there had been a further average loss of 0.3 m (Bullen 1993).

Fig. 17.7 Aberavon seawall

In January 1991, as a result of many internal voids that had developed within the seawall, a section collapsed in a not-very-severe storm (Bullen 1993). The south-eastern part was built seawards of high-water and toe erosion is on-going due to wave reflection and turbulence. This is exacerbated by the adjacent tidal harbour which interferes with northerly longshore transport and starves the fronting beach of sediment. Conversely, the north-western half of the seawall was constructed landwards of high-water. Here the fronting beach is outside the tidal harbour's influence and receives sediment recharge from the south, resulting in dry sand being blown against and over the wall. Therefore, adjacent to the tidal harbour beach levels have continued to fall resulting in exposure of sheet piles at the seawall toe and this has led to rock armour protecting the seawall which in turn is protecting the promenade (Fig. 17.7).

Margam Sands (OS ref: SS768855), south-east of the tidal harbour (Fig. 17.6), was once an extensive dune line and is now a 3 km length of slag (waste from steel-making processes) and rock armour revetment (Fig. 17.8; SBCEG 1999). The Tata Steel frontage had experienced erosion over many years and between 1975 and 1991 there was a maximum beach level fall of 1.1 m (Bullen 1993). The 1991 aerial survey showed a back beach/dune line recession of between 20 and 40 m and Bullen (1993) concluded that during this 16 year period there had been significant sediment loss, equating to 950,000 m^3 from back beach recession and eroding foreshore. This was equivalent to an average beach level fall of 0.7 m and parallels were drawn with

Fig. 17.8 Blast furnace slag revetment

similar losses experienced between 1830 and 1870 due to disruption caused by the construction of Port Talbot docks. The slag revetments which protect the steelworks and other infrastructure such as access roads (Fig. 17.8) are themselves susceptible to erosion and sections have been reinforced with stone armour revetments. Consequently, this particular shoreline was recommended for priority monitoring (SBCEG 1999).

17.2.4 Llanelli (OS Ref: SS625530)

At the beginning of the nineteenth century, Llanelli had one small dockland while above the high tide mark, housing was built on copper slag banks (Llanelli Community Heritage 2008). Industries included agriculture, wool and pottery, ironworks, steelworks, tinplate, copper smelting and coal and lead mining. However, it became an important centre for coal exports and in 1834 these led to the construction of New Dock. The commencement of shipbuilding in 1860 required the building of North Dock to provide larger ship facilities (Rees 2006). Transport systems were centred on these industries with additional roads, tramways, railways, canals and docks being developed in response to economic growth. By the mid twentieth century, industry had started to decline and thousands of jobs were lost.

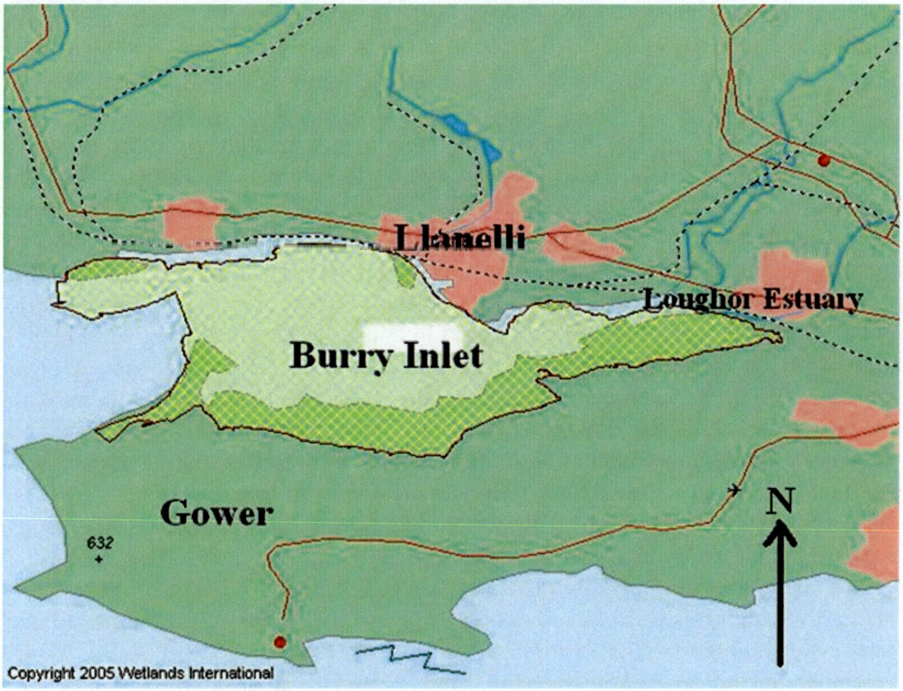

Fig. 17.9 Burry Inlet RAMSAR site (Wetlands International 2005)

The docks fell into disrepair and land was unusable due to high amounts of contaminants, by-products of various local power stations, metal industries and mining. Towards the end of the twentieth century, Llanelli's beach was surrounded by decommissioned industrial sites, docklands and part of the town. Consequently, Llanelli had high unemployment and was not an attractive prospect for developers.

Geologically, Llanelli (Fig. 17.1) and environs date from the Carboniferous with extensive Holocene coastal floodplains (Bowen 1980). The Burry Inlet is a RAMSAR site and lies between Llanelli, Burry Port and the northern edge of the Gower Peninsula (Fig. 17.9). Since medieval times there have been considerable coastal morphology changes and seawalls prompted by coastline change protect the land. It is hard to ascertain through cartography, due to human intervention, whether there have been recent local accretional or erosional patterns (James 2006). The area has a spring tidal range of 8 m and the predominantly southwest-facing coastline is characterised either by low-lying wetlands straddling tidal rivers, or fronting sand dunes (CBCEG 1999). Storm surge and high tide combinations can cause inundation beyond normal tide lines, causing problems for coastal lowlands, which have seen intensive and dramatic land-use change (James 2006). Seawalls, groynes and beach nourishment currently protect existing and new developments, including an important rail link, while Llanelli beach has a sandy foreshore that attracts tourists. Llanelli's seawall is approximately 2 km long and joins the southeast dune system. The seawall has been locally undermined resulting in the addition of rock armour (CCCESC 2004) while Burry Port has two westward groynes to prohibit sand movement. These structures restrict sand movements which consequently affect the coastline (Llanelli Star 2006).

Therefore to facilitate regeneration, funding from the National Lottery, Carmarthenshire County Council and the Welsh Government enabled >800 Ha (2,000 acres) of contaminated industrial wasteland to be converted into the Millennium Coastal Park (MCP, Fig. 17.10). The development objective was to improve housing, commercial accommodation and recreational facilities, including construction of the 22 km Millennium Coastal Path. There are 75,000 people living in the surrounding areas (BBC 2004a) and in 2004, the area received over 500,000 tourists (BBC 2004b). The MCP gives views over the Burry Inlet and Gower Peninsula (Fig. 17.9) and a signature structure, the Earth Sculpture 'Walking with the Sea', was specifically designed to be seen from land, sea and air. It is located on the coastal path between Burry Port and Llanelli (Fig. 17.10) and previously, this coastal section had no sea defence. Constructed from Pulverised Fuel Ash (PFA), an extremely fine waste product from the old coal-fired power stations, it extends seawards into the coastal environment (Fig. 17.10). The use of PFA was seen as an environmental benefit i.e. reuse instead of disposal. Beaches were incorporated into the Earth Sculpture's western and eastern flanks while silt from Burry Port Harbour and sewage sludge facilitated rapid natural re-vegetation (UKQAA 2003). Redevelopment included new houses and seafront apartments adjacent to Llanelli Beach, such as the Millennium Quay (Fig. 17.10). High-quality housing has been constructed on floodplains to the east and further housing and office buildings have been located around the path.

Prior to redevelopment, Llanelli's coastline had not been extensively studied due to its past contaminated status and limited recreational importance. Unfortunately, in the redevelopment strategy, there is no evidence of due consideration being given to prevailing coastal processes. There had been evidence of previous problems when maintenance of river training walls in the Loughor Estuary (Fig. 17.9) ceased in the 1950s following the decline of Llanelli docks. Siltation following collapse of these walls caused nearshore channel migration and since 1970, there has been considerable damage to Llanelli beach (James 2006; CBCEG 1999). Subsequent to the MCP development, there have been a number of problems along the coastal frontage.

Burry Port is a busy recreational harbour which requires storm protection and a clear navigation channel. There are two adjacent beaches, and the eastern one is eroding while West Beach, backed by extensive sand dunes and short seawall, is relatively stable. Longshore eastward sediment transport (CBCEG 1999) is interrupted by groynes and a seawall directly west of the harbour mouth. Sand bar formations subsequently block the harbour entrance requiring sea and land dredging, and sediment is subsequently removed from the system, causing beach erosion to the east (James 2006). In 2006, a 50 m section of the path near the harbour was damaged and in March 2007, following a severe storm, a 400 m length of the coastal path was lost (BBC 2007, Fig. 17.11). Prior to its construction, the

Fig. 17.10 Millennium coastal park

Fig. 17.11 Coastal footpath failure

beach was several metres seaward of this path, illustrating the consequences of stabilising this shoreline. The path has subsequently been repaired and protected by rock armour.

The Earth Sculpture is subject to westward river currents at low tide and eastward sea currents on incoming tides. Its construction seaward of the shoreline is compounded by its PFA composition. Consequently, it is vulnerable to wave attack and since construction, the structure has eroded together with the newly constructed beach on its eastern side, while PFA shows as black streaks on along the coastline. On completion there had been gentle gradients from the hinterland onto the beach but these are now steep. In the March 2007 storm, parts of the Earth Sculpture retreated by >10 m leaving in places >2 m vertical falls to the beach. (CCCESCTFG 2007a, b). Concrete fence posts erected to prevent people falling over an ever increasing cliff edge, now lie at intervals along the beach, some 2 m from the sculpture base. Rock revetments have been constructed to protect the coastline but without these, more of the coastal path would be lost (Fig. 17.11). Retaining walls defending the important rail link have been repaired several times and are also protected by rock armour (Fig. 17.12). Residential and commercial buildings have been constructed within 8 m of the Llanelli seawall (Fig. 17.13) and are often damaged in severe storms (SWEP 2007). These have sometimes required evacuation of residents (Llanelli Star 2005). Severe high tides and storms in March 2006 and March 2007 led to considerable seawall and promenade damage (Fig. 17.13, CCCESCTFG 2007a). The seawall has been further protected by rock armour at its base and as it joins the dune system to the east, sand levels rapidly increase. These dunes are in effect the only sustainable part of Llanelli's coastline.

Fig. 17.12 Rock armour protection of rail link

Fig. 17.13 Redevelopment adjacent to Llanelli beach

17.3 Discussion

Historically, industrial development has affected to some degree the equilibrium of South Wales's coastline, and the Institute of Oceanographic Sciences concluded that human intervention, including port developments and seawall construction, had been the main beach erosion mechanism (Bullen 1993). Similar consequences have been documented elsewhere, including Marina di Massa, Tuscany, where following port developments, beach erosion led to every kilometre of coastline being protected by 1.4 km of hard structures (Cipriani et al. 1999). Later assessments at Marina di Pisa showed 2.2 km of hard structures protecting every kilometre of coastline (Cipriani et al. 2004). Therefore, coastal erosion has been locally aggravated by some of the very strategies implemented to reverse the pattern (Gillie 1997; Weerakkody 1997). Erosion is currently affecting many EU member states with coastlines retreating on average between 0.5 and 2 m $year^{-1}$ and by 15 m $year^{-1}$ in a few extreme cases (Europa 2007a). Of the 875 km of European coastlines that started to erode within the past 20 years, 63% are located less than 30 km from coastal areas altered by recent engineering works (Europa 2007b). The Welsh Government's (WG) coastal erosion policy is to economically and sustainably reduce risks to people and the environment. However, commercial and residential development in coastal zones has led to increased infrastructure along the edge of beaches (Jędrzejczak 2004). This in turn has created a need for coastal defences to protect facilities, exemplified by these Welsh case studies. Therefore, regeneration and economic policies are conflicting with sustainable shoreline management.

During the initial industrialisation of South Wales, there was little knowledge and even less incentive to understand shoreline response to infrastructure development. However, beach consequences of seawalls and groynes were clearly recognised with publication of the 1911 Royal Commission on Coast Erosion and Afforestation (HMSO 1911). Throughout the decision-making process at Llantwit Major, there was no evidence of coastal processes being given due consideration when rebuilding the Surf Life Saver Club and revetment at a cost of approximately £750,000. In reality, hard engineering was not the correct option but the then local authority allowed its policies to be driven by replacing what was lost, without due consideration of why it was lost. Not for the first time, ignoring natural processes had economic implications for local communities. Fortunately, albeit after the event, managed retreat will be the future strategy at this location. The example of Porthcawl shows that once a shoreline is stabilised, more and more investment is required to build bigger and more substantial seawalls. Coastal processes continue irrespective of intervention and while not an aesthetic solution, Porthcawl's bitumen revetment has remained structurally sound for >25 years and consequently, can be considered a success. Industrial developments at Port Talbot have historically caused local beach and dune system loss. The tidal harbour, necessary for maintaining economic steel production, has affected both up- and down-drift adjacent coastlines. It is a substantial structure that is here to stay, and according to

SBCEG (1999) it receives >12 million tonnes per annum of iron ore and other raw material for steel production. Coastal defence measures necessitated by this structure include rock armour revetments to protect concrete and blast furnace slag seawalls at Aberavon and Margam respectively. Tata Steel needs a coastal location and it is argued that despite beach consequences, stabilising this shoreline on socio-economic grounds can be justified.

Initial development at Llanelli was undertaken in ignorance of potential coastline impacts, and following industrial decline, the need for socio-economic regeneration resulted in the MCP vision. The County Council's planning system and Unitary Development Plan (UDP) supported this development, and despite previous seawall failures and beach losses, it appears little consideration was given to prevailing coastal processes. Although the Environment Agency (EA) was consulted about flooding issues during the MCP planning stage, neither it, nor respective local authorities were further involved (CCCESCTFG 2007a). It is important for all coastal professionals to understand causes and consequences of coastal erosion and flooding. This includes sea level rise, global warming, natural and anthropogenic land changes, reduction of river sediment supply and increased coastal development (Doornkamp 1998; Coudert 2005). Beach adjustments usually involve erosive and sedimentary processes and these are mainly a response to natural and anthropogenic change (Anfuso et al. 2000). The whole MCP shoreline is experiencing post-development problems and these have caused additional pressures on sea defences. Resolving these will be difficult, not least because of cost. An MCP funding clause specified that the coastal path must be maintained and remain in place for 75 years and if lost, construction costs would have to be repaid (CCCESCTFG 2007a). Between 1994 and 2004 inclusive, £50,000 was spent on coastal protection and maintenance (BBC 2004a; CCCESC 2004), while in 2006/2007, Llanelli and Burry Port received a joint budget of £75,000 for sea defence repairs. Although this was a £17,000 increase on the previous year, it was not enough and repairs within the first month used the entire budget (CCCESCTFG 2007a). An additional combined budget of £47,550 was given to fill and repair coastal defences and the Earth Sculpture toe, but the local authority wanted this doubled for storm protection and emergency contingency funding (CCCESC 2006). In 2007 following severe storms, the WG and Carmarthenshire County Council gave £83,000 and £400,000 respectively to repair Llanelli's seawall (CCCESCTFG 2007a). Considering estimations of approximately £15–£20 M were needed to improve and repair coastal defences between Llanelli and Burry Port (Wales on Sunday 2004), the funding deficiency is clear. This is a major issue, as the coastal path could be lost if this is not addressed (CCCESC 2006).

Infrastructure at Burry Port is defended with rock armour and coastal structures. Harbour defences result in sediment loss from the system and continuation of current practices will cause further adjacent beach depletion. Therefore dredged material should be used to recharge beaches to the east. The low-lying coastal path is too close to the sea and original design problems are accentuated during storms. Its location should have been higher and further from the coastline and if it is not moved, further failure, as shown in Fig. 17.11, is inevitable. The Earth Sculpture needs both structural and aesthetic restoration with current erosion rates exposing PFA layers. Fundamental theory regarding headland and bay formation should have alerted developers to the folly of extending seaward from the shoreline. Being built from PFA in a high-energy wave environment compounded this error. Direct and refracted wave attacks combined with strong estuary currents are eroding the sculpture from both the west and east, and management responses include gabion cages and rock armour. Currently Llanelli beach has many hard engineering structures that are incapable of total storm protection, while defences in place before the MCP development now protect additional infrastructure (Fig. 17.13). It would no longer be practical to remove Llanelli's seawall due to the height difference between promenade and foreshore. The majority of the seawall is >3 m above the beach and near the railway line, reaches a maximum height of 3.92 m. By protecting the seawall base with rock armour, which is not uncommon in the UK (Simm 1996), wave impacts are reduced. However, for effective protection greater volumes are needed. The lack of financial backing has left Llanelli's coastal defences inadequate and the local authority has to choose the least expensive instead of the most effective long-term option.

It is traditional for Anglo-Welsh coastlines to be managed through hard-engineering holding-the-line practices (Turner et al. 2007). Seawalls are constructed to protect coastal assets and these are subsequently vulnerable to wave attack and prevailing coastal processes. Therefore, seawalls themselves require protection and rock armour toe protection is a common response. Stamski (2005) suggested developing a long-term regional-scale erosion response and armouring plan to predict erosion hotspots. This would reduce the need for emergency works and increase defence capability. However, shoreline stabilisation has been cited as a costly measure that often aggravates recession (e.g. Gillie 1997), but Basco (1999) believed there are many misconceptions about seawalls. They are positioned to defend coastal infrastructure and therefore, development policies and not shoreline stabilisation are the root cause of erosion problems.

17.4 Conclusions

Shoreline stabilisation in South Wales began with the development of its metal and coal industries, especially during the nineteenth century. Development legacies have proven environmentally and financially costly at many locations; and even when early in the twentieth century seawall and groyne impacts were recognised, economic imperatives meant coastal defence strategies still followed previous practices. Following industrial decline, socio-economic regeneration has been underpinned by coastal location. However, lessons from the past have not been learned and damages to new developments have resulted from inappropriate siting of infrastructure and advancing the line without due consideration of coastal processes. Existing seawalls have been undermined and are now themselves protected by rock armour. This is becoming critical as there is a lack of funding to undertake effective remedial measures. New strategies which consider prevailing coastal processes are needed for long-term management, including a rethink of policy. Otherwise, paving the shoreline may be the only sustainable future for the Welsh coastline.

References

Anfuso G, Gracia FJ, Andres J, Sanchez F, Del Rio L, Lopez-Aguago F (2000) Depth of disturbance in mesotidal beaches during a single tidal cycle. J Coast Res 16:446–457

Basco DR (1999) Misconceptions about seawall and beach interactions. In: Ozhan E (ed) Land-ocean interactions: managing coastal ecosystems. Proceedings of the Medcoast-EMECS joint conference, vol 3. Medcoast, Ankara, pp 1565–1578

BBC (British Broadcasting Corporation) (2004a) Coastal park takes top award. http://news.bbc.co.uk/1/hi/wales/south_west/3990489.stm. Accessed 11 June 2008

BBC (British Broadcasting Corporation) (2004b) £20 m bill to protect coastal park. http://news.bbc.co.uk/1/hi/wales/south_west/4066219.stm. Accessed 11 June 2008

BBC (British Broadcasting Corporation) (2007) £400,000 to repair park erosion. http://news.bbc.co.uk/1/hi/wales/south_west/6405479.stm. Accessed 11 Mar 2011

Bowen DQ (1980) The Llanelli landscape: the geology and geomorphology of the country around Llanelli. Llanelli Borough Council, Llanelli, 280 pp, ISBN 10: 0906821010, ISBN 13: 9780906821015

Bullen (1993) Coastline response study: worms head to penarth head: final report, vol 1. Bullen and Partners, Mold, Wales, 285 pp

Carter RWG (1988) Coastal environments: an introduction to the physical, ecological and cultural systems of coastlines. Academic, London, p 617

CBCEG (Carmarthen Bay Coastal Engineering Group) (1999) Shoreline management plan (8c). Management unit no. 6/3. Llanelli Beach to Machynys. www.pembrokeshire.gov.uk/objview.asp? Language=&object_id=1676. Accessed 28 Dec 2010

CCCESC (Carmarthenshire County Council Environment Scrutiny Committee) (2004) Coastal protection: the management and maintenance of coastal assets. http://www.carmarthenshire.gov.uk/. Accessed 11 June 2009

CCCESC (Carmarthenshire County Council Environment Scrutiny Committee) (2006) Coastal protection: the management and maintenance of coastal assets. http://www.carmarthenshire.gov.uk/. Accessed 11 June 2009

CCCESCTFG (Carmarthenshire County Council Environment Scrutiny Committee Task and Finishing Group) (2007a) Coastal erosion along the coastline of carmarthenshire. http://www.carmarthenshire.gov.uk/. Accessed 31 Oct 2008

CCCESCTFG (Carmarthenshire County Council Environment Scrutiny Committee Task and Finishing Group) (2007b) Coastal erosion – progress update – executive summary. http://www.carmarthenshire.gov.uk/. Accessed 31 Oct 2008

Cipriani E, Pelliccia F, Pranzini E (1999) Beach nourishment with nearshore sediments in a highly protected coast. In: Ozhan E (ed) Land-ocean interactions: managing coastal ecosystems, proceedings of the Medcoast-EMECS joint conference, vol 3. Medcoast, Ankara, pp 1579–1590

Cipriani LE, Wetzel L, Aminti DL, Pranzini E (2004) Converting seawalls into gravel beaches. In: Micallef A, Vassallo A (eds) Management of coastal recreational resources – beaches. Yacht marinas and coastal ecotourism. ICoD, Malta, pp 3–12

Coudert E (2005) The prospective analysis by blue plan as a base for a sustainable Mediterranean coastal strategy. In: Ozhan E (ed) Medcoast 05: proceedings of the seventh international conference on the Mediterranean coastal environment, vol 1. Medcoast, Ankara, pp 103–114

Doornkamp JC (1998) Coastal flooding, global warming and environmental management. J Environ Manage 52:327–333

Europa (2007a) Coastal zone policy. http://ec.europa.eu/environment/iczm/home.htm. Accessed 4 Jan 2008

Europa (2007b) European commission puts spotlight on coastal erosion http://ec.europa.eu/fisheries/press_corner/press_releases/archives/com04/com04_21_en.htm. Accessed 4 Jan 2008

Gillie RD (1997) Causes of coastal erosion in Pacific Island Nations. J Coast Res Special Issue 24:173–204

HMSO (His Majesty's Stationery Office) (1911) Royal commission on coast erosion and afforestation: third (and final) report. HMSO, London, 177 pp

James H ed (2006) Where sea meets land. http://www.terra-demetarum.org.uk/articles.htm. Accessed 11 June 2009

Jędrzejczak MF (2004) The modern tourist's perception of the beach: is the sandy beach a place of conflict between tourism and biodiversity? Coast Rep 2:109–119

Ketchum BH (ed) (1972) The waters edge. MIT Press, Cambridge, 393 pp

Llanelli Community Heritage (2008) Seaside. http://www.llanellicommunityheritage.org/HistoryFiles/LCH_HistoryFile_Seaside.htm. Accessed 12 Feb 2008

Llanelli Star (2005). Gran evacuated as storm batters coast. http://moreresults.factiva.com/results/index/index.aspx?ref=LLANLL0020050114e11d0001f. Accessed 11 June 2009

Rees D (2006) Carmarthenshire: the concise history. University of Wales Press, Cardiff, 145 pp

SBCEG (Swansea Bay Coastal Engineering Group) (1999) Swansea Bay Shoreline management plan. Sub-cell 8b: worms head to lavernock point. Shoreline Management Partnership and H.R. Wallingford, 73 pp

Simm JD (ed) (1996) Beach management manual. CIRIA Report 153. CIRIA, London, 448 pp

Stamski R (2005) The impacts of coastal protection structures in California's Monterey Bay National Marine Sanctuary. US Department of Commerce. http://sanctuaries.noaa.gov/science/conservation/pdfs/stamski.pdf. Accessed 31 Oct 2008

SWEP (South Wales Evening Post) (2007). Homes at risk as waves wreck coastal defences. http://moreresults.factiva.com/results/index/index.aspx?ref=SWEPOS0020070104e31300014. Accessed 11 June 2009

Turner RK, Burgess D, Hadley D, Coombes E, Jackson N (2007) A cost-benefit appraisal of coastal managed realignment policy. Glob Environ Chang 17:397–407

UKQAA (United Kingdom Quality Ash Association) (2003) Case study, millennium coastal path, Llanelli, South Wales. www.ukqaa.org.uk/Publications.html. Accessed 12 Dec 2010

Wales on Sunday (2004) £20 m bill to save walk (5th Dec 2004). http://www.highbeam.com/doc/1G1-125825409.html. Accessed 11 June 2009

Weerakkody U (1997) Potential impact of accelerated sea-level rise on beaches of Sri Lanka J Coast Res Special Issue 24:225–242

Wetlands International (2005) Ramsar site information service: Ramsar report for Burry Inlet. http://www.wetlands.org/RSPB/Default.htm. Accessed 11 June 2009

Williams AT (2002) Coastal cliff engineering: case studies from Wales, UK. In: Esin C (ed) Coastal engineering, vol 6. Tubitak: Turkey Scientific and Technical Research Council, Antalya, pp 44–59

Williams AT, Davies P (1990) Coastal cliff protection measures in South Wales, U.K. In: Quelennec RE, Ercolani E, Michon G (eds) Littoral 1990. RGM, Marseilles, pp 699–703

Williams AT, Davies P, Ergin A, Balas, CE (1998) Coastal recession and the reliability of planned responses: Colhuw Beach, the Glamorgan Heritage Coast, Wales, UK. J Coast Res Spec Iss 26:72–79

Williams AT, Davies P, Ergin A (2002) Coastal engineering and erosion along the Glamorgan Heritage Coast, UK. In: Smith JK (ed) Coastal engineering 2002: solving coastal conundrums. Proceedings of the 28th international conference, vol 3. World Scientific, London, pp 3539–3551

Chapter 18
Coastal Stabilization Practice in France

Edward J. Anthony and François Sabatier

Abstract The expansion of coastal urban fronts, leisure ports and tourism in the course of the twentieth century has been the main driver of large-scale modification of the coastal zone in France. The development pressures generated by mass tourism and the economic boom of the 1960s have had their strongest effects in the Mediterranean, but pressures have also been important on the coast of Normandy and Picardy. In the Mediterranean, where large-scale planned development involving joint state and private capital ventures was implemented, this situation has, in many cases, exacerbated coastal instability, while endangering coastal ecosystems, and the growth of urban fronts has commonly led to a drastic reduction in beach width and to dune degradation. Coastal sediment budgets have also been seriously affected by updrift stabilization of cliffs and beaches, especially in Normandy and Picardy. In France, some of the causes of, and the responses to, shoreline destabilization have been essentially a matter of 'hard' engineering, for both historical and cultural reasons, although the situation has been changing over the past decade. A brief overview of shoreline stabilization procedures and structures highlights the overwhelming predominance of seawalls and groynes. Recent practices have tended to move closer to beach nourishment, which is gaining ground in France.

Four case studies briefly highlight the benefits and disbenefits of shoreline stabilization in France. These are: (1) a sandy city-front beach protected by breakwaters on the macrotidal southern North Sea coast downdrift of the large

E.J. Anthony (✉)
Aix-Marseille Univ, Institut Universitaire de France, CEREGE, UMR CNRS 6635,
Europole Mediterraneen de l'Arbois, B.P. 80, 13545 Aix en Provence Cedex, France
e-mail: anthony@cerege.fr

F. Sabatier
CEREGE, UMR CNRS 7330, Aix Marseille Univ, Europôle Méditerranéen de l'Arbois,
B.P. 80, Aix en Provence 13545, France
e-mail: sabatier@cerege.fr

J.A.G. Cooper and O.H. Pilkey (eds.), *Pitfalls of Shoreline Stabilization: Selected Case Studies*, Coastal Research Library 3, DOI 10.1007/978-94-007-4123-2_18,

industrial port of Dunkirk, (2) the regularly renourished and heavily groyned gravel barrier of Cayeux in Picardy, the largest coastal gravel barrier in France, (3) a strongly eroding sector of the Rhône delta shoreline where various combinations of coastal stabilization projects have succeeded each other for over a century and a half, and (4) the regularly renourished gravel beach of Nice, on the French Riviera. These examples show that the emphasis on coastal stabilization at whatever cost that has underpinned coastal management practice needs to be reconsidered, as stabilization will become costlier in the future, as pressures from coastal development increase, as sea level rises and as sediment stocks diminish. Openings in this regard are coming from larger environmental awareness, the recognition of the failure or poor performance of many coastal stabilization projects, and the diversification of the actors involved in coastal management and planning. These developments are progressively generating a new logic of wider concert, on the basis of a more prospective, upfront and long-term approach to coastal management, instead of the logic of a 'stabilization-only' and a commonly one-shot immediate response to storm erosion problems that had tended to prevail in the past.

18.1 Introduction

Large-scale occupation of the coastal zone in France, especially since the beginning of the twentieth century, has been essentially a product of urban, port and tourism development, resulting sometimes in dramatic and chronic shoreline destabilization. The development generated by mass tourism and the economic boom of the 1960s have left their imprint on virtually all sectors of the coast of France, although pressures have been very variable, attaining their maximum in the Mediterranean, the only sector where large-scale planned development involving joint state and private capital ventures was implemented. This situation has been characterized, especially along this Mediterranean coast, by sometimes misguided economic development that has exacerbated coastal instability while endangering coastal ecosystems. The most direct and most important development effect has been a drastic reduction of beach width due to the growth of urban fronts, while coastal erosion has become commonplace due to updrift stabilization of cliffs and beaches. This situation is especially typical of two 'hotspot' regions of coastal stabilization: the Mediterranean coast and the Normandy-Picardy coast (Fig. 18.1). On the Mediterranean coast, urbanization, railways, and coastal routes have all been squeezed into the narrow sand barrier fringing the Languedoc-Roussillon coast and on the small narrow coastal plains of the French Riviera.

In many ways both the causes of, and the response to, shoreline destabilization have been essentially a matter of 'hard' engineering (Fig. 18.2), for both historical and cultural reasons, although the situation has been changing over the past decade. Following a brief overview of shoreline stabilization procedures and structures in France, we will present four examples from various parts of the French coast, from the North Sea to the eastern corner of the Mediterranean (Fig. 18.1), that illustrate the benefits and disbenefits of shoreline stabilization.

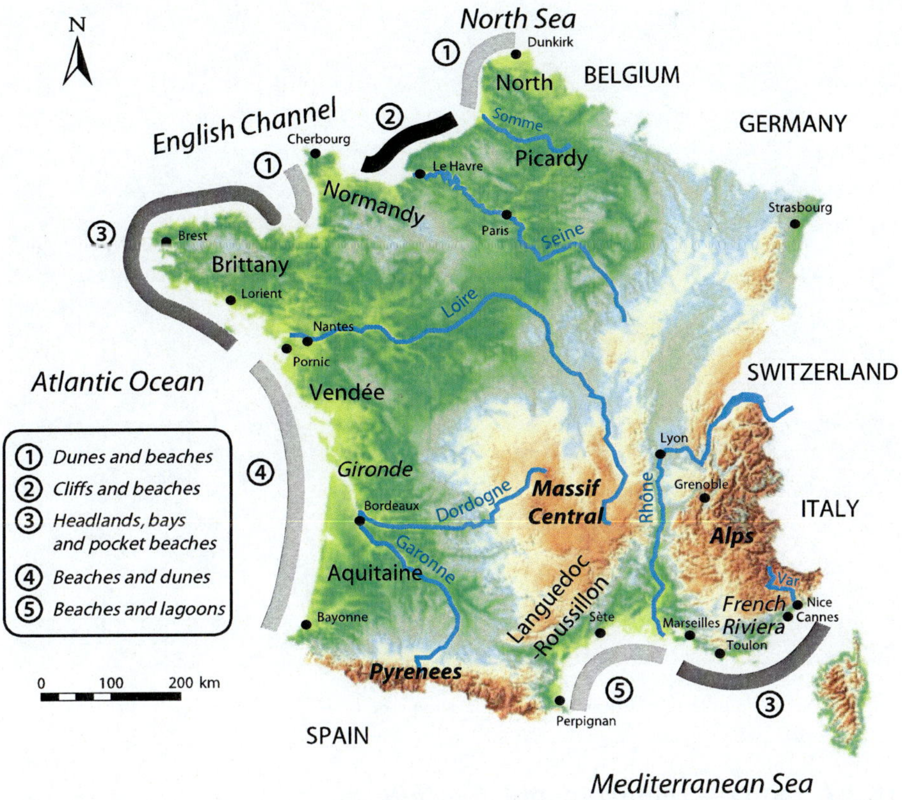

Fig. 18.1 Map showing the coast of France. Coastal stabilization structures show a relatively high density on: (1) the coast of Normandy and Picardy where eroding cliffs that supply sediments to beaches on many coastal resorts have been stabilized, inducing, in turn beach erosion and the construction of numerous rock revetments, seawalls and groynes. (2) the Mediterranean coast where strong development pressures have affected shoreline stability, generating heavy recourse to seawalls, groynes and breakwaters

18.2 Shoreline Stabilization Practices

France has a long tradition of civil engineering that dates back to the Napoleonic when specialized and prestigious higher education schools of mining and engineering were created. Coastal defence and stabilization in France have been the *chasse gardée* of coastal engineers and were, for decades, hinged on a 'hard structural' approach involving the construction of rock armouring and revetments, seawalls, groynes and breakwaters. This approach has been changing over the last decade as hard engineering structures have had rather mixed fortunes in terms of stabilizing the coastline, while the price paid has sometimes been high both in socio-economic and environmental terms.

Fig. 18.2 An example of multiple coastal protection structures on a coastal resort in Normandy

18.2.1 Rock Armouring and Seawalls

The earliest forms of shoreline stabilization in France, as in many other countries in Europe, are rock armouring and seawalls. Rock armouring is still widely used in France in sectors where urban seafronts are absent, especially for cliff base protection, but in many areas it has been replaced by masoned seawalls. Seawalls have become much more popular than rock armouring because they are more aesthetic and, once enlarged, are generally backed by a seafront promenade, but some seawalls are still fronted by rock armouring, especially in northern France and in Normandy. In many coastal resorts, the seafront promenade on the seawall serves as a hallmark of quality, important in attracting tourists. This advantage complements the defensive role of many seawalls in France. The construction, however, of seawalls accommodating promenades has, on many sectors of coast throughout France, been carried out to the detriment of dunes, and has also generally resulted in beach narrowing. Other effects include destabilization of the beach face and of the back-barrier water table. There are also, examples, such as in Wissant Bay, near Dunkirk in northern France (Fig. 18.1), where a deficient sand supply has resulted in beach lowering in front of the seawall (Sedrati and Anthony 2008). The effects of seawalls are also sometimes hard to isolate when these structures are associated with a defence system including groynes and breakwaters.

18.2.2 Groynes

The pressure on beachfronts in France has been matched by a significant increase in the number of implanted groynes, which is nearly close to 1,000, although their distribution is quite variable, depending on the density of coastal resorts. They are generally constructed in sets of two or more groynes, to stabilize beaches by preventing longshore drift of sand or gravel. Groyne fields tend to extend downdrift, as erosion ensues where updrift sediment trapping has been efficient. The largest number of groynes (about 548 in 2011) is found along the coast of Normandy and Picardy, where a high concentration of coastal resorts exists (Fig. 18.1). The density decreases dramatically in Britanny where embayed beaches prevail. Groynes become numerous once again along the central Atlantic coast between Pornic and the Gironde estuary, and in the Mediterranean. Over 95% of the groynes in France are the usual long or short orthogonal groynes. T-groynes are much less common, and are generally associated with artificial beaches. Groynes are also used to canalize urban runoff to the sea.

Groyne construction has stagnated over the last decade as stricter controls are imposed by the DREAL (a state agency responsible for overseeing, among its missions, coastal protection in each region). This situation also reflects the fact that groynes on many coastal sectors have not solved the problem of beach erosion and where dense groyne fields have been built, erosion has been displaced downdrift, creating problems for other communes. This is illustrated by example #2 in Sect. 18.3 below. Groyne dimensions have also been a problem in places, especially in the Mediterranean, where certain short groynes constructed in the early 1970s turned out to be totally inadequate in stopping sand from drifting alongshore over longitudinal bars, or were destabilized and became detached from the beach as a result of chronic erosion. Along many seafront beach sectors, especially on the coast of Normandy, the groyne field generally ends downdrift with a long terminal groyne or a jetty, which is an efficient way of depriving the neighbouring commune downdrift of sand, but many jetties are used to canalize small river outlets. Jetties enclosing leisure ports have been responsible for the erosion of beaches in a whole host of tourist resorts throughout the Mediterranean.

18.2.3 Breakwaters

Breakwaters are much less commonly used in France for beach protection, no doubt due to their much more expensive implementation costs, especially along the southern North Sea/English Channel and Atlantic coasts where large tidal ranges require massive breakwater configurations. A few breakwaters have been constructed on the high-energy mesotidal beaches of the Atlantic coast. Not surprisingly, breakwaters are much more common in the Mediterranean.

18.2.4 Submerged Geotextiles

Although submerged geotextile tubes aimed at wave attenuation have been implemented on beaches for over a decade, the method is still relatively novel in France and has been experimented on a couple of beaches in the Mediterranean (Isebe et al. 2008). The most recent project on Marseillan beach, near Sète (Fig. 18.1), implemented in 2010, concerns a 1 km-long stretch of shore on which a 2.5-high and 6-m wide géotextile tube has been implanted at a distance of 350 m from the beach and a depth of 4.5 m below mean sea level. The beach has also been recharged to the tune of 600,000 m^3. A study to assess the efficiency of the operation is scheduled in autumn 2011.

18.2.5 Dune Rehabilitation

Dune rehabilitation is widely practised in France, and has gained ground as environmental protection groups have gained impetus. Various sand-trapping designs have been implemented. These include fences, geotextiles and plantations, the efficiency of which has been demonstrated in many case studies (e.g., Anthony et al. 2007). In certain cases, these operations are combined with significant sand nourishment and beach and dune profile modifications aimed at optimizing sand immobilization.

18.2.6 Beach Nourishment

There is a long history of beach nourishment in France, inadvertently practised on many beaches which served as sites for materials excavated during building operations, especially of hotels, such as in Cannes and Nice. Such waste disposals on beaches were banned in 1926. Beach nourishment in France has traditionally been coupled with groynes and breakwaters, used as supporting measures to minimise sand losses and maintenance. In many projects, nourishment is a by-product to get rid of sand dredged from a nearby harbour (Hanson et al. 2002). Beach nourishment material is generally dumped on the beach by trucks in the case of shingle and pebbles, and hydraulically in the case of dredged sand. In several cases, in situ tests were performed to check the design, but monitoring after nourishment has not been systematic in many cases.

Beach nourishment is now tending to supersede other forms of beach protection in France, and many communes or communities are now resorting to this practise, especially in Britanny and in the Mediterranean. Over the last decade, beach nourishment has emerged as the most commonly employed form of beach protection and the French coast is replete with examples conducted under the auspices of numerous beachfront communes that have tended to move away from hard structures such as groynes and breakwaters. The case of the Mediterranean beach of Petit Travers, near Sète (Fig. 18.1) is a fine example of the recent massive scale of beach nourishment in the Mediterranean. The beach is situated downdrift of a field of breakwaters and groynes and chronic beach erosion has directly threatened the coastal road behind the beach (Sabatier et al. 2009a), a now classic situation on the Languedoc-Roussillon coast. More than 1 million m^3 of sand was recharged on these beaches from January to April 2008. The monitoring carried out by the DREAL in the year following this operation has highlighted progressive loss of the recharged material. Although beach nourishment is considered as a 'soft' approach to management, nourishment operations are financially costly. There are also strong ecological concerns in the Mediterranean regarding nourishment, especially on the shoreface, as such operations can damage sea grass colonies of *Posidonia oceanica* that play a role in wave energy dissipation, and, therefore, beach protection.

18.2.7 Beach Drainage

Beach drainage has been practised in France over the last two decades, via a single operator and a procedure called *Ecoplage* (*Ecobeach*). Drainage comprises perforated drainage pipes buried below the upper beach surface, and connected to a pump and discharge system. The concept is based on the principle that sand beach accretion is favoured by enhanced permeability due to an artificially lowered water table. The drainage devices are largely buried and therefore the technique has no visual impact. Results have been rather mixed, partly as a result of poor site selection, such as in macrotidal areas, inadequate design, and lack of management. While installation costs are relatively low, both maintenance and management costs are high, and communes have tended to overlook this point in several of the cases where the system has been installed.

18.3 Examples of Shoreline Stabilization

18.3.1 Case #1: The Sandy Macrotidal Beach of Dunkirk

Following the extension, by several kilometres, of the western pier of the port of Dunkirk (Fig. 18.1) in the 1970s, France's third most important industrial port, erosion threatened the urban seafront of Dunkirk in the North Sea, near the Belgian border. On this macrotidal coast (5.6 m at spring tides), sediment supply is essentially governed by the proximity of nearshore tidal sand banks that may be driven onshore by storm waves (Anthony et al. 2010). The erosion was thus likely due to the fact that the port pier and permanent dredging operations carried out to maintain the shipping channel prevented sand from such tidal banks from attaining the coast. The southern North Sea is a storm-dominated short-fetch sea. Considerable refraction and dissipation of waves occur over the numerous shallow nearshore tidal sand ridges that make up the shallow shoreface, resulting in modal inshore wave heights less than 0.5 m high more than 80% of the time.

Two offshore breakwaters were built in 1978 spaced 500 m apart and the beach nourished to the tune of 200,000 m^3 with sand dredged from nearshore sand banks. The structures were designed to stabilize the macrotidal barred beach and protect the residential and resort centre located downdrift of the port (Fig. 18.3). In the face of the rather poor performance of the first two breakwaters, a third was added in 1984, and the project accompanied by a nourishment operation of 160,000 m^3 in 1988. The breakwaters are submerged at high tide, and the overall operation has promoted beach stabilization without significantly impacting the beaches further downdrift. Oblinger and Anthony (2008) showed that the breakwaters are extremely efficient in completely attenuating impinging waves during a large part of the rising and falling limbs of the tide. This generates, however, a wave shadow zone associated with muddy sedimentation in the lee of the breakwaters. The study highlighted an original and rather complex situation, wherein the opposed longshore flows that lead to long-term stabilization of the beach behind a breakwater are due essentially to wind forcing, aided by tides, rather than to waves. In wave-dominated microtidal settings such as in the Mediterranean, this stabilization is commonly generated as a result of the existence of wave-exposed and wave 'shadow' zones that reflect breakwater disorganisation of the wave-generated longshore current, thus leading to local sediment accumulation. In the case reported here, the longshore current in the breakwater sector appears to be only indirectly influenced by waves through longshore compensatory flows that balance strong, exclusively offshore, wave-induced currents. The structure of these currents behind the breakwater and the wave energy dissipation engendered by the breakwaters below MSL both lead to a long-term balance that generates only mild morphological change and maintains sand in situ. In this regard, the breakwaters have been efficient. However, this efficiency is also favoured by both tidal translation rates associated with this tidal environment and the wave energy conditions. The overall fetch-limited low-to-moderate energy context (due also largely to refraction and dissipation by the nearshore sand banks) and the relatively high tidal translation rates of wave fronts imposed by the large tidal range, especially in the mid-beach zone, inevitably lead to low morphological and sediment budget changes.

18.3.2 Case #2: The Gravel Spit of Cayeux in Picardy

The erosion of the gravel spit of Cayeux (Fig. 18.4a) along the coast of Normandy and Picardy was initially triggered by both the stabilization of cliffs and the construction of several jetties that formed staunch drift cell boundaries for gravel

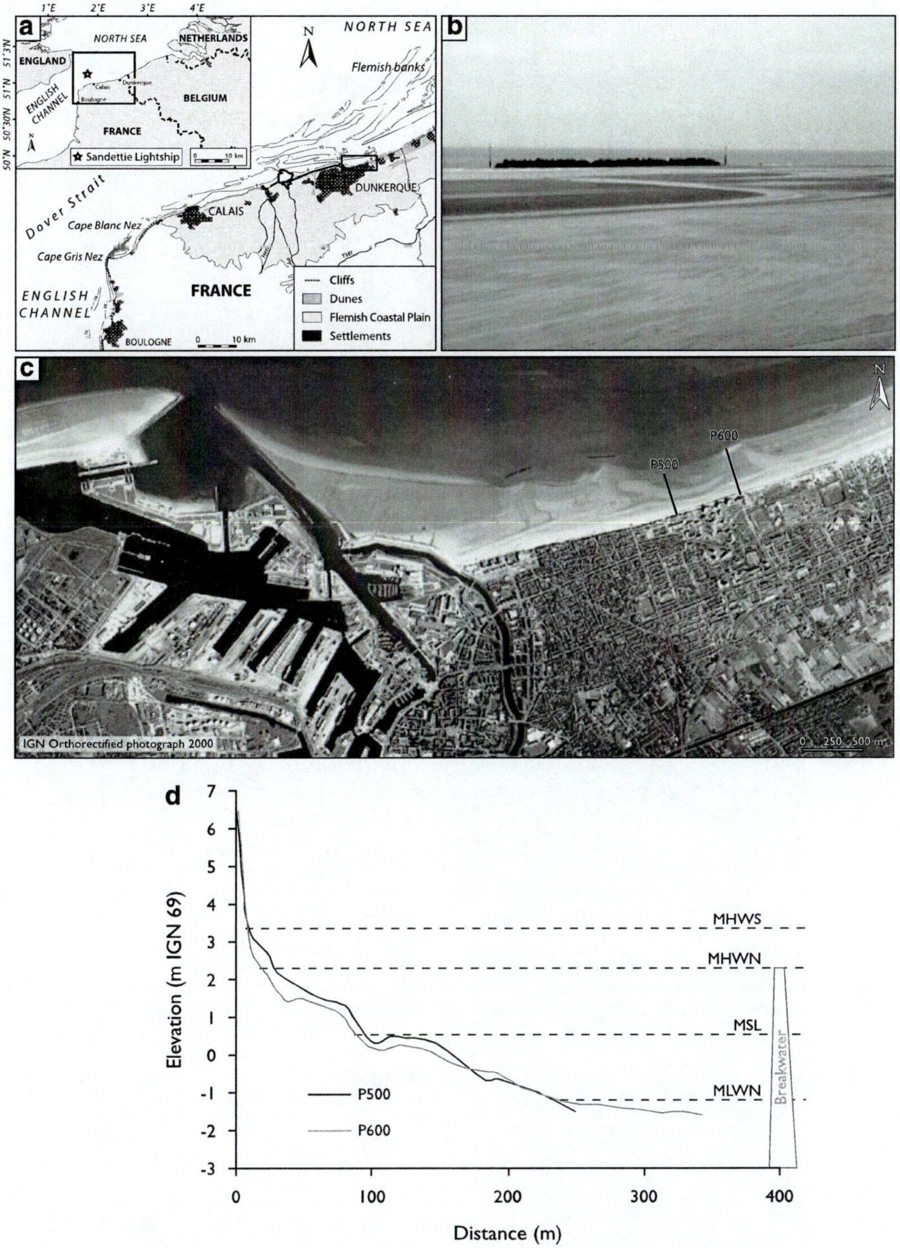

Fig. 18.3 The breakwaters on the macrotidal beach of Malo-les-Bains, in Dunkirk. Panels (a), (b), (c) show respectively geographical location (note the numerous nearshore sand banks in (a), an aerial photograph of the breakwaters (including profile and ADCP currentmeter locations in the course of the study by Oblinger and Anthony (2008)), and ground photograph of one breakwater. The lower panel shows typical beach profiles and tidal levels behind a breakwater. These breakwaters have been beneficial in stabilizing the beach, due largely to the relatively low wave energy conditions on this macrotidal beach

Fig. 18.4 Cayeux spit in Picardy, the largest gravel barrier in France. (**a**) Oblique aerial photograph of the distal part of the spit and the polders behind. (**b**) Erosional and accretional sectors of the spit, the former comprising a field of no less than 83 groynes between the proximal hinge point at Onival and the now-threatened seafront town of Cayeux-sur-Mer. (**c**) Offsets caused by the groynes near the proximal hinge point of the spit

Fig. 18.4 (continued)

transported downstream for spit growth. As the jetties were constructed and the cliffs on this coast stabilized, especially in the 1970s to protect cliff-top urbanization and farmland, there has been a tenfold drop in the gravel supply to the spit, from 20,000 to 30,0000 m^3 prior to cliff stabilization, to 2,000–3,000 m^3. The spit has served as protection for centuries for polders on the Picardy coast (Fig. 18.4a), and the dramatic decrease in gravel supply has been responsible for chronic erosion of its proximal part over the last 30 years (Fig. 18.4b), thus weakening its protective role (Anthony and Dolique 2001). The severe storm that hit western Europe in February 1990 led to massive erosion of the spit over a stretch of 800 m, resulting in flooding of the polders. The solution implemented to counter erosion has consisted of the construction of an impressive succession of groynes (Fig. 18.4c), the number of which has steadily grown over the years to attain a total of 83 groynes in 2011 on the 8 km-long southern stretch of the gravel spit. The downdrift spread of the groyne field has been matched by erosion which now attains the central sector of the spit, threatening the urban front of Cayeux-sur-Mer where rock revetments have been installed to consolidate the spit. In addition to these works, the eroding sector is nourished in gravel, to the tune of about 30,000 m^3 per annum, by lorry transport from the downdrift accretion sector at the approaches to the Somme estuary. The total cost to date (2011) of this stabilization operation has been 40 M€, and the annual maintenance operations run at 350,000 € (Bastide 2011).

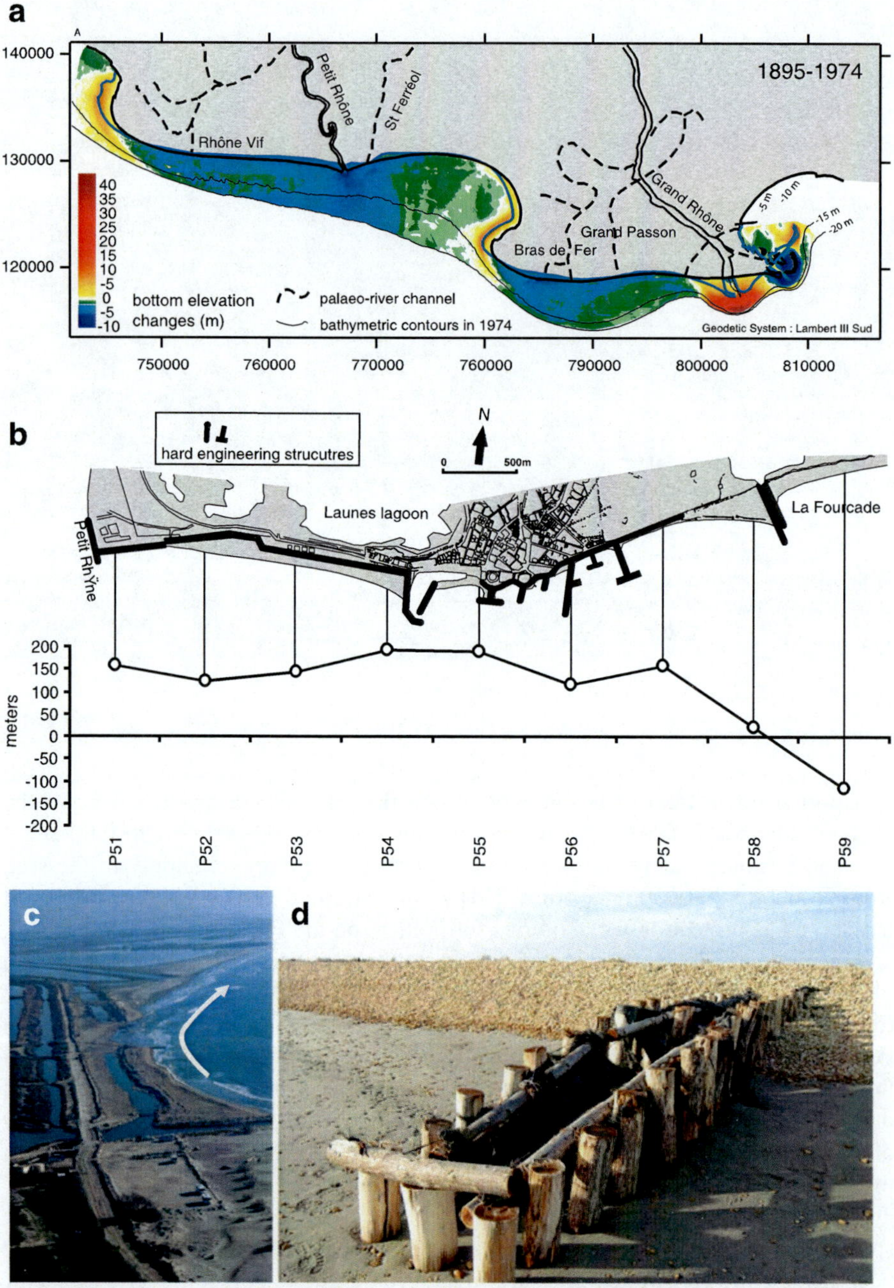

Fig. 18.5 Changes in the shoreface of the Rhône delta between 1895 and 1974 showing erosional and depositional sectors (**a**) associated with variations in sediment supply from the distributary mouths of the Petit and Grand Rhône, with *arrowing* showing longshore drift directions (Modified after Sabatier et al. 2009a). (**b**) Simulation of the efficiency of hard engineering structures in

18.3.3 Case #3: The Rhône Delta Shores

The seafront town of Saintes-Maries-de-la-Mer was in the nineteenth century a postcard village located up to 1 km inland amidst dunes on the sandy shores of the Rhône delta (Sabatier et al. 2006), the most important delta in the western Mediterranean. The mechanisms of retreat or advance of the sandy shores of the delta depend essentially on changes caused by major floods in dominant sediment supply-routing associated with distributary mouths. Shores adjacent to distributary mouths that no longer route adequate quantities of sediment are thus subject to rapid erosion that feeds downdrift accumulation sectors characterized by spits (Fig 18.5a). As the Grand Rhône distributary increasingly captured catchment liquid discharge and sediment supply, correlative dwindling of sediment supply from the mouth of the Petit Rhône distributary, just updrift of Saintes-Maries-de-la-Mer, led to erosion of the delta shores (Sabatier et al. 2009a). Saintes-Maries-de-la-Mer is thus in the rather unfortunate situation of being located on an updrift erosional shore that feeds one of the three main sand drift termini on the shores of the Rhône delta.

In order to limit sea flooding, a backshore seawall was constructed in 1859, initiating a long chain of successive shoreline stabilization measures that is still on-going (Fig. 18.5b). Wooden palisades were constructed on the beach at the start of the twentieth century and groynes in the 1940s in order to counter erosion, followed by rock armouring. After over 60 years of massive armouring, the shore fronting this once picturesque village is now largely artificial. Simulations of shoreline retreat rates by Sabatier et al. (2009a), assuming no stabilization structures, have shown that the town of Saintes-Maries-de-la-Mer would have completely disappeared by the early 2000s (Fig. 18.5c). It may, therefore, be considered that the stabilization structures have actually ‘saved’ this town. This situation has, however, created two major constraints.

The first concerns the shoreface bathymetry which has been rendered steeper by this shoreline stabilization. The upper part of the shoreface (down to -12 m) exhibits slopes that have increased from 0.4% (in 1872–1895) to 1.2% (in 2005), inducing in turn higher incident wave energy levels on the shore. The breaking wave height and the setup for a 100-year wave (Hs $=$ 7.2 m) are 20% larger in 2005 than in 1872. The impact of this increase in shore-incident wave energy on longshore sediment transport and on the stability of the coastal stabilization structures has not been evaluated, but must, in conjunction with sea-level rise,

Fig. 18.5 (continued) stabilizing the shoreline of Saintes-Maries-de-la-Mer in 2000 (After Sabatier et al. 2009b). (**c**) Oblique aerial photograph, taken in 2007, showing the marked manipulation of the shoreline in Saintes-Maries-de-la-Mer involving revetments, groynes, artificial dune reconstruction and beach nourishment using gravel, which is not native to the Rhône delta shoreline. Note the storm breach just updrift of the groyne in the middle of the photograph. *Arrow* shows drift direction. The 4,000 m^3 of gravel recharged over a 300-m section of the shore had migrated by 400 m downdrift in only 2 years, bypassing groynes (**d**). Photographs **c**, **d** by courtesy of Parc Naturel Régional de Camargue pose a renewed threat to the town of Saintes-Maries-de-la-Mer. The second concerns downdrift erosion of the eastern part of the town seafront which is now being exacerbated by this stabilization, resulting in increasing protuberance of the stabilized zone in the west. Between 1895 and the 1960s, the eastern sector retreated at rates of 2–4 m per annum that increased to 5–9 m per annum following the construction of revetments in the western sector and jetties channeling urban runoff. In order to counter erosion and possible flooding of a nearby sewage treatment plant, the beach was artificially nourished with 4,000 m^3 of gravel in 2007 over a distance of 300 m, and an artificial dune created backshore of the nourished zone. The nourishment, the grain size of which ($D_{50} = 200$ mm) had nothing to do with the natural beach sand ($D_{50} = 0.2$ mm), has, in no way, solved the problem. Three years after this nourishment, the shoreline had retreated to its pre-nourishment position. Moreover, the gravel had migrated over 400 m downdrift (Fig. 18.5d) and the artificial dune it was meant to protect was breached by a storm (Fig. 18.5c).

18.3.4 Case #4: The Gravel Beach in Nice

This case concerns the urban front of Nice, in the French Riviera in southern France (Fig. 18.6), a leading world tourist destination, especially attractive for its sunny coast. Rapid socio-economic development in the Nice area has brought considerable pressure to bear on the gravel beaches bounding the Baie des Anges, generating considerable beach narrowing (Fig. 18.6b). Other effects that have impacted the beaches include the reclamation and extension of the Var delta plain through infill, and armouring of the shoreline for the construction of Nice-Riviera airport (Fig. 18.6a), thus completely cutting off the former natural gravel supply from this river (Anthony 1994; Anthony and Julian 1999). The present beach sediment budget is one of zero natural sediment inputs, resulting in chronic beach erosion (Anthony et al. 2011). The 4.5 km-long beach has been artificially nourished since 1976 to the tune of 558,000 m^3, representing an input of 123 m^3 of gravel per metre of beach throughout the Nice seafront, making this long-term operation one of the most important in France, and certainly one of the most significant for gravel beaches in the world. The nourishment gravel is provided at little cost essentially from the numerous housing and construction sites on the Plio-Pleistocene puddingstone geological formations that constitute the hilly foundations of much of the city, and which are rich in rounded gravel clasts. Nourishment has been nil in certain years (1979, 1980, 1983–85, 2001–2002) but reached a peak of over 97,000 m^3 in 2000.

The beach shows no significant change in net width (Fig. 18.6c) over a monitored 30-year period (1976–2006), despite the massive nourishment (Anthony et al. 2011). In fact the two sectors where beach width has shown a net decrease received over 211,000 m^3 of nourishment. Analysis of wave climate records show no significant change in significant wave heights over the nourishment period. Since

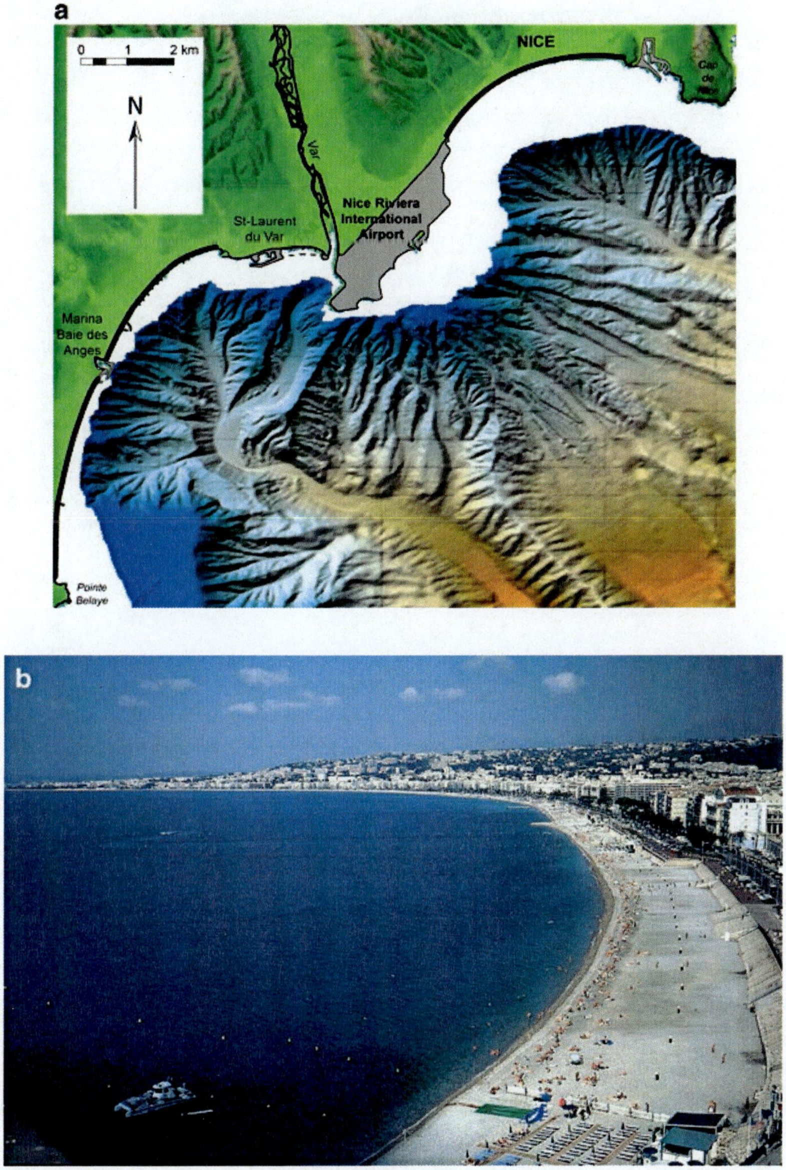

Fig. 18.6 The steep French Riviera margin in the Nice area (**a**). The figure shows an assemblage of the topography (Courtesy of IGN) and bathymetry (Courtesy of IFREMER). Photograph of Nice beach in early summer at the start of the tourist season (**b**). The seawall supports the prestigious *Promenade des Anglais*, the construction of which in the nineteenth century led to considerable narrowing of the original gravel barrier bounding the bay, bringing its seaward face close to the steep shoreface. Mean beach width has hardly varied over the period 1976–2005 in spite of massive nourishment (**c**) (After Anthony et al. 2011)

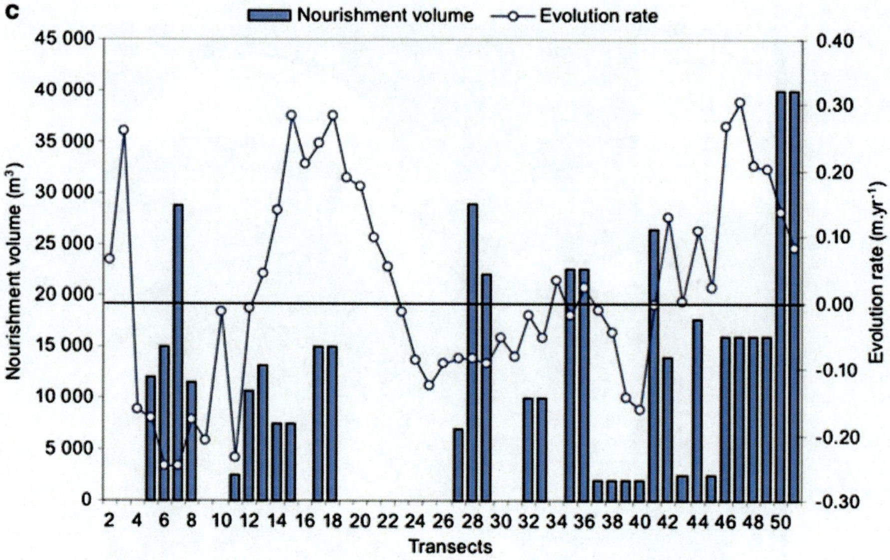

Fig. 18.6 (continued)

there are no possibilities for gravel leakage alongshore from Nice beach, and that only a small fraction of recharge may have been stored permanently through aggradation of the upper beach platform, much of the >550,000 m^3 of nourished gravel has been lost to the steep inner shoreface. This situation raises two questions regarding: (1) the reasons for the relative 'inefficiency' of nourishment in significantly enhancing beach width, and (2) the fate of the beach without nourishment. The steep shoreface of the French Riviera (Fig. 18.6a) is an important environmental factor in regulating the sediment budgets of the beaches in the area. Nourishment on Nice beach is generally practised in spring, and involves spreading out of the recharged materials and flattening of the bermed profile in order to extend the width of the beach, often by several metres, and enhance its comfort and 'carrying' capacity during the summer tourist season. Both nourishment and beach profile restructuring operations are solely aimed at obtaining as wide a beach as possible, fulfilling the recreational role of this beach in summer. Morphodynamically, the combination of nourishment and profile flattening eliminates the natural beachface, induces a lower, more uniform intertidal profile that merges with the steep inner shoreface, where beach gravel may be irreversibly evacuated downslope during storm conditions associated with surge levels and enhanced energy reflection. Recharged gravel is redistributed alongshore, and offshore leakage is probably enhanced where small narrow submarine canyon heads impinge on the beach. Mean beach width shows an oscillating alongshore pattern that may be due to the influence of these canyons as pathways of gravel loss offshore. An example is that of 25,000 m^3 of nourishment in a localized sector of the beach in winter 1992 that gained a dramatic 30 m in width.

The following week, a storm with a significant maximum offshore wave height of 3.48 m led to a reduction of the beach width by 20 m. This reduction was attended by steepening of the beachface, thus suggesting storage of some of the nourishment material through landward beachface retreat, but the bulk of the nourishment material went down to the inner shoreface.

18.4 Discussion

The four examples briefly presented above illustrate the benefits and disbenefits of shoreline stabilization in France. The Dunkirk example clearly shows that the breakwater design, initially accompanied by significant sand nourishment, has been basically successful, keeping the sand budget relatively constant, probably because of the relatively low wave-energy conditions and the large tidal range. The breakwaters do not seem to have impacted the sandy coast further downdrift to the east because the sediment supply to this sector, which is located far enough from the Dunkirk port breakwater extension, is not dependent on longshore supply from the breakwater sector, but is essentially controlled by nearshore tidal sand banks. In contrast, the cases of the gravel spit in Picardy and of the sandy shore of Saintes-Marie-de-la-Mer on the Rhône delta are examples that are particularly illustrative of the vicious circle of shoreline stabilization. The Cayeux spit and back-barrier polders are a French epitome of land-use conflicts and of the lobbying capacity of farmers and wildfowl hunters (French President François Mitterand visited the site and promised total reconstruction following the February 1990 storm flooding). Proponents of the stabilization of the emblematic Cayeux spit will always have good reasons for maintaining this spit and the empoldered farmlands and campsites it is meant to protect, but at what cost to tax payers? The heavy casualties and damage to coastal communities on empoldered lands in Vendée (Fig. 18.1) caused by Storm Xynthia in February 2010 have brought considerable attention to the problems faced by coastal defence and stabilization in France, including the necessity of abandonment and set-back lines (Sabatier et al. 2009b). This is a lesson that has not been learnt in the case of the Rhône delta town of Saintes-Maries-de-la-Mer where the town authorities have consistently sought to fix the shoreline, to the extent that the protected sector of the town is protruding farther and farther offshore as the updrift and downdrift shores retreat. The overall cost of this 'town-saving' project has not been estimated, but it must be running into tens of millions of euros, provided by taxpayers at the local, regional, national and European levels, with funding no doubt obtained through a combination of astute political lobbying and spurious engineering expertise predicting so-called beneficial stabilization, with no signs of the vicious circle in which this venture is now firmly locked being called into question.

The massive nourishment in Nice has been beneficial in that it has at most contributed to beach stabilization, especially when balanced against the negligible cost of obtaining the recharged sediment. It is not clear, however, to which extent Nice beach would or would not subsist without nourishment. Gravel beaches are resilient features and headland-bound gravel beaches in particular may be expected to contain their gravel stock under conditions of a stable sea level through overwash and overtopping processes during storms (Orford and Anthony 2012). The effects of these processes on Nice beach are probably annihilated by the beach-flattening practice and the cross-shore redistribution of gravel. A measure of stability and resilience without nourishment is given by sectors of the other, longer, bay beach south of the Var delta (Fig. 18.6a), which is subject to the same complete depletion of natural gravel supply from the Var river, and which has not benefited from the least artificial nourishment in the past. At some stage, cheap nourishment gravel for Nice beach will start running out as the building operations on the hills overlooking the city that provide the gravel will become rarer or more distant from the city.

18.5 Conclusion

The emphasis on coastal stabilization at whatever cost that has underpinned coastal management practice in France needs to be reconsidered, as stabilization will become costlier with time, as pressures from coastal development increase, as sea level rises and as sediment stocks diminish. Larger environmental awareness and the recognition of the failure or poor performance of many coastal stabilization projects based on 'hard' and costly engineering structures, represent openings in France that have gone hand in hand with a diversification of the actors involved in coastal management and planning, leading to greater checks and balances, especially under the impetus of pressure from environmental groups. The new logic prevailing in this regard is one of wider concert, on the basis of a more prospective and upfront approach to coastal management, instead of the logic of a 'stabilization-only' and a commonly one-shot immediate response to storm erosion problems that had tended to prevail in the past. In this regard, it is interesting to note that French administrative regions with a seaboard are individually setting up a 'Sea and Coast Forum' aimed at prospective multi-party management of the coast and of marine resources, especially with the perspective of sea-level rise. This approach is a forerunner to the setting up of a National Coastal Observatory, probably within the next 2 years.

References

Anthony EJ (1994) Natural and artificial shores of the French Riviera: an analysis of their interrelationship. J Coast Res 10:48–58

Anthony EJ, Dolique F (2001) Natural and human influences on the contemporary evolution of gravel shorelines between the Seine estuary and Belgium. In: Randall RE, Barnes RSK, Neal A, Packham JR (eds) The ecology and geomorphology of coastal shingle. Westbury Academic and Scientific Publishers, Otley, UK

Anthony EJ, Julian M (1999) Source-to-sink sediment transfers, environmental engineering and hazard mitigation in the steep Var River catchment, French Riviera, southeastern France. Geomorphology 31:337–354

Anthony EJ, Vanhée S, Ruz MH (2007) An assessment of the impact of experimental brushwood fences on foredune sand accumulation based on digital elevation models. Ecol Eng 31:41–46

Anthony EJ, Mrani Alaoui M, Héquette A (2010) Shoreface sand supply and mid- to late Holocene aeolian dune formation on the storm-dominated macrotidal coast of the southern North Sea. Mar Geol 276:100–104

Anthony EJ, Cohen O, Sabatier F (2011) Chronic offshore loss of nourishment on Nice Beach, French Riviera: a case of over-nourishment of a steep beach? Coast Eng 58:374–383

Bastide J (2011) Morphodynamique et enjeux d'aménagement des franges littorales d'un estuaire macrotidal tempéré: la Baie de Somme, Picardy, France. Thèse de Doctorat, Université du Littoral Côte d'Opale

Hanson H, Brampton A, Capobianco M, Dette HH, Hamm L, Laustrup A, Lechuga A, Spanhoff R (2002) Beach nourishment projects, practices and objectives – a European overview. Coast Eng 47:81–111

Isebe D, Azerad P, Bouchette F, Ivorra B, Mohammadi B (2008) Shape optimization of geotextile tubes for sandy beach protection. Int J Numer Methods Eng 74:1262–1267

Oblinger A, Anthony EJ (2008) Wave attenuation and intertidal morphology of a macrotidal barred (ridge-and-runnel) beach behind a breakwater: Dunkerque, northern France. Zeits für Geomorphol 52(Suppl 3):167–177

Orford JD, Anthony EJ (2012) Coastal gravel systems. In: Sherman DJ (ed) Morphodynamic systems. Elsevier (in press)

Sabatier F, Maillet G, Provansal M, Fleury TJ, Suanez S, Vella C (2006) Sediment budget of the Rhône delta shoreface since the middle of the 19th century. Mar Geol 234:143–157

Sabatier F, Samat O, Brunel C, Heurtefeux H, Delanghe-Sabatier D (2009a) Determination of setback lines on eroding coasts. Example of the Gulf of Lions (French Mediterranean Coast). J Coast Conserv 13:57–64

Sabatier F, Samat O, Ullmann A, Suanez S (2009b) Connecting large-scale coastal behaviour with coastal management of the Rhône delta. Geomorphology 107:79–89

Sedrati M, Anthony EJ (2008) Sediment dynamics and morphological change on the upper beach of a multi-barred macrotidal foreshore, and implications for mesoscale shoreline retreat: Wissant Bay, northern France. Zeits für Geomorphol 52(Suppl 3):91–106

Index

J.A.G. Cooper and O.H. Pilkey (eds.), *Pitfalls of Shoreline Stabilization: Selected Case Studies*, Coastal Research Library 3, DOI 10.1007/978-94-007-4123-2,

O

P